Field Techniques in Glaciology and Glacial Geomorphology

Field Techniques in Glaciology and Glacial Geomorphology

Bryn Hubbard
Neil Glasser
Centre for Glaciology
University of Wales, Aberystwyth

John Wiley & Sons, Ltd

Other Wiley Editorial Offices

John Wiley & Sons Inc., 111 River Street, Hoboken, NJ 07030, USA

Jossey-Bass, 989 Market Street, San Francisco, CA 94103-1741, USA

Wiley-VCH Verlag GmbH, Boschstr. 12, D-69469 Weinheim, Germany

John Wiley & Sons Australia Ltd, 33 Park Road, Milton, Queensland 4064, Australia

John Wiley & Sons (Asia) Pte Ltd, 2 Clementi Loop #02-01, Jin Xing Distripark, Singapore 129809

John Wiley & Sons Canada Ltd, 22 Worcester Road, Etobicoke, Ontario, Canada M9W 1L1

Wiley also publishes its books in a variety of electronic formats. Some content that appears in print may not be available in electronic books.

Library of Congress Cataloging in Publication Data

Hubbard, Bryn.
   Field techniques in glaciology and glacial geomorphology / Bryn Hubbard, Neil Glasser.
      p.   cm.
   Includes bibliographical references and index.
   ISBN-13 978-0-470-84426-7 (cloth : alk. paper) — ISBN-13 978-0-470-84427-4 (pbk. : alk. paper)
   ISBN-10 0-470-84426-4 (cloth : alk. paper) — ISBN-10 0-470-84427-2 (pbk. : alk. paper)
   1. Glaciology—Field work. I. Glasser, Neil F. II. Title.
   GB2402.3.H64 2005
   551.31′072′3—dc22

                                                                                    2004028501

British Library Cataloguing in Publication Data

A catalogue record for this book is available from the British Library

ISBN-13 978-0-470-84426-7 (HB)   978-0-470-84427-4 (PB)
ISBN-10 0-470-84426-4 (HB)   0-470-84427-2 (PB)

Typeset in 10/12pt Sabon by Integra Software Services Pvt. Ltd, Pondicherry, India

Contents

Preface

All geography and Earth science students carry out fieldwork during their undergraduate degree, either on supervised field-courses or independently in the form of an extended project such as a dissertation. Students and researchers of glaciology and glacial geomorphology embarking on field-work, many of them for the first time, currently do so without a standard text informing them of accepted and practicable techniques for addressing their chosen research topics. Currently, students therefore obtain such information either by word of mouth or by searching the methods sections of journals and research papers, neither of which is entirely adequate. At the same time, readers of glaciological texts may be unaware of exactly how a certain field data set was generated. In this book we provide information on the techniques currently used to study the glacial environment. Our aim is to provide an accessible text on *how field glaciology and glacial geomorphology are done* rather than one on the *theory of glaciology and glacial geomorphology*, which is adequately covered by existing texts. Thus, some level of understanding of glaciology and glacial geomorphology is assumed throughout the text. In providing a text on how aspects of glaciology and glacial geomorphology are studied in the field we hope to provide information that is relevant to two user groups: those who wish to carry out such investigations themselves and those who wish to find out how information was collected by others.

This book is designed primarily for glaciological investigations in high latitudes rather than low latitudes. Thus, for example, our use of the term 'winter season' may be taken to mean 'wet season' at low latitudes. Conversely, our use of the term 'summer season' may be taken to mean 'dry season' at low latitudes. We use the term *glacier* in its broadest sense, that is to describe any substantial ice mass including valley glaciers, ice caps and ice sheets. We also make the distinction between *glacierized*, describing those areas still covered by glacier ice, and *glaciated*, describing formerly glacierized areas (i.e. that are not currently glacierized).

By necessity, the discussion of field techniques is somewhat selective and (unintentionally) biased towards the authors' particular research areas and operational scale. Thus, we limit our discussion to approaches and techniques that are available to most researchers at a reasonable budget – and many are explicitly included because of their availability to undergraduate researchers. Thus, specialist, logistically demanding topics such as satellite-based data collection and oceanography are not covered in any detail.

We also do not focus on snow investigations *per se*, for which excellent method-based texts already exist (e.g. Gray and Male, 1981). This said, we hope that we have covered the main techniques available to the majority of today's glaciologists and glacial geomorphologists.

Some of the methods described in this book are relatively simple; some are much more complex. Some of the techniques are old; some are new. Readers should be aware that change is constant in this rapidly developing field. Technological advances will inevitably occur and many of the techniques included in the text will change over time. Indeed, we are resigned to the fact that some of the techniques included in the text will have been superseded by the time of publication. Readers should not be afraid to amend the methods outlined or to experiment with new methods. As one reviewer of our original book proposal put it: 'The best outcome of books like this are that they attract newcomers and raise the level of standard practice; the worst outcome is that they entomb the science and deaden initiative.' We hope to achieve the former without doing the latter.

Acknowledgements

We thank the following people for providing comments on parts of this book at various stages in its completion: Matthew Bennett, Paul Brewer, James Etienne, Mike Hambrey, Duncan Quincy and Laurence Fearnley.

Thanks are due to many people who kindly shared with us information about specific techniques in their research area: Sven Lukas for information on geomorphological mapping; Duncan Quincy for information on remote sensing techniques; and Becky Goodsell for information on mapping glacier structures.

Neil Glasser wrote large portions of this book whilst on study leave in the Department of Geography, University of Canterbury, Christchurch, New Zealand. Thanks are due to everyone in Christchurch, in particular Wendy Lawson and Ian Owens, for their hospitality there.

We would also like to thank all those people with whom we have spent time in the field over the years: Matthew Bennett, Kevin Crawford, James Etienne, Urs Fischer, Becky Goodsell, Dave Graham, Stephan Harrison, Richard Hodgkins, Alun Hubbard, Dave Huddart, Bernd Kulessa, Krister Jansson, Peter Jansson, Wendy Lawson, Doug Mair, Ben Mansbridge, Peter Nienow, Tavi Murray, Anne-Marie Nuttall, Nick Midgley, Martin Sharp, Martin Siegert, David Sugden, Richard Vann, Richard Waller, Jemma Wadham, Charles Warren and Ian Willis.

We acknowledge the support and advice of all colleagues (both past and present) at the Centre for Glaciology, University of Wales, Aberystwyth as well as Ian Gulley and Antony Smith (Institute of Geography and Earth Sciences, University of Wales, Aberystwyth) for drawing many of the diagrams used in this book.

Bryn Hubbard and Neil Glasser
Aberystwyth

1

Introduction

1.1 AIM

The aim of this book is to provide students and researchers with a practical guide to field techniques in glaciology and glacial geomorphology. Many books and papers have been written about glaciology and glacial geomorphology, but nearly all of these present the *results* of glaciological or geomorphological studies rather than describing the *methods* by which these results were achieved. We have written this book with three principal audiences in mind: (1) undergraduate fieldtrip and dissertation students who may be conducting fieldwork independently and for the first time, (2) undergraduates studying a standard theoretical course in glaciology or glacial geomorphology, whose understanding may be enhanced by knowledge of the techniques used to achieve various theoretical outcomes, and (3) postgraduate research students and professionals who may be designing field projects and equipment and perhaps implementing them for the first time.

1.2 THE SCOPE OF THIS BOOK

Glaciology and glacial geomorphology are essentially field sciences and the emphasis of this book is therefore on fieldwork. We recognize that not all problems can be solved by field research, partly because of the complexity of glaciological and geomorphological problems in nature and partly because not all problems lend themselves readily to investigation in the field. Some properties of glaciers are difficult or time-consuming to measure in the field (e.g. patterns of spectral reflectance, temporal changes in altitude or velocity,

Field Techniques in Glaciology and Glacial Geomorphology   Bryn Hubbard and Neil Glasser
© 2005 John Wiley & Sons, Ltd

mapping of large-scale surface structures such as crevasse patterns), and these are more readily investigated via other methods such as remote sensing. Fieldwork is seldom a stand-alone task and other non-field techniques are therefore sometimes required to solve glaciological and glacial geomorphological problems. These non-field techniques are especially useful as a precursor (e.g. aerial photograph interpretation) or follow-up (e.g. laboratory analysis) to field investigations. Such non-field techniques are included in this book where they relate closely to field sampling and field sample preparation, but due to space constraints we do not elaborate in this book on laboratory-based chemical or physical analytical techniques, glaciological theory, modelling or large-scale remote sensing in detail. These techniques are introduced only at a basic level where relevant to the glaciological and glacial geomorphological techniques presented, and references are supplied for those wishing to explore the techniques in more detail.

Any book such as this necessarily provides only a snapshot of technology and techniques that are continually changing: new and improved versions of some of the equipments presented in this book are doubtless currently being developed and are possibly already in use elsewhere. Such omissions cannot be avoided. Indeed, it is our intention that one consequence of this review will be to stimulate such developments further, perhaps particularly by appealing to skilled workers in technical fields outside glaciology, who may be unaware of the glaciological problems and technologies involved.

1.3   BOOK FORMAT AND CONTENT

This book is divided into ten chapters. Chapters 1 and 2 cover the theory and practice of fieldwork. In addition to introducing the book, Chapter 1 discusses briefly the role of field data acquisition in the broader disciplines of glaciology and glacial geomorphology. In particular, we investigate the relationships between fieldwork, modelling and remotely sensed information. Chapter 2 is concerned with preparation for fieldwork and conducting that fieldwork. Chapters 3–7 cover the range of field techniques commonly used in glaciology with chapters on ice, meltwater, ice drilling and borehole sensors, ice radar, and mass balance and glacier velocity. Finally, Chapters 8–10 cover field techniques in glacial geomorphology with chapters on glacial sediments, glacial landform identification and mapping, and reconstructing glacier fluctuations. Throughout the book, we have used text boxes of individual case studies to illustrate the theoretical outcomes that can be achieved using these techniques.

We conclude each chapter with an indication of the manner in which the specific field techniques presented in the chapter might be applied to undergraduate fieldwork or dissertation research projects. These project ideas should not be viewed as site-specific to individual glaciers or field areas, since this would be of limited value to a widely distributed readership. Rather they are general and lasting ideas, in which we have attempted to identify examples of investigations where a student can test the interrelationships between groups of variables.

1.4 THE ROLE OF FIELDWORK IN GLACIOLOGY AND GLACIAL GEOMORPHOLOGY

Fieldwork is an essential component of both glaciology and glacial geomorphology because field measurements and field observations yield the data that allow us to formulate, test and improve theories within these subject areas. It follows that the ability to measure and describe successfully in a consistent, rigorous and precise manner is fundamental to both disciplines and that these procedures are adequately documented. If we define a technique simply as a *measurement procedure* (Goudie, 1994), then this book would be concerned only with how glaciologists and glacial geomorphologists actually measure variables in the field. However, many variables also have properties that require not only measurement but also careful description in the field (such as the description of sediments, landforms and glacier structures). We therefore prefer a broad definition of fieldwork, which involves (to a lesser or greater degree) the following tasks:

- Hypothesizing and predicting
- Designing an investigation
- Observing
- Describing
- Measuring
- Classifying
- Collecting and organizing data
- Analysing
- Interpreting
- Evaluating
- Formulating models.

In general terms, we can identify at least four reasons for carrying out field-based data collection:

1. To provide routine information relating to the size or status of a physical system (e.g. ice surface temperature, ice albedo, ice-mass dimensions, moraine structure and morphology);
2. To provide information relating to directions and rates of change in the size or status of a physical system (e.g. long-term ice-mass size and short-term glacier fluctuations);

3. To identify the dominant processes operating in a system (e.g. ice-mass motion, surface melting; debris entrainment, transport and deposition);
4. To provide information relating to rates of process operation and the controls over those rates (e.g. ice-mass surface velocity).

The role of fieldwork in glaciology is to provide, wherever possible, measurement (in quantitative terms) or description (in qualitative terms) of variables that can be used to generate and/or validate other types of models, whether conceptual, numerical or physical. In turn, these thought models, numerical models and laboratory models can often help to generate and to direct effective fieldwork programmes. The controlled conditions that can be achieved in models are useful for studying processes that cannot be measured accurately in the field (Slaymaker, 1991). These experimental conditions have been applied to fluvial geomorphology (Dietrich and Gallinatti, 1991), drainage basin studies (Walling, 1991) and mass movements (Okuda, 1991), but to a lesser extent in glaciology and glacial geomorphology.

1.5   THE RELATIONSHIP BETWEEN FIELD GLACIOLOGY AND GLACIOLOGICAL THEORY

Glaciological fieldwork neither exists in a vacuum nor represents an end in itself. It is directed by theory and it, in turn, directs theory. The nature of this interplay between glaciological data acquisition and theoretical development can be illustrated using three examples drawn from the literature. These examples are intended to demonstrate how data gained through fieldwork relates to wider glaciological knowledge and where relevant how that knowledge can direct field data acquisition.

1.5.1   Determining the mass balance of the Antarctic Ice Sheet

If melted, the East and West Antarctic Ice Sheets would raise the level of the Earth's oceans by 55 m. The question of whether the mass balance of the Antarctic Ice Sheet is positive or negative is therefore of great interest to glaciologists (Jacobs, 1992). Mass balance is traditionally measured in the field by glaciologists using accumulation and ablation stakes. However, these measurements are problematic in Antarctica for two main reasons. First, the precipitation measurements required to estimate accumulation rates and the ablation measurements required to estimate ice loss on the ice sheet are sparse and difficult to make because of the size (over 13 million km^2) and remoteness of the Antarctic continent and its climatic extremes. Second, in the coastal areas, where precipitation rates are relatively high, measurements are problematic because of the strong winds. Standard precipitation gauges fail, so that surface snow accumulation is used as surrogate for precipitation (Schlosser *et al.*, 2002). An alternative is therefore to compare precipitation records to measured snow accumulation rates but this requires detailed precipitation records and snow accumulation measurements. A second approach is to measure mass balance manually over a number of years using traditional accumulation and ablation stake methods. However, the Antarctic Ice Sheet is huge and it is therefore difficult to ensure that the chosen glaciers are representative of the ice sheet as a whole. Instead, glaciologists from numerous countries have concentrated on obtaining measurements from a few locations, mainly along transects or in locations in close proximity to their respective Antarctic research establishments. For example, Schlosser *et al.* (2002) point out that at only one site on the entire Antarctic continent has surface accumulation been measured weekly at an array of stakes at a location where meteorological information is also available. Even these data date back only to 1981. The huge practical and logistical difficulties of obtaining reliable estimates of precipitation, snow accumulation and accurate ice velocities on the ice sheet mean that other approaches must be adopted. The problem has therefore been approached by using firn cores to estimate past snow accumulation rates (Stenni *et al.*, 2000), by inferring the physical properties of the ice sheet from satellite imagery (Bindschadler and Vornberger, 1998; Frezzotti *et al.*, 2000), by estimating ice surface elevations from ERS-1 satellite radar altimetry (Rémy *et al.*, 1999), by producing digital elevation models from satellite imagery (Fricker *et al.*, 2000), by predicting ice sheet behaviour through numerical modelling studies (Huybrechts, 1993; Naslund *et al.*, 2000), and by using geomorphology to assess changes in the vertical and horizontal extent of the ice sheet in the past (Sugden *et al.*, 1995, 1999). The problem of determining the mass balance of the ice sheet illustrates neatly that field measurements are but part of a suite of methods including also remote sensing and modelling.

1.5.2 Developing and calibrating models of valley glacier motion

Various stages of integration of field data with other data sources are involved in developing and applying numerical models to reproduce ice-mass motion. In the following sections, we summarize in a very simplified manner these linkages in terms of (i) theory, (ii) modelling and (iii) field data, following the example of an extended research programme at Haut Glacier d'Arolla, Switzerland.

1.5.2.1 Theory

Most of the theoretical developments underpinning the motion of ice masses by ice deformation were made in the 1950s, culminating in Glen's flow law for ice (Nye, 1953; Glen, 1955). In tensor notation with i, $j = x$, y, z, the three axes of the Cartesian coordinate system, this law takes the form:

$$\dot{\varepsilon}_{ij} = A\tau_e^{n-1}\tau_{ij} \tag{1.1}$$

here, $\dot{\varepsilon}_{ij}$ is the strain rate tensor, A is a rate factor that reflects ice hardness (mainly considered as solely temperature-dependent), τ_e is the effective stress, that is a measure of the total stress state of the ice, τ_{ij} is the imposed stress tensor and the exponent n is a constant.

This basic theory of ice deformation is supplemented by that relating to basal motion, whereby ice can slip over its substrate (the so-called basal sliding) and/or move with a substrate that is itself deforming (the so-called bed deformation). Limited theoretical developments have been made in this general field, with most ideas guided by specific field studies. However, in general it is accepted that motion by both basal sliding and bed deformation (i) occur effectively only in the presence of meltwater and (ii) are enhanced by higher (or increasing) water pressures. Since basal motion is sensitive to the presence and pressure of meltwater, Shreve's (1972) theory to explain and predict the passage of meltwater through an ice mass is also relevant to modelling valley glacier evolution. According to Shreve (1972) water within ice-walled channels is driven by two forces: (i) gravity acting on that water, forcing it downwards, and (ii) the pressure gradient exerted on the water by the channel walls, driving the water along a gradient dictated by the ice-surface slope. The resulting water-pressure potential (ϕ) is given by:

$$\phi = \rho_{ice}\, g\, \alpha_{surface} + (\rho_{water} - \rho_{ice})\, g\, \alpha_{bed} \tag{1.2}$$

where ρ is density, g is gravitational acceleration, and α is slope angle with the horizontal. This pressure potential may be differentiated with respect to distance over real ice-mass geometries to define equipotential surfaces down which pressurized englacial and subglacial water will be driven.

1.5.2.2 Modelling

Spatially distributed numerical models of ice-mass evolution are based on calculating changes in ice thickness through time over a spatial scheme representing an actual ice mass. The most advanced of these models are three-dimensional (3D) and, where necessary, thermomechanically coupled to cater for temperature-driven variations in A in equation (1.1). Such models generally solve mass balance, temperature, stress, strain and ice thickness iteratively on arrays of cells that are usually fixed and pre-defined, termed finite differences. In the simplest solutions, the zero-order approximation considers motion to be driven only by local shear stress calculated from local slope, ice thickness, density and gravity. However, at valley glaciers characterized by spatially variable boundary conditions and relatively high stress gradients, significant stresses can be transferred longitudinally or laterally. Calculating the inherited longitudinal stresses requires a higher level of numerical solution, carried out by first-order approximation models. For vertical or bridging stresses to be calculated, higher-order terms again need to be solved, carried out by full solution models. However, these full solution models are computationally demanding and have, to date, only been solved on fixed boundary, finite-element geometries (e.g. Gudmundsson, 1999; Cohen, 2000). Although these bridging stresses are unlikely to be critical at valley glaciers, longitudinal stresses are. Evidence from Haut Glacier d'Arolla, Switzerland, for example, indicates that such stresses can account locally for up to half of the total stress field (Hubbard, 2000).

In modelling Haut Glacier d'Arolla, Switzerland, Hubbard *et al.* (1998) first assumed the value of n in equation (1.1) to be 3 (which is generally accepted; Hooke, 1981), then tuned the ice hardness parameter A by minimizing the sum of differences between the glacier's computed surface velocity field and that measured at the glacier. Next, the authors supplemented this internal motion field with a temporally and spatially distributed basal motion component. The nature of this component was determined on the basis of fieldwork to be centred along the axis of a major subglacial channel established during the late spring (resulting in a major temporary speed-up, the so-called spring event) and active through the summer, melt season. The result of this 3D modelling was a greatly improved fit between the (annually averaged) modelled vertical velocity field measured at the glacier and that generated by the model relative to the no-sliding model.

Before the motion and evolution of Haut Glacier d'Arolla were modelled, Shreve's (1972) theory was coded into a separate model and run for the 3D geometry of the glacier. This was done in order to predict the locations of major subglacial drainage channels to direct a future hot-water drilling programme (section 1.5.2.3). Results of this model (Sharp *et al.*, 1993) predicted the existence of two major subglacial channels in the ablation area of the glacier. One of these channels was chosen as the focus of the subsequent hot-water drilling programme, which successfully intersected a major, multi-year subglacial channel (section 1.5.2.3).

1.5.2.3 Fieldwork

Fieldwork is central to the development, testing and calibration of the theory and models outlined in sections 1.5.2.1 and 1.5.2.2, and those theories and models have, in turn, directed various stages of fieldwork at Haut Glacier d'Arolla. Several key fieldwork stages may be identified, along with their links to theory and modelling, and to techniques covered in other sections of this text.

- Initially, the surface and bed topographies of the glacier were measured by optical survey (section 9.8) and ice surface radar (section 6.5) respectively (Sharp *et al.*, 1993). These data were first used as boundary conditions for the model of Shreve's (1972) pressure potential, which was run to predict the locations of major subglacial drainage channels at the glacier (Figure 1.1). The same data sets were later used as boundary conditions for the 3D first-order numerical model of the glacier's flow (Hubbard *et al.*, 1998; Hubbard, 2000).
- The surface velocity field of the glacier was measured during the winter (when significant sliding is assumed not to occur) by optical survey in

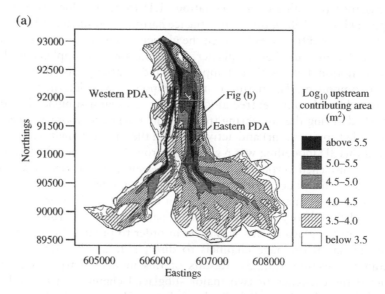

Figure 1.1 Subglacial drainage channel prediction by applying Shreve's (1972) hydraulic equipotential analysis to the geometry of Haut Glacier d'Arolla, Switzerland: (a) shaded map of the upstream area contributing meltwater to each cell calculated by assuming meltwater flows down Shreve's equipotentials (each predicted channel is marked as a preferential drainage axis (PDA)); and (b) expanded map of the Eastern PDA in the glacier's ablation area, which was selected on the basis of this analysis for a hot-water borehole drilling research programme. Figure reproduced from Mair *et al.* (2001), after Sharp *et al.* (1993), with the permission of the International Glaciological Society

(b)

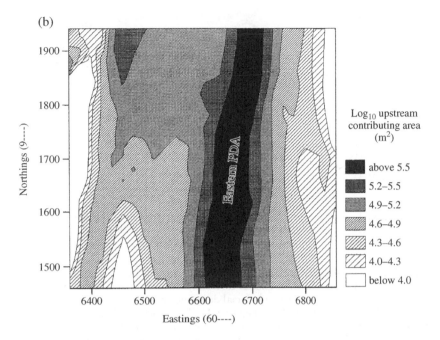

Figure 1.1 (Continued)

order to tune the value of the ice hardness parameter A in the numerical model of the glacier's ice flow.

- A borehole drilling programme was undertaken in an area of the glacier predicted by the equipotential theory and model to encompass a major subglacial drainage channel. The resulting borehole records of water pressure (section 5.4) indicated that a major channel was located in the predicted position and that the pressure variations initiating at the channel evolved through the summer (Gordon *et al.*, 2001) and propagated for some tens of metres either side of the channel (Figure 1.2; Hubbard *et al.*, 1995).

- Associating borehole water-level data with repeated surface velocity measurements (section 7.4.2) indicated that subglacial drainage became pressurized and led to unstable basal motion with its initial seasonal disruption, causing the so-called spring event (Figure 1.3; Mair *et al.*, 2001, 2002).

- Repeated dye-tracing experiments (section 4.6) from moulins distributed over the glacier indicated that, following the spring event, subglacial drainage channels progressively opened and extended up-glacier from the terminus through the summer melt season (Figure 1.4; Nienow *et al.*, 1998).

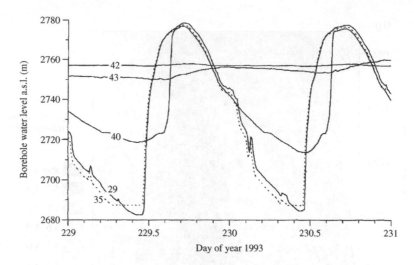

Figure 1.2 Water-level time series recorded in a set of boreholes aligned in a transect across the ablation area of Haut Glacier d'Arolla, Switzerland, in the summer of 1993. Borehole 35 is located above the subglacial drainage channel that flows approximately north–south at northing ~91 800 (Figure 1.1). The other boreholes are located progressively further away from the channel, towards the glacier centreline, according to: 29 = 14 m west, 40 = 31 m west, 42 = 52 m west, 43 = 68 m west. The amplitude and timing of these diurnal water-pressure cycles indicate that pressure waves are generated at the channel and propagate laterally away from it, forcing pressurized water across the glacier bed and inducing sliding there. Reproduced from Hubbard *et al.* (1995) with the permission of the International Glaciological Society

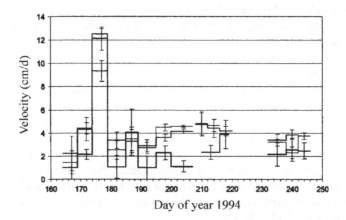

Figure 1.3 Time series of along-glacier velocities, averaged over five-day periods, recorded across the borehole transect at Haut Glacier d'Arolla, Switzerland (Figure 1.2), throughout the 1994 melt season. Note the very high velocities associated with the spring event between 170 and 180 days and the generally higher summer velocities measured following the spring event relative to those recorded immediately before it. Reproduced from Mair *et al.* (2001) with the permission of the International Glaciological Society

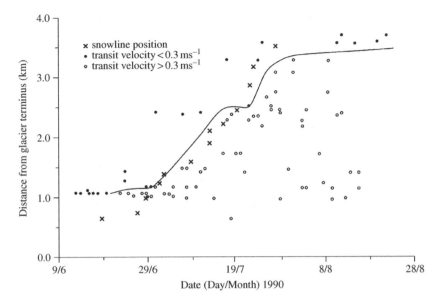

Figure 1.4 Plot of the straight-line distance of individual dye tracer tests from the terminus of Haut Glacier d'Arolla, Switzerland, against the date on which the test was carried out. Tests with a net transit speed of faster than $0.3\,\mathrm{m\,s^{-1}}$ are plotted as dots and those with a net transit speed of slower than $0.3\,\mathrm{m\,s^{-1}}$ are plotted as circles. The boundary between the two domains is marked by a solid line. The seasonal retreat of the glacier's surface snow line is plotted as crosses. This study illustrated clearly that a rapid subglacial drainage network progressively extended from near the glacier terminus in early June to approaching the glacier's headwall by September. Reproduced from Nienow *et al.* (1998) with the permission of Wiley

- The internal velocity field of the glacier was measured within the borehole array by repeating borehole inclinometry (section 5.4.1.3) and borehole tilt cells (section 5.4.1.4) in order to investigate the relationships between subglacial drainage conditions and the glacier's 3D velocity field (Figure 1.5; Harbor *et al.*, 1997). Temporal changes in the resulting velocity field were used to direct and evaluate the 3D model of the glacier's full motion field, including temporally and spatially distributed basal motion (Hubbard *et al.*, 1998).

1.5.3 Investigations on surge-type glaciers in Svalbard

Surge-type glaciers are those that exhibit cyclical instabilities related to changes intrinsic to the glacier rather than external forcing factors (Dowdeswell *et al.*, 1991). The dynamics of surge-type glaciers are characterized by cyclic periods of fast velocity separating longer periods of slow flow. Surge-type glaciers occur in many locations around the world, but

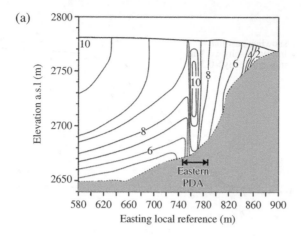

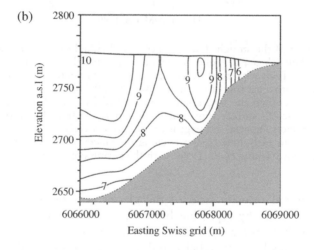

Figure 1.5 Contoured plots of along-glacier ice velocities in glacier half-sections: (a) velocities (m/a) recorded over a 1-year period in a set of boreholes aligned in a transect across the ablation area of Haut Glacier d'Arolla, Switzerland (Figures 1.1 and 1.2), showing the location of the subglacial preferential drainage axis (PDA); (b) modelled velocities produced by a 3D first-order approximation model of ice flow supplemented by periods of basal motion at the PDA directed by field velocity data (Figure 1.3); and (c) velocities measured in Athabasca Glacier, Canada, where there appeared to be no disruption of the internal velocity field caused by spatially distributed basal motion. Reproduced from: (a) Harbor *et al.* (1997) with the permission of the Geological Society of America; (b) Hubbard *et al.* (1998) with the permission of the International Glaciological Society; and (c) Raymond (1971) with the permission of the International Glaciological Society

(c)

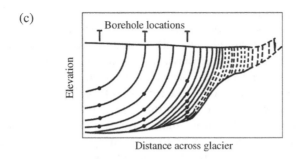

Figure 1.5   (Continued)

they are particularly concentrated in Svalbard (Norwegian High Arctic), where between 13 and 90% of the glaciers are surge-type (Lefauconnier and Hagen, 1991; Hamilton and Dowdeswell, 1996; Jiskoot *et al.*, 1998, 2000). These classifications are based mainly on analysis of remotely sensed images, with surge-type behaviour indicated by the presence of looped medial moraines, formed as fast-flowing, active-phase surge-type glaciers flow past less active or stagnant neighbours and deform the medial moraines between them, potholes formed on the glacier surface during the quiescent phase and a heavily crevassed surface indicative of a glacier in the active phase of a surge cycle (Dowdeswell and Williams, 1997). Surges have also been described from field observations (e.g. Hagen, 1987; Glasser *et al.*, 1998) and by analysis of multi-temporal remote-sensing data (e.g. Jiskoot *et al.*, 2001). However useful these classifications are in identifying glacier surges, they unfortunately provide little detailed information about the physical processes responsible for glacier surges (the surge mechanism) or the trigger for surges. Field investigations of the dynamics of surge-type glaciers are therefore crucial. Indeed, some of the measurements of the physical properties that control the dynamics of surge-type glaciers can only be made in the field. Thus field studies are required of long-term mass balance (Dowdeswell *et al.*, 1995), the nature of glacier thermal regime (Hagen and Saetrang, 1991; Murray *et al.*, 2000b), their hydrological characteristics (Björnsson, 1998; Bennett *et al.*, 2000), the physical properties of the bed that underlies them (Porter and Murray, 2001), the rate of surge-front propagation, velocity and basal shear stress (Murray *et al.*, 1998), together with field descriptions of their structural glaciology (Hambrey and Dowdeswell, 1997) and overall debris structure (Hambrey *et al.*, 1996). Thus remotely sensed data, although useful in identifying and classifying surge-type glaciers, cannot provide a physical explanation for surge-type behaviour. Field measurement and monitoring of these physical processes are the only means of identifying surge mechanisms.

A note on terminology
In this book we use certain terms that may best be defined clearly at the outset. We use the term 'glacierized' throughout to refer to a basin or area that currently contains an ice mass. In contrast, we use the term 'glaciated' to refer to a basin or area that formerly contained an ice mass but no longer does so. We also use the term 'glacier' in its broadest sense to refer to any ice mass, apart from where definitions are being discussed or where the distinction between different ice-mass types is important. In the latter case, the term ice mass is used to refer to any lasting body of ice located at the Earth's surface, and more specific terms are used according to their more rigorous definitions. Finally, our experience in researching predominantly high-latitude glaciers (in the strict sense) may be reflected in our choice of seasonal descriptions as warm/summer as opposed to cold/winter. We hope researchers of low-latitude glaciers will forgive us this simplification and assume that, in such cases, our reference to cold/winter generally implies a season of positive mass balance and that our reference to warm/summer generally implies a season of negative mass balance. However, we are well aware that even these distinctions are too simplified for the complex mass balance patterns that characterize some tropical glaciers.

2

Planning and conducting glaciological fieldwork

2.1 AIM

The aim of this chapter is to provide an overview of the preparations that may be necessary for planning and conducting fieldwork. The following topics are covered: research design and fieldwork preparation (e.g. planning field data acquisition within the framework of a broader project), and data manipulation (e.g. methods of data recording; automated data logging, downloading and power requirements; analysis of accuracy and errors in fieldwork and subsequent representation).

2.2 DESIGNING AND PLANNING FIELD-BASED RESEARCH

Research design is an important element of every fieldwork project. Thorough planning and preparation for research can save hours or even days of wasted fieldwork effort. A properly formulated research project will therefore have well-defined aims and objectives: the data collected in the field will be designed to address those aims. In this case, it is important that the methods used in the field relate to the objectives of the study and that the objectives of the study address the aim of the study. In distinguishing

Field Techniques in Glaciology and Glacial Geomorphology   Bryn Hubbard and Neil Glasser
© 2005 John Wiley & Sons, Ltd

between these terms, a useful definition of an aim is that it states the overall purpose of the study (e.g. to investigate . . . or to study . . .), reflecting the solution of a general problem that may not be fully achievable. Objectives, on the other hand, are a set of specific tasks (each of which can be fully completed) that are designed to go as far as logistically possible to addressing the aim. In preparing a research programme or report the aim is usually presented before a list of objectives is introduced by something along the lines of: 'In order to meet this aim the following specific objectives will be addressed:' After this, a list of objectives may be presented, followed by a list of methods that will be used to achieve the objectives (Box 2.1).

Box 2.1   Field data collection and defining the aim, objectives and methods of an example glaciological investigation: Meltwater-suspended sediment concentrations at a temperate glacier

Research projects, particularly those carried out by undergraduate students, are most effectively planned, executed and reported according to a rigorous protocol constructed around the definition of an aim, objectives and methods. This protocol may be best illustrated through an example. While this example begins with the aim of the study, it would in most cases be driven by research hypotheses that arise from an initial review of the literature. This review and the hypotheses would normally be presented before the aim.

Aim – The aim of this study is to investigate the relationships between suspended sediment concentration (SSC) and discharge in a proglacial meltwater stream over different time periods.

Objectives – In order to address this aim the following specific objectives will be undertaken at a suitable field site:

(i) The discharge of a suitable meltwater stream will be measured once every 4 hours between 8 am and 6 pm for a period of 2 weeks. During this time period, discharge will be measured more intensively, each hour, over two periods of 24 hours. Errors in discharge measurements will be evaluated.

(ii) At the same time as discharge is measured the SSC of the stream will also be measured. Errors in SSC measurements will be evaluated.

(iii) SSCs and discharges will be characterized quantitatively and the relationships between the two properties will be investigated.

Methods – In order to achieve these objectives the following methods will be used:

(i) The *discharge* of the meltwater stream will be measured by establishing a rating curve for the sampling location by salt-dilution gauging and thereafter recording stage against a fixed datum to be identified in the field. Salt-dilution gauging will be carried out repeatedly, as conditions allow, throughout the study to evaluate errors in the rating curve.

(ii) SSC will be measured by collecting water samples of between 250 and 1000 ml by USDH sampler and vacuum filtering them in the field through 0.4-μm pore-size cellulose nitrate filter papers. The volume of filtrate for each sample will be measured in the field to a precision of 1 ml by measuring cylinder. The filter papers will be removed from the filtration chamber by tweezers and stored in sealed plastic bags. In the laboratory, the filter papers will be dried overnight in an oven at 45 °C and ashed at 450 °C for 3 hours. The resulting sediment samples will be weighed to a precision of 0.01 g.

(iii) SSCs and discharges will be characterized quantitatively through the use of descriptive statistics and the relationships between the two properties will be investigated by plotting the quantities against each other and against time and through the use of statistical analysis of their time series.

The aims and objectives of a research project also dictate the geographical area of investigation, the sampling strategy used, the number of samples or the types of measurements to be collected and the nature of follow-up analyses (e.g. laboratory work) to be carried out. In reality, all glaciologists and glacial geomorphologists have different working methods, and no two research projects are identical. However, there is general agreement that research projects should follow a number of logical steps (Figure 2.1). The progression is from initial problem identification, through hypothesis formation and experimental design, to identifying a suitable site, data collection, organization, analysis and presentation. The results of the research are considered and the initial hypothesis is either accepted or rejected. If the hypothesis is rejected on the basis of the data collected, a new or modified hypothesis must be proposed. If the hypothesis is accepted on the basis of the data collected, then a list of predictions of the hypothesis can be made in

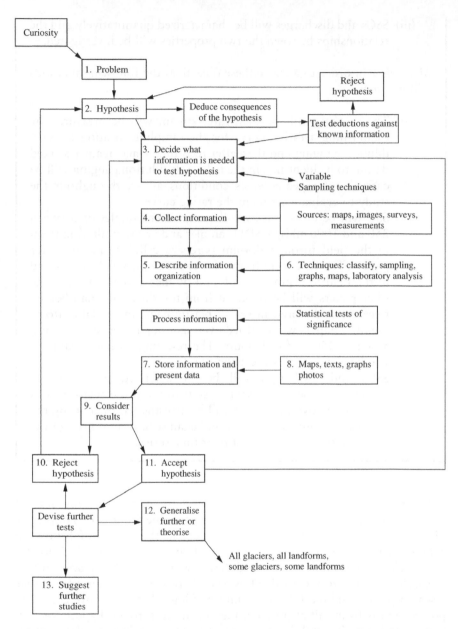

Figure 2.1 Scientific method and the role of fieldwork in experimental design. Based on ideas presented by Brockie (1972)

order to produce generalizations and theories. Finally, in all cases it ought to be possible to make suggestions for further studies that develop the line of enquiry.

2.2.1 Designing a field research project – a worked example

Kennedy (1992), and Parsons and Knight (1995) provide further guidance on research planning for the uninitiated. In this section we outline the logical steps in a research project using as an example the compilation of a glaciological map for a glacier in a remote area that has not yet been visited by the researcher (i.e. mapping the extent of the glacier and its margins, its principal surface features and debris structures).

1. *Define the aims of the mapping and delimit the areal extent the map is to cover.* It is essential at the outset to have a well-defined aim and a well-defined field study area. Once a 'target' glacier has been identified it is important to define precisely the aim of the mapping. For example, is the aim to produce a map that can be compared to previous maps to calculate changes in glacier extent, thickness and volume over time, or is the aim to map structural glaciological features to compare with the distribution of sediment types across the glacier? The precise aim will depend on whether or not published maps and literature already exist, on the availability of aerial photography and satellite imagery, boreholes and other sub-surface data, the time available for the survey and budget constraints. A glacial geomorphological study, concerned for example with mapping landforms and sediments in the proglacial area, would also need to address the issues listed above.

2. *Conduct a desk-based study before embarking on any fieldwork.* Preparatory work is essential to avoid waste of time and effort in the field and to avoid duplication of effort. A desk-based study should start with the analysis of any existing topographic or other thematic maps of the proposed field area, as well as a study of the available literature for the area. Literature searches can be carried out fast and efficiently using the Internet and other electronic literature databases. These allow keyword searches to be made using title keywords, the names of specific geographical areas, authors names or journal titles. For further information on literature searches and how to develop these into a literature review, see Hay (1995). In particular, it is important to check if the area you intend to map or adjacent areas have already been mapped, as this will provide an indication of the glacier surface features and landforms likely to be present. A preliminary glacier map or geomorphological map could be produced from the visual interpretation of aerial photographs and satellite imagery, if these are available.

3. *Field research in the map area.* If it was produced, the preliminary interpretative glacier map or geomorphological map should be checked in the field and, where necessary, corrected. The map can also be supplemented by further detailed field mapping and description, field measurements, by sketching the relationships between individual glacier surface features and landforms, by constructing topographic profiles and surveying and by collecting samples for subsequent laboratory analysis. It is also advisable to take photographs of the field area to provide documentation of the range of surface features and landforms present.

4. *Laboratory analysis.* This involves the laboratory analysis of the physical or chemical properties of any samples collected. In some projects this will be a major component, in others it will not.

5. *Compilation of final map.* The final stage is the compilation of a publication-quality glacier or glacial geomorphological map using standard symbols in order to make the map understandable and quickly accessible to the reader. Ideally the map would be produced digitally, using one of a number of software packages, to allow for corrections and modifications during the final stages. The final map should be accompanied by a written description and interpretation documenting the types of surface features and landforms present, and their spatial distribution. If possible some attempt should be made to account for the genesis of the features described on the map. This documentation should also include some discussion of the relationship between the new map and any previously published studies in the field area. If you are unsure how to write such a report, see guidance in Parsons and Knight (1995).

2.3   LOGISTICAL PREPARATIONS FOR FIELDWORK

Most glaciological research is undertaken in remote locations that are not within easy reach of major habitations. Thus, logistical preparations need to be made well in advance of fieldwork in terms of both carrying out the research and living on site. Although it is difficult to generalize in terms of what is needed under each of these categories, research and personal, some general considerations that may be valid to both emerge. These are considered briefly in the following sections.

2.3.1   Research logistics

Equipment failure or breakage can be a major problem if it occurs during glaciological fieldwork. Clearly, any equipment that is to be used in fieldwork should be fully tried and tested before each fieldwork campaign, even if the equipment has been in careful storage and even if it functioned perfectly when it was last used. This is particularly the case for equipment involving electrical components as corrosion can occur during storage, particularly if the equipment or storage location is damp. Consideration should be given to the possibility of equipment failure and to a backup in the event of a partial or complete failure, such as taking, or ensuring ready access to, a backup instrument. Depending on the sophistication of the instrument concerned, it may be possible to prepare for carrying out field repairs on the problems most likely to arise. If possible, wiring diagrams and structural plans of key equipment should therefore be taken into the field. In any case, a small field tool kit often proves indispensable to field researchers; it is surprising how often simple wrenches and screwdrivers prove invaluable in the field. Equipment left in the field, even for very short periods, should be weatherproofed with full consideration given to possible weather conditions whilst left unattended. For example, equipment should be boxed and sealed and caches marked with flags if left unattended at a location where snow is a possibility (Figure 2.2).

Providing and ensuring sufficient, uninterrupted power to instruments can also prove a challenge under field conditions. Depending on the scale of the project and the logistical support available to it, field power supply can vary between helicopter-lifted generators supplying hundreds of watts and solar-powered chargers that can trickle-charge equipment batteries at some tenths of a watt. Between these two scales, sealed lead acid batteries or 'gel cells', trickle-charged by solar panels, are probably the most commonly used type of power supply to field equipment, although small, portable generators are now available. The advantage of solar panels is that they have no tangible fuel demand and are very robust and portable. They do, however, provide limited power and require sunlight to operate. Generators require fuel and are less portable, but are capable of providing very high power. Some generators may need carburettor adjustment to operate at high altitude, and this should be investigated before departing for fieldwork. Finally, external power supply may not be the only consideration as many items of equipment have internal 'clock' batteries that need infrequent replacement. These should also be checked and replaced as necessary prior to fieldwork. Ideally, spare internal batteries should be carried with the equipment. It is also worthwhile to remember that battery power is reduced at low temperatures (particularly alkaline batteries) and plenty of spares may be needed. For example, alkaline batteries powering digital cameras can run down at an alarming rate at cold temperatures. In this case, it may be worth considering using a camera that uses commonly available batteries or one that can be charged from a small solar panel. Power drain can be reduced on digital cameras by switching the LCD screen off during use.

Finally, it is essential to consider closely the retrieval of data from field equipment. First, ensure a robust means of retrieving data from field instruments. Where such data transfer is automated, ensure robust connections,

(a)

(b)

Figure 2.2 Equipment cached on a glacier surface: (a) hot-water drilling and associated equipment as left overnight in poor weather. Note the numerous flags used to demarcate the area of the cache; and (b) digging the same cache out following a snow storm. Note that the entire cache was buried by fresh snow and that none of the cache was visible, other than the flags, upon returning to the site. This example illustrates graphically why it can be essential to take a GPS reading of the positions of material located in the field

a consistent transfer (baud) rate and operation under field conditions. Download a dummy data set as part of the equipment testing protocol prior to departing for fieldwork. Second, ensure suitable data backup procedures and store backup data separately from original data to ensure that one set is saved in the event of a mishap that is site-specific or computer-specific.

2.3.2 Personal logistics

Despite different glaciological fieldwork projects involving different living conditions, from permanent dwellings providing professional catering and on-site laboratories (Figure 2.3a) to 'backpackable', tent-based accommodation (Figure 2.3b), some general recommendations are applicable to any field-based living.

In planning a research programme, full consideration should be given to the health and safety issues, whether they relate to carrying out the research itself or more generally living and moving about the environment concerned. In the case of deep field Antarctic research, for example, such considerations commonly form part of a formal procedure that all researchers are obliged to participate in. In such cases, researchers may attend intensive field training both prior to departure to the base station in the field and at or near to that station prior to departure to the deep field camp. Such inductions will involve testing all the clothing and equipment likely to be necessary at the field camp, as well as familiarization with field communications and communication protocols. Danger from animals may also pose a threat in certain environments.

Adequate provision should be made to avoid sickness and to cure or reduce sickness if it should arise. Evidently, researchers embarking on fieldwork should be in good health and free from potential ill health. For example, a full dental check well before departure is recommended. If travelling abroad, then the requirement for vaccinations should be investigated up to a year in advance of departure. If working at altitude, suitable provision for acclimatization (and the requirement to return to lower altitudes in the event of sickness) should be factored into fieldwork itineraries. More generally, a full first-aid kit should always be located in field camps, as should a means of communicating with the authorities in the event of an emergency. Smaller, individual kits should be carried by field teams operating away from base. Indeed, field researchers operating off base should always notify others (and agree a return time and plan of action should it lapse) and carry standard emergency items including spare, warm clothing, a survival bag and a first-aid kit. Ideally, research off base should always be carried out in company and a hand-held global positioning system should also be carried to enable relocation of the base in the event of bad weather. Additional protection may be needed in certain areas, depending on

(a)

(b)

Figure 2.3   Different styles of field base: (a) a view of Kongsvegen from the window of the rest room located above the dining room, and near to well-equipped laboratories, at Ny Ålesund base in the Arctic; and (b) a field camp under canvas following recent snow at Taylor Glacier in the Antarctic. Field conditions such as these should be factored into expectations regarding the nature and duration of any field-based glaciological research programme

local conditions. In the Arctic, for example, it is recommended (indeed required in most cases) that, off base, a rifle is carried at all times in the event of a polar bear attack.

In a related vein, personal insurance, including adequate provision for helicopter rescue, should always be taken out. The policy should be checked to ensure that it covers fieldwork activities, especially if those activities could be considered as mountaineering or skiing, which often invoke supplementary insurance premiums. Finally, if leading other researchers then considerations of legal responsibility should be investigated and clearly defined before departure. In advising students on independent fieldwork on or near glaciers, in addition to standard health and safety assessment forms (which most institutions provide), it may be helpful to issue guidance relating to rules of conduct and personal equipment requirements (Box 2.2).

Box 2.2   Guidance on rules of conduct and personal logistical provisions for unsupervised undergraduate fieldwork at an Alpine glacier

In advising undergraduate students participating in independent field-based research guidance such as that below may be provided. The following set of rules must be adhered to at all times while you are conducting fieldwork.

- Do not walk on the glacier alone. If you do walk on the glacier, ensure you are roped to another person and that you have adequate training. Be particularly wary of glacier-walking after snowfall, when narrow surface crevasses may be covered by fresh snow.
- Always let somebody else know your plans in advance and, where possible, work in pairs. Arrange, and let others know, your fall-back procedure in the event of a problem arising.
- Do not participate in dangerous activities, including rock climbing and mountaineering, during the period of your unsupervised research.
- You must bring the following personal clothing and equipment:

  (a) a strong tent, preferably with room to provide a safe shelter for cooking;
  (b) food, utensils and (individually or to share) cooking apparatus;
  (c) a sleeping bag, or combination of bags, to a minimum warmth rating of four seasons;
  (d) a sleeping mat;
  (e) a waterproof coat and over-trousers;

Box 2.2   (Continued)

  (f) a pair of sturdy boots (plastic or water-proofed leather – you
       may be working in rain or snow);
  (g) warm trousers (not jeans) and tops;
  (h) a warm hat;
  (i) good quality sunglasses;
  (j) sun cream;
  (k) several pairs of spare, thick socks;
  (l) a water bottle; and
  (m) a compass and a whistle.

● Personal insurance, including mountain rescue cover (helicopter
  rescue to £5000 recommended), must be taken out for the full
  period of research.
● Always carry warm clothing, a first-aid kit (including an emergency thermal bag) and food with you in the event of either a
  change in the weather or immobilization.

In addition to the above requirements, you may find it useful to bring
the following:

  ● a torch (head torches are very useful) and spare batteries;
  ● a novel/personal music for rainy days;
  ● a pair of light trainers or espadrilles for wearing around camp;
  ● a sun hat and lip salve; and
  ● a bottle of your favourite spirit for cold nights.

2.4   FIELDWORK DATA

2.4.1   Some golden rules

Data, whether quantitative or qualitative, are the principal reason for
conducting fieldwork, and they represent the most important end-products
of that research effort. Their importance is paramount and some golden
rules of data collection and handling may be suggested:

● Back data up, on different media if possible, and store the data sets
  separately.
● Check data in the field to ensure that items of equipment and measurement procedures are operating correctly. In addition to the general

magnitude of the values recorded, check for trends and the frequency of measurement. Ensure you can answer 'yes' to the following questions: (i) are the data values realistic? (ii) is the variable changing? (iii) is the variable changing broadly as it is anticipated to and, if not, are the observed changes feasible? (iv) is the instrument recording as often as it should? and (v) is the instrument downloading the full data set?

- Design data collection carefully so that it relates to the objectives of the study (section 2.2).
- Annotate data and collection procedures carefully in a waterproof note-book in waterproof ink or a pencil. Look after the notebook and always carry a spare pen or pencil. Back this information up whenever possible.

2.4.2   Data accuracy, precision, sensitivity and error

It is important to be aware of concepts of accuracy, precision, sensitivity and error in dealing with and in reporting quantitative data. Working definitions of each of these are presented in the following sections.

2.4.2.1   Accuracy

Accuracy is the degree to which data match their correct values. The accuracy of a data set is therefore central to its quality (and the uses it can be put to), and is dictated by the measurement and recording proced-ures used to collect the data. Inaccuracies contribute towards error (section 2.4.2.4).

2.4.2.2   Precision

Precision is the degree to which individual measurements of the same quantity agree. Thus, it represents the repeatability of a measurement, and is usually expressed as a variance or standard deviation of a set of repeated measurements of a constant quantity. Precision is therefore quite different from accuracy and it is important to appreciate that, no matter what the precision of a data set is, those data can still be inaccurate. Imprecision, like inaccuracy, contributes to overall error (section 2.4.2.4).

2.4.2.3   Sensitivity

Sensitivity is the change in the magnitude of a causal variable or an input value that is needed to produce a specified change in a response variable or an output value. Most commonly, sensitivity is expressed as the smallest recordable change in the property being measured. For example, the sensitivity of a length measurement made by a total station may be 10^{-3} m (a millimetre). This property relates closely to the resolution to which field data may be expressed. As a simple rule, no value should be expressed to a greater number of significant figures than the sensitivity at which it was measured. For example, a length measured by total station with a sensitivity of 10^{-3} m could be expressed as 123.45 m and not as 123.450 m.

It is often advisable to consider at an early stage what data sensitivity is sufficient to address the objectives of a particular study as increasing the resolution of field data can be demanding and involve a trade-off with data coverage and/or the number of samples collected.

2.4.2.4   Error

As a result of inaccuracy and imprecision, field data should normally be expressed with two components: (i) a numerical value giving the best estimate possible of the quantity being measured (in a specified system of units) and (ii) the degree of uncertainty associated with that estimated value. This degree of uncertainty is the error and is commonly expressed as a $\pm$ value. For example, the measurement of distance to a velocity prism might yield a result expressed such as 123.45 ± 0.1 m. This error term can include both systematic (sometimes referred to as determinate) errors and random (sometimes referred to as indeterminate) errors and these can themselves include instrumental error, operator error and errors arising from, for example, natural variations in the property being measured. In field investigations it is particularly effective to evaluate total error empirically by repeatedly measuring a single representative property and analysing the resulting data set. Once such a data set has been recovered, it can be used to express error in one or more of several different ways. Amongst others, these alternatives include expressions of (i) maximum error (calculated as half the difference between the maximum and the minimum values measured), (ii) probable error (calculated as the range that includes 50% of the values measured), and most effectively (iii) standard deviation (calculated as the square root of the sum of the squares of deviations from the mean value of the data set, divided by the number of observations less one) or variance (calculated as the sum of the squares of deviations from the mean value of the data set, divided by the number of observations less one; i.e. the standard deviation squared).

It is important to recognize that errors will generally increase as they are compounded by linking several stages of analysis, each with an associated error. For example, transforming two measured distances into a velocity by dividing the difference between the two distances by the time interval separating the measurements involves the compounded error of twice that of a single-distance measurement plus that associated with the time measurement. Such compounded random error calculations form a major branch of quantitative data analysis and involve procedures that are covered in some detail in specialist texts such as Taylor (1997).

3

Glacier ice: Character, sampling and analysis

3.1   AIM

The aim of this chapter is to describe the main ice types, or facies, that make up glaciers and to provide guidance on how to identify, sample and analyse ice for its physical properties. We begin by defining different types of ice mass and by summarizing the principal defining characteristics of surface, englacial and basal ice facies. We then describe field ice sampling methods, including a description of small- and medium-sized ice coring equipment, and the transport and storage of the ice samples they yield. The chapter concludes with a brief description of some of the field-based analytical techniques applied to ice.

3.2   ICE MASSES AND ICE FACIES: PRINCIPLES, DEFINITION AND IDENTIFICATION

There are many different types of ice mass present on the Earth's surface, from the giant East Antarctic Ice Sheet, which has a surface area of over 10 million km^2, to the smallest hanging glaciers, corrie glaciers and glacierets, which may be no more than some hundreds of metres across. These different names for ice masses are generally defined on the basis of the size, shape and situation of the mass concerned. As a rule of thumb, glaciers have an exposed headwall and are bounded by rock or sediment on at least three of their four

Field Techniques in Glaciology and Glacial Geomorphology   Bryn Hubbard and Neil Glasser
© 2005 John Wiley & Sons, Ltd

sides (the terminus can end in water, below). In contrast, ice sheets and ice caps have no exposed headwall and submerge the underlying topography, with the exception of isolated rock pinnacles known as nunataks. They are therefore generally unconstrained by topography and consequently have a domed shape. Ice motion is multi-directional, flowing radially out from the centre of one or more such domes. The largest ice masses, defined as having a surface area of >30 000 km^2, are termed 'ice sheets', of which the larger is the Antarctic Ice Sheet (which may be subdivided into the smaller West Antarctic Ice Sheet and the larger East Antarctic Ice Sheet) and the smaller is the Greenland Ice Sheet. Ice caps are of similar morphology to ice sheets, but smaller, having a surface area of <30 000 km^2.

Ice thins as it approaches the edges of an ice sheet, and in these locations it may be directed by local topography to form outlet glaciers or by surrounding slower-moving ice to form ice streams. Both frequently drain large areas of the interior of ice sheets and may be characterized by very rapid surface velocities. Also both usually terminate in the sea, with outlet glaciers some-times flowing along fjords. In general, any glacier that terminates in seawater may be referred to as a tidewater glacier. In contrast, where outlet glaciers terminate on dry land they may be termed 'piedmont glaciers'. The classic glacier form is the Alpine-type valley glacier, contained within discrete rock headwall and marginal boundaries. Such glaciers may also include smaller cirque glaciers that are contained entirely within small armchair-shaped hollows. Finally, hanging glaciers may form on rock faces as steep banks of ice, frequently supplying ice by avalanche to valley glaciers located beneath them. In some areas, very small valley glaciers are termed 'glacierets'.

Although this morphological classification of ice masses provides us with a useful descriptive vocabulary, the distinctions are rooted in form rather than in process. A more fundamental classification may therefore be made on the basis of the thermal properties of the ice mass concerned, with a fundamental distinction being made between ice that is at the melting point (termed 'temperate' or 'warm') and ice that is below the melting point (termed 'cold'). Three major ice-mass classifications have been derived on the basis of this distinction: (i) high polar ice masses are characterized by perennially and ubiquitously cold ice; (ii) temperate ice masses are charac-terized by ice that becomes ubiquitously warm for at least part of the year; and (iii) sub-polar or polythermal ice masses fall between these two extremes, experiencing melting at some time of, or throughout, the year, which does not affect the entire ice mass.

Temperature is not the only physical property that can be used to define and to classify different types of ice. Indeed, virtually all of the physical properties of ice vary within and between the Earth's ice masses. For example, even to the untrained eye, snow is visibly different from firn, which is, in turn, different from clean glacier ice, and all are very different from the debris-rich ice that is located near the bases of glaciers. Further, upon closer examination, other general ice types emerge, many of which can be split into sub-types defined on the basis of a recurrent suite of physical similarities that distinguish each type from all other types. This distinction allows individual ice facies and sub-facies to be interpreted under the assumption that the physical differences that make each facies unique indicate a correspondingly unique combination of processes influencing the origin and history of that ice. A variety of ice facies have been defined on the basis of studies carried out at glaciers of all thermal regimes. Although this does not provide an exclusive classification, ice facies are probably best summarized according to their location at an ice mass, yielding surface ice facies, englacial ice facies and basal ice facies.

3.2.1   Surface ice facies

The surface of ice masses can be divided systematically into elevation zones principally on the basis of their thermal characteristics. Five such zones are commonly identified, each of which is associated with a distinctive snow or ice facies (Figure 3.1).

3.2.1.1   Dry snow facies

At the highest elevations and in very cold environments, the snow that nourishes ice masses experiences no melting at all, irrespective of season.

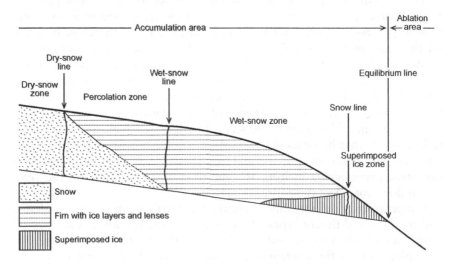

Figure 3.1 Thermal zonation of surface ice facies from the head of a glacier to the top of the ablation area (Based on Paterson, 1994; with permission)

This snow is dry and powdery, and firnifies slowly by being compressed by newly deposited, overlying snow. This process occurs in the absence of meltwater over depths of several tens of metres.

3.2.1.2   Percolation facies

The percolation facies is located immediately below the dry snow zone, separated from it by the dry snow line. The percolation zone is characterized by some surface melting in the summer, but not enough to raise the temperature of the snow to 0 °C. Thus, all meltwater generated refreezes again within the snow pack. The facies can therefore be identified by its composition of snow that can contain some liquid water during the summer and ice lenses formed from its refreezing. As a rough guide, Benson (1961) identified the lower extent of the dry snow zone, the wet snow line, in Greenland as corresponding to the −5 °C summer isotherm.

3.2.1.3   Wet snow facies

The wet snow facies is characterized by sufficient surface energy input and surface melting to raise its temperature throughout to the melting point by the end of the summer. However, not enough melting occurs to remove the entire snow pack. The wet snow facies can therefore be identified by its well-advanced metamorphism, the effects of which include the rounding, compaction and clustering of individual crystals. The lower boundary of the wet snow zone, which is the point at which no snow survives the melt season, is termed the 'firn' or 'snow line'.

3.2.1.4   Superimposed ice facies

Although none of the previous season's snowfall survives *in place* below the snow line, it is possible for mass to survive in the form of refrozen meltwater. In this case, meltwater from the wet snow zone flows down-glacier and survives the following summer having refrozen on the ice surface. This process is particularly important at Arctic ice masses and the Greenland Ice Sheet in particular.

3.2.1.5   Glacier ice facies

The area below the snow line (or the superimposed ice zone if one exists) is characterized by local net mass loss. Thus, all mass here is composed of ice

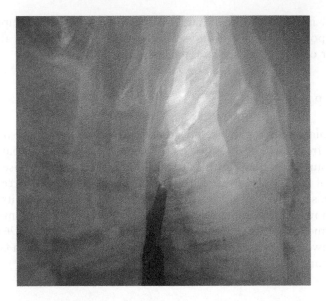

Figure 3.2 Bubble-foliated and debris-free glacier ice exposed in a crevasse

that has formed, and flowed from, up-glacier. This glacier ice can be distinguished from firn by the fact that all of its air passageways have been sealed off to form isolated bubbles. This process places a fairly well-defined lower boundary on the density of glacier ice at $\sim$830 kg m^{-3}. This represents a lower boundary because the density of ice can rise to $\sim$910 kg m^{-3} at depth as included gas bubbles are compressed under pressure. Glacier ice is not therefore of a uniformly opaque, white colour. Instead, it can often be identified by its bubble foliation, or stratification, at a scale of centimetres to decimetres, with bubble-rich layers formed during the cold or dry season separating bubble-poor layers of clearer ice formed during the wet or warm season (Figure 3.2). This facies commonly contains very low concentrations ($\sim$0.1 kg m^{-3}) of well-sorted, fine debris that is probably of aeolian origin.

3.2.2   Identifying surface ice facies

All of the surface ice facies summarized in section 3.2.1 can be identified by close inspection on the ground. However, most commonly these facies need to be identified at the relatively large, ice-mass scale at which remote-sensing techniques are well suited. Radar is particularly useful in surface facies discrimination as radar wave scattering and propagation are sensitive to liquid water and grain size (section 6.2). For example, Hall *et al.* (1995) associated specific scattering characteristics with individual facies as follows:

(a) *Dry-snow facies*: low backscatter because of the small grain size of snow.
(b) *Percolation facies*: high backscatter because of the presence of ice lenses and pipes.
(c) *Wet-snow facies*: Variable backscatter because of the variability of snow wetness.
(d) *Superimposed ice zone*: often not detectable in radar imagery particularly after a heavy ablation season.
(e) *Glacier ice facies*: temporally variable backscatter, depending on whether the glacier ice is bare (fully exposed) or partially masked by snowfall.

Potential sources of error in mapping glacier facies from satellite images include discriminating ice from snow, discriminating ice from marginal water bodies, and the identification of debris-covered ice (Williams *et al.*, 1991; Sidjak and Wheate, 1999).

In detail, the grain size and the liquid water content of snow and ice can be detected by optical, microwave and radar sensors. Techniques such as spectral classification and band-ratioing may be used to differentiate between ice and snow types and between glacierized and non-glacierized terrain, replacing traditional mapping techniques such as manual digitization. Li (Li *et al.*, 1998), for example, successfully combined both manual digitization and a supervised classification to delineate ice and snow from the surrounding terrain. Band-ratioing exploits trends between spectral responses from different wavelengths to differentiate between surface types. Landsat TM is particularly appropriate for band-ratioing because the positions of the imaging bands are sensitive to fluctuations in water, ice and vegetation characteristics (e.g. Huggel and Kääb, 2002). Bronge and Bronge (1999) used these features to distinguish between blue ice and snow facies (TM Band 3/TM Band 4) and to investigate variations in snow grain size (TM Band 3/TM Band 5). More complex ratios include the normalized difference water index ($(TM4 - TM1)/(TM4 + TM1)$), which Huggel and Kääb (2002) used for the accurate delineation of water bodies in the Swiss Alps.

3.2.3   Englacial ice facies

Once glacier ice is incorporated into the body of an ice mass its physical properties can change further in response to the temperature and stress conditions imposed on it. In order to distinguish this ice, formed by firnification near the glacier surface and not influenced by direct contact with the glacier bed, from basal ice (section 3.2.4), it may usefully be termed 'englacial ice' (Lawson, 1979). Lawson (1979) further distinguished between an englacial diffused facies and an englacial banded facies. The former is standard, firnified ice (section 3.2.1), forming the bulk of the glacier, and the latter contains coarse angular debris derived supraglacially from the valley sides.

Further, various researchers have noted from ice cores (Vallon *et al.*, 1976) and direct observations beneath glaciers (Theakstone, 1966; Hubbard and Sharp, 1995) that bubbles may have been excluded from ice located towards the beds of glaciers. As a result of these observations Hubbard *et al.* (2000) and Tison and Hubbard (2000) divided englacial diffused facies at temperate Tsanfleuron Glacier, Switzerland, into two ice zones. Here, the Upper Zone is composed of standard englacial diffused ice, and the Lower Zone is composed of highly metamorphosed englacial ice from which practically all bubbles have been removed. The latter facies is considered to form by the metamorphism of pre-existing Upper Zone ice as it experiences enhanced strains during deformation over the glacier's rough bedrock substrate. This facies extends for 10–20 m above the glacier bed, and is the same as clear facies basal ice (section 3.2.4), previously identified at the glacier bed.

3.2.4   Basal ice facies

Lawson (1979) was the first to define and present a systematic classification of ice facies formed at or near the glacier bed, the so-called basal ice facies. This classification (Table 3.1) was based on field studies at polythermal Matanuska Glacier, Alaska. Although subsequent studies have built on Lawson's scheme, it has frequently been necessary to refine, extend or amend it in the light of observations at other glaciers. Consequently, a degree of overlap has arisen between different schemes and no universally applicable basal ice classification scheme currently exists. In the following sections, we summarize Lawson's original scheme. Knight and Hubbard (1999) provide a summary of the various other schemes that have been published (Tables 3.2 and 3.3).

Table 3.1 Lawson's (1979) basal ice facies classification identified from research at Matanuska Glacier, Alaska

Zone	Facies	Sub-facies	Postulated origin
Supraglacial			Surface processes
Englacial	Diffused		Surface firnification
	Banded		Surface firnification with inclusion of valley-side debris
Basal	Dispersed		Surface firnification and basal refreezing
	Stratified	Discontinuous	Basal refreezing
		Suspended	Basal refreezing
		Solid	Basal refreezing

Table 3.2 Hubbard and Sharp's (1995) basal ice facies classification identified from research at numerous Alpine glaciers

Zone	Facies	Sub-facies	Postulated origin
Basal	Clear		Stress-induced metamorphism of englacial ice near the glacier bed
	Laminated		Regelation (Weertman, 1957, 1964)
	Interfacial	Layered	Freezing of flowing basal meltwater
		Continuous	Freezing of static basal meltwater
	Dispersed		Incorporation and transport of interfacial facies
	Solid		Freezing of water-saturated sediment
	Stratified		Any basal facies tectonically thickened and interstratified with Englacial facies ice
	Planar		Closure of ice-marginal fracture planes incorporating aeolian dust

Table 3.3 Knight's (1987) basal ice classification identified from research at Russell Glacier, Greenland

Zone	Family	Sub-family	Postulated origin
Basal	Clotted		Basal entrainment in the interior of ice sheet
	Banded	Clotted ice between debris bands	Tectonic intercalation of Clotted and Solid families
		Debris bands	Basal entrainment (Solid family) and folding/thrusting
	Solid	Stratified	Regelation (Weertman, 1957, 1964)
		Frozen till	Adfreezing and overriding of frozen ice-marginal sediments
		Old snow	Overriding of old ice-marginal snow

3.2.4.1 Basal dispersed facies

According to Lawson, the basal dispersed facies is mainly characterized by the inclusion of a uniform distribution of debris of clay- to pebble-size. The debris is present in concentrations from 0.04 to 8% by volume (with a mean of 3.8%), and the overall facies ranged in thickness from 0.2 to 8.0 m along the terminus of Matanuska Glacier.

3.2.4.2 Basal stratified facies

This facies is composed of bands of debris-rich and clean ice of variable thickness and lateral extent. Lawson split the facies into three sub-facies:

1. *Basal stratified suspended sub-facies*: This sub-facies contains randomly orientated, suspended particles and aggregates ranging from 0.02 to 55% by volume and 'includes the relatively pure ice that occurs throughout the stratified facies' (Lawson, 1979; p. 11).
2. *Basal stratified solid sub-facies*: This sub-facies is composed of well-defined layers of sediment-rich ice (>50% by volume), commonly characterized by the preservation of internal structures.
3. *Basal stratified discontinuous sub-facies*: This sub-facies contains irregular aggregates of fine-grained debris. The vertical thickness of the sub-facies was observed to vary laterally between 0.05 and 2.0 m, and separate units overlapping to form more continuous debris-rich zones.

3.3   SAMPLING GLACIER ICE

Although ice is a surprisingly strong material to the uninitiated, there are numerous ways of sampling it. In the following sections, we summarize the most common of these methods in order of increasing sophistication.

3.3.1   Axe, hammer and chisel

In the absence of power tools, ice samples can be recovered through the use of a hand axe or hammer and chisel (ideally a lump hammer and a bolster-type chisel). A 40- or 50-cm-long climbing or technical axe provides a very suitable hammer. The best method for removing an ice block using these types of tools is to select a suitable location, ideally one that is already characterized by a surface convexity, and to cut progressively around the required sample block by chipping a channel of ice away from it. Once most of the block has been separated from the main body of the glacier, it will be possible to fracture the remaining contact through a sharp blow from a bolster or log-splitting wedge. Care should be taken not to fracture the ice block itself, particularly during the latter stages, and it may be necessary to chip away more of the boundary of the block than initially anticipated.

Ice located close to the glacier surface has commonly been influenced by a high degree of surface and internal melting. This creates a layer, typically some centimetre to tens of centimetre thick, of rotted ice whose physical properties have been altered by surface melting. These effects typically include recrystallization, and the presence of influent surface debris and influent surface meltwater. The layer is readily identified in the field by its poor coherence, allowing the crystals to fall apart easily. This layer of rotten ice should be removed in the field, prior to storage. This can be achieved by chipping it away to the depth at which the ice becomes structurally sound.

Once separated, the ice block should be isolated from operator-borne and atmospheric contamination by wrapping it in cling film or sealing it in specialist plastic bags, or both. Protection from melting can be achieved through the use of freezers in the field, through storage in cold snow, or, for a period of some hours, through packaging in insulated 'cold boxes'. If the latter option is chosen, the block should also be encased in ice chips within the box, preferably along with endothermal cold packs. These packs contain separated chemicals such as urea and water which are mixed in the field, reducing the pack's temperature to several degrees below 0 °C for several hours. In our experience, a 30-cm^3 ice block that is well packed in ice-cold ice chips (in this case recovered from a sub-freezing basal cavity) can be stored safely in a cold box for ~6 hours. With the addition of endothermal cold bags, this can probably be extended to >12 hours, although factors such as external air temperature and radiative heating of the box should also be considered.

3.3.2 Ice screw

In instances where relatively small ice samples are required, such as for stable isotope analysis, ice can readily be sampled with a threaded ice screw, commonly available off the shelf. It is better not to use hammer-in screws for this purpose as the ice can be more difficult to remove from the screw after sampling. The length of screw is chosen in the light of the quantity of sample required. An ice screw of internal radius (r) and length (L) yields a maximum volume (V) of:

$$V = \pi r^2 L \tag{3.1}$$

Thus, with a typical internal radius of ~8 mm, an ice screw approaching a length of 200 mm would be required to obtain a 30-ml sample. Sampling involves inserting the screw by rotating it into the ice to be sampled. As the screw rotates and advances, its teeth fracture the ice into fine shards which pass through the screw and exit its outer end as the screw advances. These shards can be gathered directly into a sample bottle placed over the end of the screw (Figure 3.3). For this reason, a bottle with an opening wider than the outside diameter of the ice screw (~20 mm) is preferable, since this allows the bottle to enclose the end of the screw, minimizing loss of material and risk of sample contamination. Once the ice screw has been fully inserted, it should be removed from the ice, warmed by hand if necessary, and gently agitated to dislodge the remaining shards into the sample bottle. It may also be helpful to compact the shards contained within the sample bottle by heating it by hand and occasionally replacing its top and shaking it vigorously. Sample storage for isotopic analysis is described in section 4.4.3.

Figure 3.3   Ice sampling for isotopic analysis using an ice screw. Note the ice fragments emerging from the top of the ice screw as its base is screwed into the face of the glacier

3.3.3   Chainsaw

Ice can be cut extremely effectively with a chainsaw. However, at the outset it must be stated that a chainsaw wound (approaching a centimetre in width) can be extremely dangerous, particularly if it is inflicted some distance from medical facilities. When using chainsaws, therefore, great care should be taken to ensure that the working environment is stable and safe, that the chainsaw is in good working order, and that the operator is professionally trained and well-equipped. Safety equipment should include hard hat, visor, ear plugs, protective gloves, protective trousers ('chaps') and protective boots. The operator should always be accompanied by a companion who is not involved directly in the chainsawing, and who is equipped with a medical kit and an emergency radio. Rapid transport to adequate medical facilities should also be arranged in case of need. Last but not least, the operator should be professionally certified. *Chainsawing of ice should not be attempted unless all of these conditions are met.*

Debris-free ice can be cut rapidly and precisely by chainsaw, making this an extremely effective method of sampling. The straightness and precision of a chainsaw cut can be used to great advantage, yielding blocks of pre-determined size. Once collected, such blocks should be treated similar to those retrieved by hand-cutting (section 3.3.1).

In addition to retrieving ice samples, chainsaws have also been used to cut tunnels into glaciers, particularly where access to the glacier bed, some distance from the margin, is required (Box 3.1). In such cases, the chainsaw is used in a similar fashion to a snow shovel in digging a snow pit (section 7.3.2.3).

Box 3.1 Tunnelling into an ice mass by chainsaw: Suess Glacier, Antarctica

With the correct training and equipment it is possible to use a chainsaw to tunnel some distance beneath glaciers. For some years, Sean Fitzsimons has tunnelled beneath various glaciers located in the

Figure B3.1 The entrance to Fitzsimons' tunnel (centre bottom; researcher for scale) through the ice-marginal apron deposits at Suess Glacier, Antarctica

Box 3.1 (Continued)

Dry Valleys area of Antarctica. For example, Fitzsimons *et al.* (1999) tunnelled a distance of ~20 m beneath Suess Glacier, a cold-based outlet glacier of the East Antarctic Ice Sheet, much of it through debris-rich basal ice using tungsten-toothed chains. Access to this artificial cavity at a temperature of ~−17 °C allowed the researchers to investigate directly the properties of the basal ice and the ice–bed interface. Here, *in situ* shear tests on the materials present indicated that the frozen sediment was almost twice as strong as the glacier ice and sediment-rich basal ice present. However, localized weaknesses within the frozen sediments were also identified. These included the presence of relatively pure ice lenses and zones of frozen sediment characterized by air-filled, rather than ice-filled, pore spaces. Thus, the authors concluded that the frozen subglacial layer at Suess Glacier was unlikely to deform pervasively, but that it was possible for that layer to contribute to motion through localized failure.

Fitzsimons, S.J., McManus, K.J. and Lorrain, R.D. 1999. Structure and strength of basal ice and substrate of a dry-based glacier: Evidence for substrate deformation at sub-freezing temperatures. *Annals of Glaciology*, **28**, 236–240.

(a)

Figure 3.4 Cutting ice using a chainsaw: (a) removing a large ice block from a tunnel cut by chainsaw in the frontal margin of Tsanfleuron Glacier, Switzerland; (b) cutting debris-rich basal ice using a tungsten chain, Taylor Glacier, Antarctica

(b)

Figure 3.4   (Continued)

Thus, the saw is used to cut a grid, whose depth is given by the length of the chainsaw bar, into the face to be cut back. Once the grid has been cut, individual blocks are removed by fracturing their attached faces by wedging them open with a lump hammer and wedge or bolster chisel. It may be easier to begin by dislodging some of the outermost blocks first, using the solid outer edge of the opening against the wedge. The optimum size of the blocks varies with the physical setting and the nature of the ice, but blocks of dimension 30–40 cm probably represent an upper limit of manageability (Figure 3.4).

One further consideration is that of the effect of debris-rich ice on standard (steel) chainsaw teeth. If debris-rich ice is present, as is common near the glacier bed, then steel tips can be blunted very rapidly – often resulting in a marked slowdown as a consequence of a single cut through debris-rich ice. Where such a situation can be anticipated, it is preferable to use chains with tungsten carbide tips, which provide a great deal of, but not complete, resistance to wear. They are also markedly more expensive than standard chains. A practical solution may be to have a supply of standard chains and a smaller number of tungsten-tipped chains that can be used only when debris-laden ice is encountered. It is also possible to sharpen standard chain tips manually with a readily available chain file.

3.3.4 Ice cores

Large-scale ice coring is a specialist activity that requires sophisticated equipment, experienced personnel and major logistical support. However, not all ice coring is large-scale, and smaller-scale coring activities are accessible to groups of two or three researchers with only limited logistical support. In the following sections, we outline the principal coring methods adopted by glaciologists. We first consider logistically light, shallow coring and then present a brief summary of the techniques involved in logistically more demanding, intermediate and deep coring.

3.3.4.1 Shallow ice coring (1–100 m)

Most shallow ice coring has been carried out using corers based on an original SIPRE (Snow, Ice and Permafrost Research Establishment (now CRREL – Cold Regions Research and Engineering Laboratory)) design (Box 3.2). These are the most basic of the mechanical drills comprising a barrel and a cutting head. The barrels of these devices are typically ∼1-m long, and have internal diameters of 3 or 4 inches. A protruding thread is located on the outside of the barrel, which serves the purpose of removing ice chips from the cutting face. The cutting head includes 2 or 3 hardened steel or tungsten carbide teeth, the protrusion of which can be adjusted to govern the depth of cut associated with each pass. In practice, this depth needs to be carefully optimized to field conditions, particularly the physical properties of the ice concerned. If this depth is set too small, then the head will skid over the ice surface rather than cut down into it. If the depth is set too large, then the teeth will become lodged in the ice, halting the progress of the drill altogether. The cutting depth is normally set in the 3–6 mm range.

Box 3.2 Shallow ice coring: Tsanfleuron Glacier, Switzerland

Tison and Hubbard (2000) used an adapted Rand corer (a variation of the SIPRE corer) to retrieve a series of eight shallow-depth ice cores from along a flow line of Tsanfleuron Glacier, Switzerland. The longest of these cores was 45 m and, of the eight cores recovered, the five located nearest the glacier's frontal margin reached its bed. The remaining three terminated englacially in the glacier's accumulation area. The authors analysed these cores for their crystallographic characteristics and identified four distinct units. Unit 1 was generally fine grained and found within ~20 m of the ice surface in the accumulation area of the glacier. The unit was composed of ice with a random fabric and its crystal size increased with time and temperature (and therefore depth) according to an Arrhenius-type function. Unit 2, characterized by the occasional presence of coarser crystals, was found beneath Unit 1 at depths equivalent to greater than ~150 years of Arrhenius growth. Where sampled in the ablation area of the glacier, the coarser crystals had developed a multiple maximum fabric, typical of high cumulative strain,

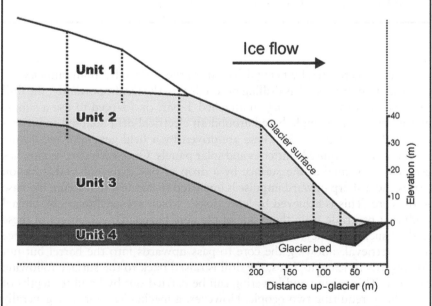

Figure B3.2 Ice crystallographic units reconstructed from eight short ice cores (dashed vertical lines) at Tsanfleuron Glacier, Switzerland

Box 3.2   (Continued)

while the smaller crystals still displayed an unchanged random fabric, implying that it had undergone little strain-induced metamorphism. Unit 3 was characterized by an abrupt increase in minimum crystal size, and occurred at a depth of ~30 m in both the accumulation area and the ablation area of the glacier. In the accumulation area, this increase coincided with the first evidence of the development of a systematic fabric, indicating the onset of dominant dynamic recrystallization. Unit 4, characterized by an increase in mean grain size, a decrease in grain-size dispersion, and an imbricated crystal structure was located within ~10 m of the glacier bed. The unit was also characterized by a strong multi-modal, girdle fabric. The minimum crystal size of this unit was consistent with a steady-state balance between Arrhenius-type growth and strain-related grain-size reduction. These changes were interpreted by the authors to result from intense, continuous deformation in the basal zone.

Tison, J.-L. and Hubbard, B. 2000. Ice crystallographic evolution at a temperate glacier: Glacier de Tsanfleuron, Switzerland. In A.J. Maltman, B. Hubbard and M.J. Hambrey (eds) *Deformation of Glacial Materials*, 23–38.

The corer is rotated by being driven from the surface by aluminium extension rods that are added as drilling proceeds into the ice. Importantly, the drill can be driven by hand, using an attached T-bar, or adapted to use a simple motor drive, for example based around an electrical drill adapted to incorporate a T-bar (Figure 3.5). These are driven by a field generator or, if only short cores are required, batteries and solar panels. Once each core section has been cut, indicated at the surface by a drop of one extension rod, rotation ceases and a sharp upward impulse is imparted to the drill to fracture the base of the core. This is achieved by 'core dogs', which are asymmetrical, hinged teeth that protrude from the inside of the core barrel. The geometry of these teeth is such that they serve both to break the core and to hold it in place during retrieval, allowing the core to pass upwards into the barrel but not downwards to leave it. Thus, the drill is raised back to the surface following each cut. This raising and lowering can be carried out by hand to depths of ~20–30 m, requiring two people. However, a mechanical winch is generally required at greater depths, typically based on a pulley and tripod system (Figure 3.6). The depth limit for this type of system is limited by the strength and flexibility of the extension rods to ~50 m.

Figure 3.5 SIPRE/Rand corer driven by a generator-powered drill with in-house-constructed T-bar

3.3.4.2 Intermediate and deep ice coring (thousands of metres)

Due to the limitations of rigid extension rods, intermediate and deep coring projects usually use drills with integral electromechanical drive motors which are suspended from the surface, and powered, by cable. Due to the cable suspension these drills have to compensate locally for the torque generated by the cutting action. This is accomplished by leaf springs attached to the upper

Figure 3.6 Generator-powered electrical pulley system being used to retrieve extension rods attached to an ice corer, Tsanfleuron Glacier, Switzerland

end of the drill barrel or sonde that protrude into the wall of the core hole (Reeh, 1984). The cutting heads on these systems are generally similar to those of the SIPRE corer (section 3.3.4.1), although barrel diameter and length, and number of cutting teeth, differ from system to system.

3.3.4.3 The intermediate-depth ECLIPSE corer

Intermediate-depth corers tend to be designed and manufactured in-house, often with specific coring projects in mind (Box 3.3). However, one commercially available intermediate-depth corer is the ECLIPSE drill, supplied by Icefield Instruments Incorporated, Canada. Named after its test site at Eclipse Dome, St Elias Range, Yukon, the system is based on several

---

Box 3.3   Intermediate-depth ice coring: High-altitude tropical glaciers

Lonnie Thompson, of the Byrd Polar Research Center, Ohio State University, USA, and co-workers have, over three decades, drilled intermediate-length cores through many of the Earth's highest and most remote ice masses. They have done this in order to reconstruct the climatic history of the areas concerned by analysing the physical and chemical properties of these ice cores. Many of these projects have been undertaken at very high altitude in remote areas, precluding transport of equipment and ice by helicopter. Thus, the drills used have been constructed to be dismantled, transported as lightweight components and re-assembled on site. Power for these drills is also provided by banks of solar panels rather than by fuel-driven generators.

An extreme example of this research is the retrieval of three cores from an altitude of 7000 m through the Dasuopu Glacier in the Himalayas (Thompson *et al.*, 2000). Analysis of these cores revealed a high-resolution record of climate change over the past 1000 years, much of which was driven by variations in the intensity of the south Asia monsoon. For example, dust and chloride concentrations in the cores indicated the presence of eight major droughts, caused by a failure of the monsoon. The worst of these was between 1790 and 1796, which historicalrecords indicate resulted in hundreds of thousands of deaths in the region. The cores also indicated major recent change, including a fourfold increase in dust and a doubling of chloride concentrations in the past century, indicating a general trend towards drying and desertification in the region. These interpretations are supported by the $\delta^{18}O$ composition of the ice, which indicates that both the last decade and the last 50 years were the warmest over the 1000-year-long record. These warming trends, interpreted as anthropogenic, have now been identified in numerous cores recovered from throughout the high-altitude tropics (Thompson *et al.*, 2003).

Box 3.3   (Continued)

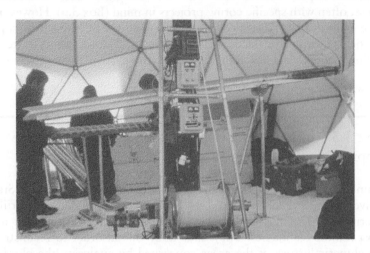

Figure B3.3   The lightweight, portable drilling system, designed, developed and tested at the Byrd Polar Research Center by Victor Zagorodnov, capable of coring to depths of ~700 m and used by Thompson and his team at Mount Kilimanjaro, Tanzania (Thompson *et al.*, 2002). The drill was powered using a combination of solar and diesel generators and can be switched from an electro-mechanical drill (used in cold, dry ice to a depth of ~180 m at Kilimanjaro) to a thermal-alcohol electric drill (used in deeper temperate ice). This image (supplied by Lonnie Thompson) shows the winch and cable, the controller units, the outer barrel resting horizontally in the tilting tray, and the inner barrel (background) being prepared for the next ice core recovery run

Thompson, L.G., Mosley-Thompson, E., Davis, M.E., Lin, P.N., Henderson, K. and Mashiotta, T.A. 2003. Tropical glacier and ice core evidence of climate change on annual to millennial time scales. *Climatic Change*, 59(1–2), 137–155.

Thompson, L.G., Yao, T., Mosley-Thompson, E., Davis, M.E., Henderson, K.A. and Lin, P.N. 2000. A high-resolution millennial record of the South Asian Monsoon from Himalayan ice cores. *Science*, 289(5486), 1916–1919.

Thompson, L.G., Mosley-Thompson, E., Davis, M.E., Henderson, K.A., Brecher, H.H., Zagorodnov, V.S., Mashiotta, T.A., Lin, P.-N., Mikhalenko, V.N., Hardy, D.R. and Beer, J. 2002. Kilimanjaro ice core records: Evidence of Holocene climate change in tropical Africa. *Science*, 298(5593), 589–593.

generations of Danish and Canadian tipping-tower designs (Johnsen *et al.*, 1980). The ECLIPSE has been developed specifically for use in remote areas, requiring that it should be hand-portable with low power consumption. Thus, the total mass of the drill is less than 200 kg, and its heaviest dismantled component (the winch drum and cable) is ~30 kg. Drilling 82-mm-diameter core (leaving a 102-mm-diameter hole) at ~75 rpm requires a maximum power of ~400 W, which is greatly increased by rapid upwinching (~40 cm s^{-1}). Power is supplied by a 24-V DC battery bank charged by solar panels or a field generator. The system has successfully drilled to a depth of 345 m.

The construction and components of the ECLIPSE drill are described by Blake *et al.* (1998) and available from Icefield Instruments Inc. (2002), which we follow closely here. The drill system has three main components: control module, drill sonde and winch. The sonde comprises the motor and anti-torque section, an inner barrel and an outer barrel. The cutter head is fitted to the lower end of the inner barrel which is itself fitted with external flights. As the inner barrel and cutter head rotate, chips are carried up the flights between the two barrels to the top of the core. The anti-torque section uses three leaf springs. The cable connection includes a hammer, thrust bearing and slip-ring to ensure the cable does not twist should the anti-torque springs lose their grip. Once raised to the surface (at a typical rate of 0.35 m s^{-1} by 24-V DC motor), the core is removed by tipping the winch tower to an approximately horizontal position (it dips slightly towards the cutting end to facilitate core removal), removing three locking pins, and extracting the inner barrel from the sonde. The core and ice chips are then removed from the top of the inner barrel. The 24-V DC winch is located on an aluminium frame consisting of an open triangular base supporting a square-faced frame (Figure 3.7).

Although most intermediate-depth drills are generally similar in design to the ECLIPSE drill, some employ electro-thermal cutting methods. Such drills use a heating element to melt an annulus around the core, and are particularly effective at coring through warmer ice (approximately above $-10\,^{\circ}$C).

3.3.4.4 Deep coring

Over the past 30 years, numerous deep drilling programmes have been (and continue to be) undertaken with the purpose of reconstructing the Earth's environmental and climatic record up to the scale of 10^5 years. Long-term environmental records are most successfully read from cores whose ice was deposited steadily through time and which have experienced minimal horizontal flow. These deep cores are therefore generally recovered from the central accumulation domes of the Earth's largest ice masses. Technological advances, particularly in mechanical drilling, has allowed progressively longer cores to be retrieved, with recently recovered cores reaching lengths of greater than 3000 m in the interior of both the Greenland and the East Antarctic Ice Sheets. The longest core is currently that recovered from

Figure 3.7   The medium-depth EPICA ice corer

Vostok Station, East Antarctica, which is 3623-m long, and which stops
~150 m above the base of the ice sheet (which itself contacts a large freshwater lake, Lake Vostok) (Box 3.4).

These deep coring programmes are logistically demanding, requiring
large groups of scientists, frequent airborne support and employing teams
of commercial drillers who work shifts around the clock. Each core length
winched to the surface is several metres long and typically of 100–140-mm
diameter. The core holes are filled with non-contaminating fluid that is of a
similar density to ice, such as *butyl acetate*, to prevent closure at depth. Due
to their large size and sophistication, there are only a small number of deep
drilling rigs worldwide, each of which includes a drilling tower of 15-m-plus
height, all, or at least the base, of which is located within a protective dome
that also houses sub-surface laboratory space for on-site core sampling and
analysis. These drills are individually designed and constructed for specific

Box 3.4   Deep ice coring: Vostok Station, Antarctica

The world's longest ice core (3623 m) was recovered as a result of Russian, American and French collaboration from Vostok Station in East Antarctica. The core is of major scientific importance for two reasons. First, the core presents a long-term record of the Earth's glacial-interglacial history. Spanning over 400 ka, the core's $\delta^{18}O$ has provided high-resolution climate information over four complete glacial-interglacial cycles (Petit *et al.*, 1999). Second, the base of the core is located ~120 m directly above a major freshwater lake of about the size of Lake Ontario, known as Lake Vostok (Kapitsa *et al.*, 1996). Analysis of the base of the Vostok core indicates that the ice located below ~3583 m is composed of refrozen lake ice (Souchez *et al.*, 2000), providing a tantalising insight into the possible chemical and biological composition of the underlying lake water, which may not have had contact with the Earth's atmosphere or oceans for millions of years. Jouzel *et al.* (1999), for example, reported the presence of colonies of microbes trapped within the frozen lake ice, indicating the presence of living forms within the lake. Since then, many researchers have combined these ice core data with geophysical and modelling studies

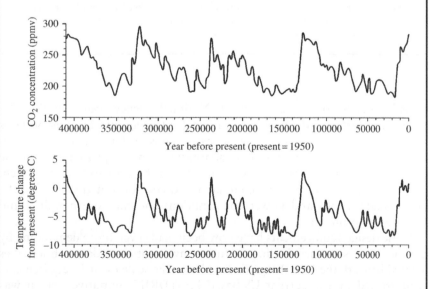

Figure B3.4   Temperature and atmospheric CO_2 record spanning four complete glacial-interglacial cycles reconstructed from the $\delta^{18}O$ composition of the Vostok ice core (Reproduced from Petit *et al.*, 1999 with the permission of *Nature*)

Box 3.4 (Continued)

to reconstruct the physical and biological processes occurring within Lake Vostok before the lake is finally accessed directly (e.g. Siegert *et al.*, 2003).

Jouzel, J., Petit, J.R., Souchez, R., Barkov, N.I., Lipenkov, V.Y., Raynaud, D., Stievenard, M., Vassiliev, N.I., Verbeke, V. and Vimeux, F. 1999. More than 200 meters of lake ice above subglacial lake Vostok, Antarctica. *Science*, **286**(5447), 2138–2141.

Kapitsa, A.P., Ridley, J.K., Robin, G.D., Siegert, M.J. and Zotikov, I.A. 1996. A large deep freshwater lake beneath the ice of central East Antarctica. *Nature*, **381**(6584), 684–686.

Petit, J.R., Jouzel, J., Raynaud, D., Barkov, N.I., Barnola, J.M., Basile, I., Bender, M., Chappellaz, J., Davis, M., Delaygue, G., Delmotte, M., Kotlyakov, V.M., Legrand, M., Lipenkov, V.Y., Lorius, C., Pepin, L., Ritz, C., Saltzman, E. and Stievenard, M. 1999. Climate and atmospheric history of the past 420,000 years from the Vostok ice core, Antarctica. *Nature*, **399**(6735), 429–436.

Siegert, M.J., Tranter, M., Ellis-Evans, J.C., Priscu, J.C. and Lyons, W.B. 2003. The hydrochemistry of lake Vostok and the potential for life in Antarctic subglacial lakes. *Hydrological Processes*, **17**(4), 795–814.

Souchez, R., Petit, J.R., Tison, J.L., Jouzel, J. and Verbeke, V. 2000. Ice formation in subglacial lake Vostok, Central Antarctica. *Earth and Planetary Science Letters*, **181**(4), 529–538.

---

tasks by organizations such as the US National Science Foundation's *Polar Ice Coring Office* (PICO), now renamed *Ice Coring and Drilling Services* and based at the University of Wisconsin-Madison. Each new design generation therefore involves several adaptations and improvements to the components of previous designs – not only to the drill sonde and winch, but also to, for example, the command and control unit (monitoring the drill's angle, depth and power consumption) and the core handling and preparation protocol.

These design requirements have now been made even more demanding by the complex issues surrounding the future investigation of deep sub-ice lakes located beneath the East Antarctic Ice Sheet. The scale of such requirements is illustrated by the current US-based FASTDRILL initiative, which was established to provide a framework to direct future deep drilling provision. In 2002, over 50 delegates from numerous international institutions and scientific disciplines attended the first FASTDRILL workshop, held at University of California at Santa Cruz.

3.4   ICE ANALYSIS

Once recovered, ice samples and cores may be analysed for their physical and chemical characteristics. These can be used to provide process-related information, for example, in correlating ice character with measures of ice softness to calibrate models of ice-mass flow. However these properties are most commonly used to reconstruct palaeoclimatic and palaeoenvironmental information. To obtain such information from ice cores, researchers first have to obtain a relationship between ice depth and age.

3.4.1   Ice core dating

If cores are recovered from the absolute spreading centres of a radial ice mass, then the age of the ice is a predictable function of depth and accumulation rate. Thus, if accumulation rate can be assumed or approximated, then straightforward densification models can be used to predict the age–depth relationship, or chrono-stratigraphy, of a core. However, reality is rarely so simple because, amongst other things, accumulation rate is not known and the locations of the spreading centres of ice masses can change over the long time periods encompassed by deep ice cores. The time period included in the Vostok core extends back from the present to over 400 000 years (e.g. Petit *et al.*, 1999) and this can be doubled for the recent EPICA core (EPICA Community Members, 2004). Thus, independent empirical dating methods are required. The simplest of these is to count individual annual layers – which are characterized by seasonal variations in bubble density, dust content, ionic chemistry and (consequently) electrical conductivity (sections 3.4.2–3.4.4). These annual layers, however, become progressively closer-spaced and less distinctive as ice becomes more compressed deeper in the ice mass and therefore along the core, eventually becoming indistinguishable. This effect is exacerbated by low initial accumulation rates. Thus, while layer-counting was not possible to any significant depth in the Vostok core (Petit *et al.*, 1999), the recent Greenland cores could be dated by layer-counting to ∼100 000 years. Alley (2000) provides a highly readable description of the acquisition and analysis of the most recent of the deep Greenland ice cores, the GISP2 core.

Other dating techniques are required below the level at which annual layers can be identified with confidence. These techniques principally involve correlating specific core properties, or suites of properties, with independently dated environmental or climatic signals identified using the 1964 AD tritium reference horizon. Volcanic ash is found in thin layers at various levels in all deep cores, and this can be associated geochemically with dated eruptions. For the Greenland ice cores, for example, the most recent large eruption that can be identified in all of the cores was the Laki fissure eruption, which occurred in Iceland in 1783–1784. Core properties can also be correlated with those of the other, better constrained cores or in the broader environmental record. These properties include patterns in the stable isotopic composition and electrical conductivity (EC) of the core. In the former case, the δ^{18}O composition of ice in an ice core depends principally on the global ice volume at the time of deposition (and to a lesser extent on local temperature at the time of deposition). Thus, δ^{18}O variations can be cross-correlated between cores because the environmental changes that caused those variations were global. One particularly useful dating method can be carried out even before a core is recovered. This involves tracing the internal reflecting horizons revealed on radar transects from a provisional or anticipated drill site to a location where the chrono-stratigraphy is known on the basis of an existing core (Box 3.5).

Although the Vostok core is physically the longest so far recovered (section 3.3.4.2), it does not cover the longest time span. The European Project for Ice Coring in Antarctica (EPICA) core from Dome C, East Antarctica,

---

Box 3.5   Tracing internal reflecting horizons to correlate age–depth relationships between separate locations: East Antarctic Ice Sheet

Internal reflecting horizons (IRHs) represent isochronous layers that may be identified by ice radar and traced for hundreds of kilometres within the Earth's largest ice masses. These IRHs are formed as a result of systematic vertical variations in ice permittivity, mainly due to changes in the impurity content of the ice and, to a lesser extent, its density and crystallography. Critically, IRHs deform passively as they move through their host ice mass. Correlating individual horizons-between sites therefore provides the opportunity to extend ice-core-derived age–depth relationships along radar profiles to un-cored locations. Siegert *et al.* (1998) applied this approach to extend the chrono-stratigraphy of the Vostok ice core to a potential drill site near Dome C, located ~200 km away. The authors linked the two locations by tracing five IRHs identified on two separate flight-lines in the 60-MHz SPRI-NSF-TUD radar database. They also bridged a gap in the continuous, Z-scope (section 6.4) radar coverage by interpolating peaks in the power reflection coefficient (PRC) in adjacent, A-scope (section 6.4) radar traces. As a result of this work, Siegert *et al.* (1998) concluded that ice at a given depth at Dome C was younger than ice at an equivalent depth at Vostok (Figure B3.5), providing important information for the future EPICA core to be drilled in the vicinity of Dome C (now approaching completion with over 3000 m of core recovered; section 3.4.1). In a subsequent, associated study Siegert and Hodgkins (2000) extended the work to link the Vostok chrono-stratigraphy to Titan Dome near the South Pole. This involved tracing IRHs for over 1000 km of radar flight-lines.

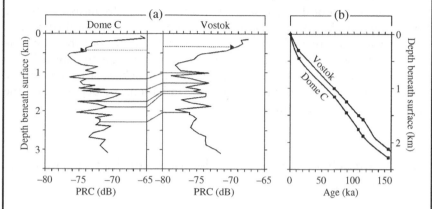

Figure B3.5 Illustration of the correlation and tracking of IRHs between Vostok and Dome C, East Antarctica: (a) correlation of five individual IRHs expressed as plots of power reflection coefficient against depth at the two sites; and (b) the resulting depth–age plots for ice at the two locations. Only the depth–age relationship at Vostok was known independently from ice core records. Reproduced from Siegert *et al.* (1998) with the permission of the American Geophysical Union

Siegert, M.J. and Hodgkins, R. 2000. A stratigraphic link across 1100 km of the Antarctic Ice Sheet between the Vostok ice-core site and Titan Dome (near South Pole). *Geophysical Research Letters*, 27(14), 2133–2136.

Siegert, M.J., Hodgkins, R. and Dowdeswell, J.A. 1998. A chronology for the Dome C deep ice-core site through radio-echo layer correlation with the Vostok ice core, Antarctica. *Geophysical Research Letters*, 25(7), 1019–1022.

when completed, will be ~3300-m long and possibly cover well over 800 thousand years (EPICA Community Members, 2004). At the time of writing 3139 m of the core have been dated (and 3190 recovered), providing an age span back to ~740 000 years before present (EPICA Community Members, 2004). Extending the core record a further 120 m to the bed of the ice sheet will extend this age span back to almost 1 000 000 years, providing a fascinating insight into the environmental and climate changes characterizing the late Quaternary Period. This task, however, will not be easy as the remaining ice to be drilled is temperate, yielding drilling difficulties and blurring the core's stratigraphy, while that stratigraphy is also more likely to be disrupted by structural complexities as the bed is approached.

3.4.2   Dust

Variations in dust within ice cores provide a strong parameter for defining annual layers because, being insoluble, dust is stable within the ice matrix. The concentration of dust particles within ice is generally measured by the degree to which the ice scatters light. The amount and type of dust in each layer is also indicative of the environment at the time that the dust was deposited. Various kinds of fallout from the atmosphere, including airborne continental dust and biological material, volcanic debris, sea salts, cosmic particles, and isotopes produced by cosmic radiation, are deposited on the ice sheet surface along with the snow.

3.4.3   Ionic chemistry and the electrical conductivity method (ECM)

Water soluble ions, such as H^+, NH_4^+, SO_4^{2-}, Cl^-, Na^+, Mg^{2+}, K^+, Ca^{2+}, are deposited at variable rates onto the surface of ice masses. These ions are derived from a variety of terrestrial, marine and atmospheric sources, both natural and, increasingly, anthropogenic. The concentration of individual ions can be measured directly from melted bulk ice samples by ion chromatography while the total ionic concentration can be approximated by measuring the bulk electrical conductivity of solid ice. This method is commonly referred to as ECM, short for the electrical conductivity method.

Bulk ice is composed of relatively pure ice crystals and relatively impure water located within veins between those crystals. Since this vein water holds the majority of the bulk ice's ionic load, ECM data represent a proxy for both the concentration of ions within these waters and the concentration of this liquid water within the bulk ice sample. This picture is complicated a little further by the fact that the inter-crystalline vein size (and therefore the concentration of water within the ice) depends on, amongst other things, the ionic concentration of the water. In ice cores, therefore, EC is principally controlled by the presence of debris and acid layers within the ice. In englacial ice, these are normally volcanic in origin, yielding internal reflecting horizons (IRHs) that can be traced laterally for hundreds of kilometres (Box 3.5). Nearer the base of the ice mass, these are normally derived from the incorporation of basal material into the overlying ice during refreezing or tectonic deformation.

The principles underpinning ECM are very straightforward: two electrodes spaced a few centimetres apart are dragged over the freshly cut surface of an ice sample. A DC voltage difference of $\sim$1200 V is established between the electrodes, and this drives a current across the core which varies with the core's EC. The current is then divided by the induced voltage to yield its conductance, which is recorded as a function of position along the core. Automated core advance and data acquisition require a specialist motorized core carriage and computer control. One disadvantage of this method is the requirement for a uniform ice surface for the electrodes to contact – variations in the ice surface texture can lead to spurious variations in the signal. This impediment has been overcome by regarding the ice core as a capacitor's dielectric, and by using curved electrodes that encircle the entire core to record its AC capacitance and conductivity (Moore *et al.*, 1989). The resulting dielectric profiling (DEP) responds to the presence of a variety of ions but is particularly sensitive to identifying the presence of volcanic acid layers. With some adaptations, the procedure can also be used to measure the permittivity of the ice – allowing direct comparison with radar wave propagation properties (Wilhelms *et al.*, 1998). Relative permittivity expresses the capacitance of a material relative to that of free air (section 6.2).

3.4.4   Gas

As ice forms from snow by firnification, the gas present within it becomes trapped in isolated bubbles. The composition of the gas within these bubbles represents that of the atmosphere between the time of initial snow deposition and the time of bubble closure. Gas is extracted from these cores by crushing ice from the centre of the core under a vacuum, and extracting the released gas into pre-evacuated sampling loops – described in detail by Barnola *et al.* (1983). The most common individual gases measured by chromatography from such samples include CO_2, CH_4, O_2 and N_2.

It is possible to measure the total gas content of a core non-invasively through proxy measures of bulk ice density. For example, the gamma densimeter uses a ^{137}Cs source to emit a fine beam of gamma radiation through the ice core. The photons are scattered by the ice in proportion to its density, and the intensity of the attenuated beam is recorded once it has passed through the core. Other things being equal, the denser the core the greater the attenuation.

3.4.5   Isotopic composition

The isotopic composition of ice, and in particular the concentration of the heavy isotopes of oxygen, ^{18}O, relative to ^{16}O, and hydrogen, ^{2}H (deuterium), relative to ^{1}H, can be used to provide valuable information about the composition of the water from which that ice formed and the temperature at the time of formation. These ratios are commonly expressed using the δ notation where, for ^{18}O:

$$\delta^{18}O = \frac{\left(\frac{^{18}O}{^{16}O}\right)_s - \left(\frac{^{18}O}{^{16}O}\right)_r}{\left(\frac{^{18}O}{^{16}O}\right)_r} \times 1000\,‰$$

where s refers to the composition of the sample and r refers to the composition of a reference (usually SMOW, standard mean ocean water, in the case of ice and water analysis). A similar equation is used to derive δD. Thus, measurements of $\delta^{18}O$ and δD in ice have been used for a variety of purposes, from identifying annual layers on the basis of seasonal temperature differences to identifying major global glacial and interglacial phases.

One additional area where the isotopic composition of ice has been used is in the field of basal ice research. Here, the spread of isotopic values and the slope of the line defined on a bivariate plot of $\delta^{18}O$ against δD of basal ice samples can provide information as to whether that ice has experienced refreezing at the glacier bed or not (Box 3.6).

Box 3.6   Co-isotopic analysis of ice to identify basal refreezing

The stable isotopes of oxygen (principally ^{18}O and ^{16}O) and hydrogen (principally ^{2}H [deuterium] and ^{1}H) fractionate to slightly different extents when water freezes. This has proved useful in studies of the origin of basal ice facies because a bivariate plot of the ratios of these isotopes, expressed as $\delta^{18}O$ and δD, against each other will yield a slightly different slope depending on whether the group of ice samples being plotted has refrozen at the glacier bed or not. This effect has been explored and utilized in several papers by Souchez and co-workers (Jouzel and Souchez, 1982; Souchez and Jouzel, 1984; Souchez and Groote, 1985). As a result of these studies the authors identified three possible situations which might be identified using this approach:

1. Where a body of water (the *initial liquid*) freezes entirely at the glacier sole, the closed system model of Jouzel and Souchez (1982) applies. In this case the ice samples will be aligned along a freezing slope which is steeper than the slope of samples of glacier ice from which the water derived. Individual samples of the basal ice can be isotopically lighter or isotopically heavier than the initial liquid from which it formed (Figure B3.6(a)).
2. Where the initial liquid is supplemented by an input of similar isotopic composition, the open system model of Souchez and Jouzel (1984) applies. In this case, a freezing slope similar to that resulting under closed system conditions is produced but with individual samples plotting only as richer in heavy isotopes than the initial liquid (Figure B3.6(b)).

3. Where lighter water is mixed with the initial liquid, the model of Souchez and Groote (1985) applies. In this case basal ice samples may be indistinguishable in terms of their slope from glacier ice (Figure B3.6(c)).

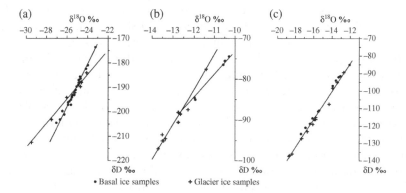

Figure B3.6 Co-isotopic plots of ice samples illustrating different basal ice freezing scenarios: (a) Aktineq Glacier, Canada; (b) Tsanfleuron Glacier, Switzerland; and (c) Gruben Glacier, Switzerland. Reproduced with the permission of the International Glaciological Society

Jouzel, J. and Souchez, R. 1982. Melting-refreezing at the glacier sole and the isotopic composition of the ice. *Journal of Glaciology*, 28, 35–42.

Souchez, R.A. and Groote, J.M.D. 1985. $\delta D - \delta^{18}O$ relationships in ice formed by subglacial freezing: Palaeoclimatic implications. *Journal of Glaciology*, 31(109), 229–232.

Souchez, R.A. and Jouzel, J. 1984. On the isotopic composition in δD and $\delta^{18}O$ of water and ice during freezing. *Journal of Glaciology*, 30(106), 369–372.

3.4.6   Ice crystallography

Ice is a crystalline aggregate and the size, shape and orientation of its constituent crystals respond to, and therefore reflect, their physical environment, particularly prevailing stress and stress history. The crystal structure of ice is particularly important because it exerts a primary control over the response of that ice to the stresses imposed on it, thereby governing the ice's softness and rheology.

Ice crystallographic analysis is usually carried out through the optical analysis of thin sections in a cold laboratory maintained at a temperature of

−20 °C or colder. Thin sections are prepared through two stages. First, a slice of ice is cut to a thickness of some millimetres, usually by using a fine-toothed band-saw. Second, that slice is frozen to a glass plate (best done by injecting liquid water onto the cold interface) and its thickness reduced to some tens to hundreds of micrometres by microtome. Although a wide range of microtomes is available commercially for biological tissue sampling, ice analysis demands a large sample size and a robust movement. The most popular microtome is therefore the SM2400 Sliding Microtome, manufactured by Leica GmBH, Germany, which is manually driven and can section a sample of lateral dimensions ∼10 cm × 10 cm. The precise procedures adopted are detailed by Langway (1958). Once reduced to a thickness of some tens of micrometres (the finer the crystals to be analysed the thinner the section needs to be), the section, still mounted on its glass plate, is removed from the microtome and viewed between crossed polarized plates illuminated from below. Because each crystal has a single optic axis along which light is transmitted, each crystal's orientation is characterized by a specific colour when viewed between crossed polaroids (Figure 3.8). This allows the outline of each crystal to be identified visually and its size to be measured. This can be achieved in a variety of ways, such as by counting crystals per unit area of thin section or by measuring the length of intersection with each crystal of one or more lines overlain onto the section. While

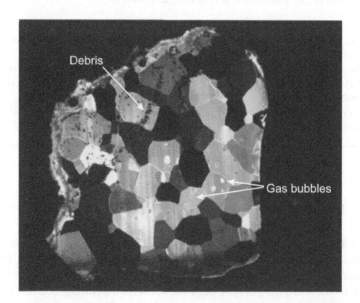

Figure 3.8   Ice thin section (∼8 cm across) viewed between crossed polarized lenses. Although most crystals can be clearly defined in greyscale, each ice crystal orientation has a different colour when viewed in full colour. Note entrained bubbles show up in outline or as lighter circles and that incorporated debris is opaque

the latter provides a generalized value for mean diameter, it should be multiplied by a correction factor to increase it to the longest diameter of each crystal. This factor is generally taken to be 1.75 (Jacka, 1984).

Ice crystal fabric expresses the bulk orientation of a crystal aggregate. To determine this fabric the orientation of a representative sample of individual crystals, preferably at least 200, is recorded. These are measured through using a universal stage (Rigsby, 1951) following procedures of Langway (1958). According to this method the thin section mounted between polarized plates is rotated on four axes, in three dimensions, within a graduated housing to the point at which the selected crystal ceases to transmit light – that is it blackens and is said to have been rotated to extinction. In this orientation the optical axis of the crystal is parallel to the line of site and the section's orientation is recorded in terms of an azimuth and a dip, similar to macro-clast fabric data (section 8.11). Ice crystal fabric data are presented in a similar manner to macro-clast fabric data, by plotting each orientation as a point on a 2D hemispheric net. This method of representation and its quantitative characterization are described in section 8.12.

3.4.7   Automated crystallographic analysis

A new generation of automated ice crystallographic analysers has recently been developed (e.g. Wang and Azuma, 1999; Wilson and Sim, 2002). These systems are based on the acquisition and analysis of multiple high-resolution digital images of thin sections (prepared as in section 3.4.6). Because these digital images are acquired simultaneously from three or more different orientations (Figure 3.9), there is no requirement to rotate the thin section to extinction. These images are then processed to produce a master image coloured according to a standard scheme, termed an *Achsen-Verteilungs-Analyse* (AVA) image. Once acquired, AVA diagrams can be analysed rapidly and automatically for size, shape and fabric by bespoke digital image analysis software. Currently, at least one automated crystal analyser is available commercially. This is constructed and distributed by the School of Earth Sciences, University of Melbourne, Australia (Wilson, 2004). The system, called *Fabric Analyser*, has an image pixel dimension of 18–20 μm over a 10 mm × 10 mm field of view. Bespoke software, *Investigator*, which can be used to analyse and characterize any AVA image, is currently available free of charge.

3.5   STUDENT PROJECTS

Unless projects can be closely affiliated to ice-coring programmes (which is an infrequent opportunity given the expense involved in acquiring such

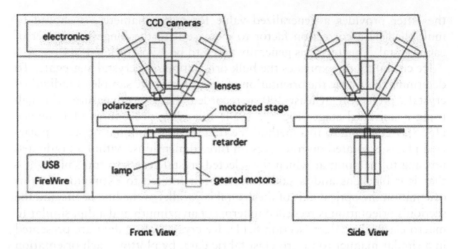

Figure 3.9 Image-analysis-based automatic ice fabric analysis instrument, fabric analyser, assembled and supplied by the School of Earth Sciences, University of Melbourne (Drawing courtesy: C. Wilson, University of Melbourne)

cores), students rarely focus on studying glacier ice in the field. However, such studies are possible for students, even with little logistical support. Some of these are outlined below:

- Identifying and mapping the presence of different ice facies at a glacier. For example, a study focusing on surface ice facies might be to investigate patterns and rates of retreat of the surface snow line. This might, in turn, be related to hydrological changes, to motion changes and to controls such as ice surface topography and meteorological data. A study focusing on basal ice facies might include applying existing facies–process relationships to reconstruct subglacial conditions in inaccessible areas of the glacier bed located up-glacier of sampling sites.
- Sampling ice, which can be melted and bottled in the field, for chemical analysis to investigate controls such as the transport of solutes from the glacier bed into clean (clear facies) basal ice located some distance above the basal interface.
- Sampling ice and melting it in the field to recover and analyse the debris incorporated within it for its concentration and sedimentology. This information could be used to investigate the characteristics of the debris associated with different ice facies and sources, which could be related to the sedimentological character of deposited proglacial landforms. Allying this sedimentological information to facies geometries and ice motion data (section 7.4) would allow reconstruction of different sediment transport rates and pathways through a glacier.

4

Glacier meltwater: Character, sampling and analysis

4.1 AIM

The aim of this chapter is to provide an overview of the techniques used to investigate the character of the meltwater supplied by glacierized basins. Focusing on proglacial meltwater streams, we provide guidance on a variety of methods to measure discharge, including velocity–area methods and dilution gauging. We also describe how to sample and analyse proglacial meltwater for its principal physical properties, including suspended sediment concentration, ionic chemistry and stable isotope chemistry. We provide an overview of the automated measurement of some of these properties through the construction and use of sensors that can be controlled by data loggers. Finally, we address the application of tracers in studies of glacier hydrology, summarizing their potential uses and providing guidance on how to inject, sample and analyse tracers.

4.2 BACKGROUND

Glaciers behave as important natural water storage reservoirs, storing water during precipitation events and releasing it by melting (section 7.2.1) during dry, warm weather. The release of water held in glacier storage therefore has a moderating influence on annual stream flows from basins

Field Techniques in Glaciology and Glacial Geomorphology   Bryn Hubbard and Neil Glasser
© 2005 John Wiley & Sons, Ltd

that contain both glacierized and non-glacierized areas. This buffering capacity is sometimes termed the compensation effect (Röthlisberger and Lang, 1987). This is one of the main reasons why farmers and hydroelectric power companies are so interested in water supplied from partially glacier-ized basins.

Runoff supplied from glaciers is often highly variable in both quantity and quality, particularly at time scales of hours to months. These variations are summarized in the following sections.

4.2.1   Regular discharge variations

Seasonal bulk glacier discharge mainly reflects air temperature and net solar radiation receipt. During the winter, glaciers are usually snow-covered (high albedo) and air temperatures remain below freezing. Surface melting is therefore minimal, and any meltwater that is produced refreezes within the sub-zero surface snow or ice layers. Thus, no surface-derived meltwater is delivered to the glacier bed and bulk outflow is comprised entirely of meltwater generated in warm areas of the bed of temperate and polyther-mal glaciers (geothermal and frictional heating in these areas produces ~10–15 mm of melt per year). Surface melting gradually increases during the spring until the low-elevation surface snowpack and ice layers have been heated up to the melting point, and surface meltwaters are delivered to the englacial and basal drainage systems. At first, this delivery results in erratic outflow patterns as drainage pathways open up, but soon large, efficient channels route these meltwaters rapidly to the glacier snout. Outflow dis-charge therefore not only rises on average, but also develops a diurnal cycle that mirrors air temperature and radiation receipt at the timescale of hours. Together, these energy inputs reach a maximum in the early afternoon, resulting in peak outflow discharges soon after this time, and fall to a minimum overnight – when surface melting commonly ceases. Note that diurnal bulk outflow actually peaks some hours after maximum energy input and reaches a minimum some hours after sunset. The latter delay reflects the time it takes for residual water to drain from the glacier once melting has reached a minimum. In this case the delay reflects water flow through the glacier's principal drainage channels as well as from the rest of the glacier to those channels. In the former case, the delay between peak melting and peak discharge generally reflects the time it takes for water melted at the ice surface to be routed through the glacier. This lag generally decreases through the melt season as the network of drainage channels extends up-glacier and as individual channel sections enlarge. The time of peak diurnal meltwater discharge therefore occurs earlier in the day as the melt season progresses (Box 4.1).

Box 4.1   Investigating drainage evolution at a temperate glacier from the magnitude and timing of diurnal discharge cycles: Gornergletscher, Switzerland

Elliston (1973) reported systematic variations in meltwater discharge from Gornergletscher, Switzerland. This pioneering study identified several important features of bulk discharge that we now know apply to most temperate and partly temperate glaciers. First, the melt season was characterized by diurnal discharge cycles that peaked in the day and were lowest overnight. These were superimposed on a base discharge that increased gradually through the melt season. Second, these cycles were disrupted by summer snow storms, which had the effect of reducing discharge to a constant low level. In some cases, the diurnal cycles did not recover fully until several days following the snowfall event. Third, the amplitude of the diurnal discharge cycles also increased through the summer, while the timing of daily peakdischarge shifted progressively earlier in the day. Thus, peak flow in late September was recorded a full 3 hours earlier in the day than in late June (Figure B4.1).

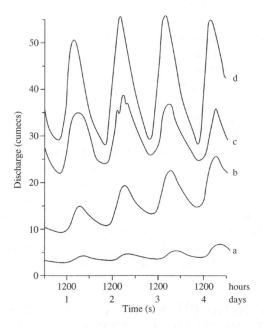

Figure B4.1   Discharge measured in one of the streams draining the Gornergletscher, Switzerland, over selected periods through 1959: (a) 17–20 May; (b) 14–17 June; (c) 23–26 June; and (d) 19–22 July. Reproduced with the permission of the IAHS

Box 4.1   (Continued)

These changes were interpreted by the author in terms of the increased efficiency of the englacial transfer of stored water – an idea that has subsequently been refined in the light of borehole-based studies of the evolution of subglacial drainage systems (e.g. Hock and Hooke, 1993; Nienow *et al.*, 1998).

Elliston, G.R. 1973. Water movement through the Gornergletscher. *International Association of Scientific Hydrology Publication*, 95, 79–84.
Hock, R. and Hooke, R.L. 1993. Evolution of the internal drainage system in the lower part of the ablation area of Storglaciären, Sweden. *Geological Society of America Bulletin*, 105(4), 537–546.
Nienow, P., Sharp, M. and Willis, I. 1998. Seasonal changes in the morphology of the subglacial drainage system, Haut Glacier d'Arolla, Switzerland. *Earth Surface Processes and Landforms*, 23(9), 825–843.

4.2.2   Irregular discharge variations

Although glacierized basins generally produce more regular discharge cycles than do non-glacierized basins, the former are occasionally characterized by highly irregular flows. Summer snow storms have the capacity to shut off surface melting and its contribution to glacier runoff. Since this water source dominates summer discharge, the diurnal cycle is essentially replaced by constant, low base flows. This pattern may persist for some days as further surface melting may be impeded by the high albedo of fresh snow, even following the onset of higher radiation and warmer air temperatures. Conversely, abnormally high bulk discharges can result from warm rain storms, particularly during late summer, when high rates of ice surface melting can be supplemented by rapid runoff in the absence of a buffering surface snowpack. Glacier outburst floods, or jökulhlaups, provide spectacularly high discharges, lasting for some hours to days. These waters may be released from ice-dammed lakes or the drainage of internal water bodies and are most extensively documented from Iceland.

4.3   MEASURING BULK MELTWATER DISCHARGE: STAGE-DISCHARGE RATING CURVES

It is not surprising, considering the value of discharge measurements, that numerous ways of measuring, or at least approximating, bulk discharge have been devised. In the first instance, any semi-quantitative indication of discharge may be preferable to leaving with no record at all. As a bare minimum, a classification scheme of low to high or 0–5 can be adopted with reference to a visual assessment of the level of the water surface. Such an approach can be supplemented with photographs of the stream reach concerned repeated over time from a fixed vantage point. However, a variety of more quantitative techniques is also available to the field researcher.

Discharge through a channel section can be calculated as the product of the flow velocity of the water and the cross-sectional area of that flow. Thus, as long as the channel shape remains the same, discharge generally scales in a repeated and predictable manner with water level or stage. However, the precise geometry of that channel shape defines the exact relationship between discharge and stage. This principle forms the basis of a weir, where a geometrically regular boundary is imposed on the outlet of a ponded channel, yielding a mathematically exact relationship between discharge and stage (termed the 'rating curve'). Most commonly such weirs have a 'V' shape, allowing a geometrical increase in discharge for an arithmetic increase in stage, thereby defining a logarithmic rating relation between the two variables. Even though such structures provide accurate and repeatable discharge data, weir installation can be extremely difficult in proglacial terrain, particularly at larger streams and where stream boundaries are mobile. Weirs are constructed in such environments, but generally only with the aid of major mechanical support which is required both to construct the weir and to ensure that the stream is routed through it.

As an alternative to a weir, field glaciologists commonly define an empirical rating relation at a fixed point on a channel where its boundary has a high depth-to-width ratio, is clearly demarcated and is considered to be stable. Once a rating relation has been defined for this location, it can be used to calculate discharge from stage data alone. In adopting this approach, it is important that the rating relation encompass the full range of discharges that might be expected. This is important because an empirical rating relation defined for an irregular channel cross-section will not be a regular mathematical function, such as a straight line or logarithmic curve, and cannot therefore be extrapolated with any degree of confidence beyond the calibration range. Further, the rating relation only holds for as long as the channel cross-section remains unchanged. Once the channel has changed, for example as a consequence of a high-magnitude event, a new rating curve must be defined. Also, it is worth checking for drift in the rating curve due to smaller channel changes by cross-calibrating with simultaneous measurements of stage and discharge whenever the opportunity arises. These data will also serve to increase the accuracy of the existing rating curve if the channel cross-section has not changed. It follows that stage data should still be collected even if logistical constraints preclude definition of a comprehensive rating relation at the outset of the research – as long as the channel boundary does not change, a rating relation can be derived and applied retrospectively.

4.3.1 Measuring stage

For a rating curve to be defined empirically, numerous measurements of stage and discharge must be made simultaneously. Stage is the simpler of the two to measure and this can be done by comparing the water surface with reference to fixed markers on the channel boundary or within the water column or by using a water depth gauge. In the former case, and at the simplest level, a ruler may be used to yield a quantitative measure of water level, although great care must be taken to ensure the base of the ruler does not move between readings – which can often be a problem in highly mobile outwash streams. Alternatively, where a channel is bounded by bedrock, the channel margin may be graduated. In either case, it is worth noting that the unit of stage graduation is unimportant – as long as the same measurement scheme is used for the calibration rating curve and the subsequent stage measurements to be transformed to discharge. Indeed, the spacing of the graduated marks need not even be uniform as long as this consistency is maintained.

Commercial stage recorders and depth gauges come in a variety of forms, foremost amongst which in field glaciological studies is a water-pressure sensor in association with a data logger. In this case, the sensor is physically shielded from the effects of particle impact and is installed at or near the base of the water column (below the lowest anticipated water level). The manner of this installation depends largely on the channel properties, such that the sensor may be placed under a retaining weight (commonly a large rock) on the base of a bedrock channel characterized by little sediment transport. However, most proglacial channels have high sediment loads, and sensors are, in such situations, attached to an arm of a metal frame constructed to extend from the stream bank into the channel. Frames are commonly constructed on site out of lengths of slotted angle such as that produced by Dexion Ltd, UK. Once installed as part of a gauging station (Figure 4.1), a water-pressure sensor may be linked to a data logger located nearby, and water height above the sensor can be recorded as frequently as the logger will permit and for as long as the system can be powered. One advantage of such continual monitoring is that the researcher need not worry about the timing of any particular discharge measurement – as long as the time of such a measurement is known it can be correlated with the stage measurement recorded by data logger at, or closest to, that time.

4.3.2 Measuring discharge

There are two principal ways of measuring stream discharge (normally expressed in $m^3\,s^{-1}$, or cumecs): (a) through measurements of water velocity and channel area, and (b) by dilution gauging.

Figure 4.1 A proglacial stream gauging station located in front of Haut Glacier d'Arolla, Switzerland, illustrating some of the difficulties in introducing a fixed structure into such turbulent and dynamic streams. The arms and uprights are constructed of lengths of Dexion angle, which are strong, portable and can be bolted together to suit individual circumstances. The sensor wires are fed along the Dexion lengths and secured to it with duct tape or cable ties (Photograph courtesy: Peter Nienow)

4.3.2.1   Velocity–area method

Channel discharge can be calculated as the product of water flow velocity ($m\,s^{-1}$) and the channel cross-sectional area (m^2) through which that water flows. Thus, if both area and velocity can be measured or approximated at a site, then discharge can be reconstructed by multiplying the two. It is important to select a survey location that is as straight and of as uniform a cross-section as possible, avoiding complex patterns of channel geometry and flow. Channel cross-sectional area can be measured by any standard survey technique. Since only a 2D survey is required (i.e. a transect across the stream expressed as height from the bed to the water surface against distance across the stream), a minimum requirement is a ruler and a tape measure (with the former lowered at regular intervals from the latter, which is extended as straight as possible directly across the stream channel), by engineer's level or by more sophisticated survey equipment. Whichever technique is used, however, it is important to remember that the more survey points the better and that the water surface also needs to be surveyed – although adequately approximated by noting the two points at which it intersects the bed (i.e. the two flow margins) and linking the two by a straight line.

Channel flow velocity can be measured by numerous techniques of vary-ing sophistication, including the use of salt tracers (measured simply by EC probes; section 4.5.1) and dye tracers (i.e. visible or measurable by fluoro-meter; section 4.6.3). However, if no dedicated equipment is available, then timing floats over a flow length is possible. Since channelized water flow velocity decreases exponentially towards the bed and banks, the most representative location for a point measurement of channel velocity is towards the middle of the channel about 40% of the way up from the bed (60% of the depth down from the surface). Since the velocity of a floating object will obviously be higher than this value, a correction factor should be applied to yield the mean channel velocity. As an acceptable approximation the mean velocity $\bar{v}$ may be determined from the measured surface velocity v_s from:

$$\bar{v} = kv_s \tag{4.1}$$

where k is a coefficient that depends on the depth of the channel (H), given by:

$$k = 0.715H^{0.072} \tag{4.2}$$

Where floats are used to estimate flow velocity, multiple tests should be conducted and averaged; each lasting some tens of seconds to get as accu-rate a mean velocity value as possible.

At-a-point stream flow velocity can also be measured by commercial velocimeters, based on rotating propellers that are faced into the flow or, more recently, Doppler shifts in transmitted acoustic waves. Indeed, the latter instruments, Acoustic Doppler Velocimeters, can measure 3D vel-ocities continuously. Since velocimeters can be used to record flow velocity at any location within the channel, at least two records should be attempted down each of many vertical profiles. These readings can then be either averaged over the entire channel or scaled for each vertical strip surveyed, providing a more accurate discharge. Correction factors similar to the value of k in equation (4.2) are available to transform point readings to mean velocity for a variety of special physical scenarios (BS3680, 1980). For example, the mean velocity $\bar{v}$ of a stream with a frozen surface of total depth <0.5 m is given as:

$$\bar{v} = 0.92v_{0.6} \tag{4.3}$$

where $v_{0.6}$ is the point velocity measured at 60% of the depth from the surface.

Finally, flow velocity can be approximated from the so-called uniform flow formulae, which recreate flow velocities theoretically from channel physical properties. These formulae, including the Manning, Darcy-Weisbach and Chézy equations, express mean velocity $\bar{v}$ ($m\,s^{-1}$) in terms of the channel's hydraulic radius R (m) (the channel's cross-sectional area divided by its wetted perimeter), the surface slope of the water S (expressed as a gradient – dimensionless) and a resistance coefficient. For example, the Manning equation has the form:

$$\bar{v} = \frac{R^{\frac{2}{3}}S^{\frac{1}{2}}}{n} \tag{4.4}$$

where n is Manning's roughness, having a typical value of 0.04 (± 0.01) for a coarse-bedded proglacial stream. There are, however, many techniques for approximating the value of n from other field data, such as channel margin particle grain size, and the reader should consult a standard hydraulic text for further details.

4.3.2.2 Dilution gauging

The accuracy of the velocity–area method of calculating discharge reduces in channels that are characterized by turbulent flow and/or a rough or irregular boundary. Unfortunately, both conditions hold in proglacial streams. However, this turbulence favours a second method of discharge calculation, that of dilution gauging. This method is also less sensitive to local flow variations and is therefore particularly well suited to measuring proglacial stream discharge. The most common form of dilution gauging uses common table salt (NaCl), although fluorescent dyes (section 4.6.1) and radioactive tracers may also be used, and in all cases the underlying principle is that of the conservation of mass. Tracer may be injected at a constant rate, involving the continuous injection of tracer into the stream, or as slugs, involving instantaneously decanting a discrete mass of tracer into the stream. In the following analysis we will deal with only the latter technique as it is easier and more commonly used in glaciological field studies, and it additionally provides a time-of-travel record for each test.

Using salt as an example, a known mass, given by the concentration of dissolved salt in a primary solution (C_P) multiplied by the volume of primary solution injected (V_P), is decanted into the stream. The injected salt pulse is then recorded as concentration against time at a site sufficiently far downstream for the solution to have mixed throughout the channel's flow width (typically some tens of stream widths in a fairly turbulent proglacial stream). The resulting salt concentration–time (or 'breakthrough') curve has the form of a plot of measured salt concentration above background

(C_m) against time (t). According to this scheme, initial salt mass input S_i is given by:

$$S_i = C_P V_P \qquad (4.5)$$

and salt mass output $(S_o$ – that passing the gauging station) is given by:

$$S_o = Q \int_0^t C_m dt \qquad (4.6)$$

Assuming no salt is gained or lost between the input and sampling sites $S_i = S_o$, and rearranging equations (4.5) and (4.6) it follows that:

$$Q = \frac{C_P V_P}{\int_0^t C_m dt} \qquad (4.7)$$

As the initial mass of salt (or C_P and V_P) is known, it remains only for the denominator in equation (4.7) (the integral of C_m with respect to t) to be measured to enable Q to be calculated. Calculating this integral is very straightforward since it is the area under the tracer breakthrough curve of the passing salt slug. Thus, in slug injection dilution tests, stream discharge is given by the product of the concentration of tracer and volume of tracer solution injected into the stream divided by the area under the plot of tracer outflow concentration against time. One slight problem here lies in the field calculation of salt concentration. This is normally done by measuring the water's electrical conductivity (EC; section 4.5.1) since, other influences (e.g. temperature) being equal, this property scales directly with salt concentration.

Putting these principles into practice requires some additional considerations that may best be illustrated through a series of stages, each accompanied by a procedural case study.

Stage I: Approximating salt injection mass and mixing length. Before conducting a gauging experiment two quantities need to be approximated: the amount of salt to be injected and the distance separating the injection site from the measurement site. First, the amount of salt to inject may be calculated from a visual assessment of the stream's discharge or cross-sectional area. A rule of thumb for the former is to use $\sim$1 litre of $\sim$20% solution for each $m^3 s^{-1}$ of estimated discharge. Kite (1993) presents a relation for the latter, whereby salt mass (M_s, kg) may be estimated from the stream's estimated cross-sectional area (A, m^2), via

$$M_s = 0.13\delta_{ec}A^{1.5} \qquad (4.8)$$

where δ_{ec} is the required rise in EC ($\mu S\,cm^{-1}$) above background (which may be taken conservatively as 50%). Second, the distance separating the injection site and the downstream monitoring site (the mixing length, L_m) should be long enough to ensure complete mixing of the solution, but not so long that the concentration of salt in the water falls too low to yield a measurable breakthrough curve. A general rule of thumb for proglacial streams is something between 10 channel widths in very turbulent channels and 100 channel widths over smoother-flowing reaches. Numerous formulae for the L_m (in m) are given in Kite (1993), the most appropriate for proglacial streams being

$$L_m = 260\sqrt{A} \qquad (4.9)$$

where A is the stream's cross-sectional area (m^2).

Case study: Stage I. The stream to be gauged is estimated to have a discharge of $1\text{–}3\,m^3\,s^{-1}$, is fairly turbulent and has a width of $\sim2\,m$. Thus, 2 litres of 20% NaCl solution are made up by dissolving 400 g of NaCl in 2 litres of water from the stream to be gauged. The salt is mixed and dissolved in one bucket, and then decanted into another before removal for calibration or gauging, since only solution (and not un-dissolved solid salt which may be present at the base of the mixing bucket) should be used. The injection and sampling sites are chosen, separated by a distance of 20 stream widths ($=40\,m$).

Stage II: Calibrating EC against relative salt concentration. Because salt concentration appears on the numerator and denominator of equation (4.7), only relative concentrations are needed for stream gauging. EC readings therefore need not be absolutely accurate but they do need to be consistent between readings of the primary solution and the breakthrough curve. The best way to ensure such consistency is to calibrate the particular EC probe to be used to record the breakthrough curve with a range of dilutions of the primary solution. If the river water is ionically pure then, theoretically, only one calibration point is needed (since a calibration curve would pass through that point and the origin $[0, 0]$ on the plot), but since stream water has a background EC, at least two additional calibration points are required – and preferably more than this. This approach has the added advantage that it is not necessary to know the exact absolute concentration of the primary solution, but only its exact volume. To construct a calibration curve of EC against salt concentration a small volume of the primary solution is removed and a *secondary solution*, of typical concentration (C_s) 0.01, is made up in a measuring cylinder. This is then gradually added to a large volume of stream water to create the various relative concentrations required for the probe calibration.

Case study: Stage II.   Ten ml of the primary solution is mixed into 1000 ml (1 litre) of river water to produce a *secondary solution* of relative concentration (C_s) 0.01. Meanwhile, a bucket is filled with 20 litres of river water (and, if possible, partially submerged in the stream to minimize temperature deviations during calibration). The secondary solution is added to the bucket in increasing amounts to yield a series of *calibration concentrations* (C_c) whose EC is measured. The EC probe is rinsed and dried between each reading. An example of the volumes added and the resulting calibration concentrations (C_c) is provided in Table 4.1. In column two of this table C_c is calculated from:

$$C_s V_s = C_c V_c \tag{4.10}$$

Here, C and V refer to relative concentration (relative to the concentration of the primary solution) and volume and the subscripts s and c refer to the secondary and calibration solutions respectively. For example, for row six in Table 4.1, where 50 ml of secondary solution is added to the 20-litre calibration bucket, $C_s = 0.01$, $V_s = 50$ ml and $V_c = 20\,050$ ml (the 20 000 ml of river water plus the 50 ml of secondary solution), yielding:

$$C_c = \frac{0.01 \times 50}{20\,050} = 2.49 \times 10^{-5} \tag{4.11}$$

The resulting calibration curve is produced by plotting each calculated relative calibration concentration (C_c) against its corresponding EC (Figure 4.2). This relationship is described mathematically by placing a best-fit straight line through the data points (in Figure 4.2 this yields a line with a slope of 2.02×10^{-7} and an intercept of -2.01×10^{-6}).

Stage III: Gauging experiment.   Once the correct volume of primary solution has been prepared, the EC probe has been calibrated in terms of

Table 4.1   EC probe and salt solution calibration table for the example salt-dilution guaging experiment described in the text

Known volume of secondary solution added to 20 litre bucket (V_s) (ml)	Calculated calibration concentration in bucket (C_c) (relative units)	Measured EC ($\mu S\,cm^{-1}$)
0	0	11
10	0.000005	29
20	0.00001	57
30	0.000015	86
40	0.00002	118
50	0.0000249	145
75	0.0000374	180
100	0.0000498	258

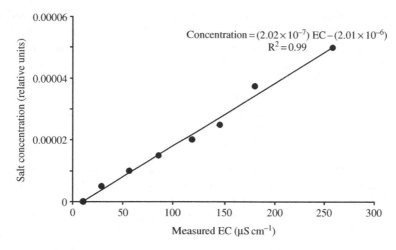

Figure 4.2 Calibration curve for the example salt-dilution gauging experiment described in the text

that primary solution, and the injection and gauging sites have been located, the experiment can be carried out. The remainder of the primary solution is injected at the injection site and EC readings are taken at the gauging site. The primary solution is poured into the main current of the stream, where the EC readings are also made. These should be taken as often as possible during the passage of the salt pulse, though longer-spaced readings may be taken as the recession limb decays. Ideally, these EC readings would be made automatically at high frequency by data logger.

Case study: Stage III.　The remaining 1.95 litres ($V_p = 0.00195$ m^3) of the primary solution is poured rapidly and smoothly into the main current of the stream. The person injecting the solution notifies a second person recording the EC at the gauging site, and EC readings are taken every 5 seconds. The initial (background) reading is 11 μS cm^{-1}, but readings rise after ~70 seconds to peak at 140 μS cm^{-1} after ~120 seconds. These then gradually tail off back to 11 μS cm^{-1} after ~350 seconds. If not recorded by data logger, all readings and times are noted in a waterproof notebook – ideally by a third person on the stream bank at the gauging station to whom readings are recited by the person holding the EC probe in the flow.

Stage IV: Calculating the stream discharge.　Once one or several dilution experiments have been completed, the data from each are used to calculate stream discharge. First, the salt breakthrough curve is plotted as an EC readings against time curve and then converted, using the equation of the calibration line, into a relative concentration–time curve. The area underneath the curve is then calculated or approximated. This can be measured by various methods, for example by using a computer programme capable of integrating a curve, or a compensating planimeter to measure the area beneath a hard copy of the plot. However, if neither is available, a decent approximation can be based on counting squares, more than half of each of which is enclosed by the concentration–time curve. Next, the product of the concentration of the primary solution injected and its volume is divided by the area beneath the curve has been calculated (and corrected for consistent units) to provide a value of discharge.

Case study: Stage IV.   The time of each EC reading and its corresponding EC value are entered into a computer programme such as MS Excel and the EC readings converted into relative concentration units by using the calibration equation. For example, an EC reading of $20\,\mu S\,cm^{-1}$ becomes $2.02 \times 10^{-6} = (((2.02 \times 10^{-7}) \times 20) + (-2.01 \times 10^{-6}))$ relative concentration units. The EC breakthrough curve is plotted (Figure 4.3a), as is the corresponding relative concentration–time curve (Figure 4.3b). The area beneath the latter is calculated, in this case by counting squares. All squares more than

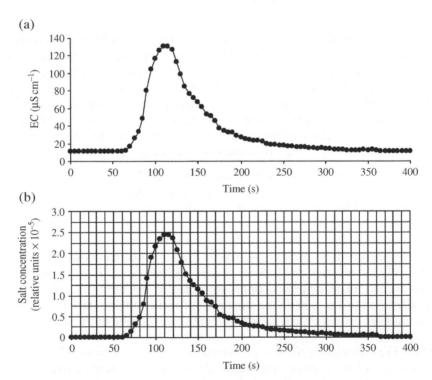

Figure 4.3   Example time-series plot of (a) measured EC; and (b) relative concentration against time during the example salt-dilution experiment described in the text. The total number of squares falling beneath the curve in (b), counted by eye, is 70

half of each of which fall beneath the curve are counted. In Figure 4.3b this yields 70 squares. The number is then transformed into concentration units in the light of the dimensions of each square; in this case, one x-axis square unit is 10 seconds and one y-axis square unit is 2.5×10^{-6} relative concentration units. This yields a corrected area of $(70) \times (10) \times (2.5 \times 10^{-6}) = 1.75 \times 10^{-3}$ relative concentration seconds (s). Remembering that the relative concentration of the primary solution is 1 and the volume of primary solution injected is $0.00195\,m^3$ (1.95 litres), equation (4.7) becomes:

$$Q = \frac{1 \times 0.00195}{0.00175} \approx 1.11 \qquad (4.12)$$

in standard units of $m^3\,s^{-1}$, or cumecs. Thus, the discharge of the gauged stream is $1.1\,m^3\,s^{-1}$.

One additional factor, however, should be considered if measurements are made at a different time or place from the calibration procedure – that of the dependence of EC on water temperature (as well as total ionic concentration). Fortunately, this is often not a problem in proglacial locations where stream water temperature varies little. However, all EC measurements made during the calibration and the gauging should ideally be corrected for temperature. Many commercial EC sensors correct automatically for temperature, providing EC values referenced to a standard temperature, usually $25\,^{\circ}C$. However, this correction can also be applied manually under the approximation of a 2% increase in EC per $1\,^{\circ}C$ rise in temperature, yielding:

$$EC_r = EC_t - (0.02(t - r)EC_t) \qquad (4.13)$$

where EC_r is the EC value at the reference temperature; EC_t is the EC value at the sampled temperature; r is the reference temperature ($^{\circ}C$), and t is the sample temperature ($^{\circ}C$).

4.4 SAMPLING AND ANALYSING GLACIAL MELTWATERS

4.4.1 Planning a sampling programme

Devising a sampling strategy will be mainly governed by programme-specific factors, such as the scientific objectives of the study and the techniques to be used. One generic consideration, however, is that of the timing of sampling. Proglacial meltwater discharge varies both annually and diurnally, reflecting temporal variations in energy input and the efficiency of flow routing across, through and beneath glaciers (section 4.2). Thus, if the purpose of a sampling programme is to characterize variability in water quality characteristics, then samples should be collected at least at the times of the extreme highest and lowest values of those characteristics. In proglacial meltwater streams, these times coincide with those of maximum and minimum discharge in the diurnal hydrograph. At many temperate and polythermal glaciers these times vary through the summer, getting progressively earlier in the day as the surface snowpack disappears and as the glacier's drainage system becomes more effective (Box 4.1). Having characterized, through a bi-daily sampling strategy, the total variation in meltwater character, it may be desirable to investigate the nature of the variation between these two extremes over a small number of more intensive sampling periods. In such situations, glacier hydrologists and hydro-chemists tend to collect samples hourly or every 2 hours over periods of 24 hours. The reason why meltwater is generally not sampled at this higher frequency throughout the period of a project is largely logistical: field meltwater samples usually require some degree of sample treatment or analysis immediately upon collection. These procedures, many of which are outlined in the following sections, require time and a successful research strategy will allow for that time (plus rest) between periods of sample collection. Further, the objectives of the study may not demand a continuous, high-frequency sampling strategy, freeing up valuable research time for other tasks.

4.4.2   Suspended sediment concentration – background

Glaciers are active erosional agents, continuously crushing and abrading the substrate over which they flow. Glacierized basins are also characterized by steep slopes and rapid meltwater transfer, typically resulting in proglacial streams with high and variable suspended sediment concentrations (SSCs). Indeed, the water in these streams is often referred to as 'glacier milk' on account of the colour it acquires as a result of this suspended sediment. SSCs in such streams typically vary between 0.1 and $6 \, \text{kg m}^{-3}$ (or g l^{-1}). Rivers draining glacierized areas therefore generally transport significantly more solid matter in suspension than do those draining non-glacierized areas.

Fundamentally, SSC reflects the availability of material for entrainment at the glacier bed and the ability of meltwaters to entrain that sediment. Between-glacier variation in SSC therefore reflects, in the first instance, the nature of the substrate: soft-bedded glaciers generally supply higher SSCs than those underlain by bedrock. However, SSCs at any given glacier are far from constant. As with bulk discharge, they vary systematically at a number of different time-scales. Since the availability of sediment rarely changes systematically, most regular SSC variability is related to the capacity of the meltwater stream to transport debris. Recorded SSCs are therefore generally positively correlated with discharge at the diurnal and seasonal time-scales. Bivariate scatter plots of SSC against discharge, however, are not linear and a number of significant deviations occur (particularly at high discharges) due to, amongst other factors, variations in the availability of sediment for transport at the glacier bed (Box 4.2). Thus,

---

Box 4.2 Interpretation of suspended sediment variations in a glacier-fed mountain stream: Place Creek, Canada

Richards and Moore (2003) measured the discharge and SSC of two streams in British Columbia, Canada, through the summers of 2000 and 2001. The main stream investigated (Place Creek) drained a partly glacierized catchment, and the other (Eight Mile Creek) drained an adjacent non-glacierized catchment. Comparison of the two data sets allowed the authors to identify different source-water periods and to isolate glacier-fed discharge during the mid-summer. This component was characterized by strong clockwise hysteresis in the relationship between SSC and discharge, at both diurnal and longer time scales (Figure B4.2). Multiple regression models, based on measures of

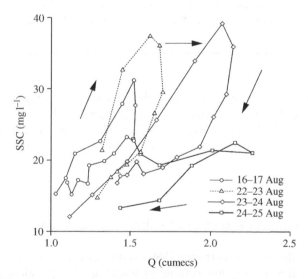

Figure B4.2 Bivariate scatter plot of SSC against discharge (Q) in glacier-fed Place Creek over four diurnal cycles through the second half of August, 2000. These data indicate nested clockwise hysteresis with diurnal loops themselves forming a longer-term clockwise loop. Reproduced from Richards and Moore (2003) with the permission of Wiley

Box 4.2   (Continued)

discharge and precipitation, were developed to predict the observed SSC variations. However, the best match with discharge during the-glacial sub-season was achieved when an 'index SSC' term was introduced. This term was defined as the SSC measured at 1500 hours each day, indicating a strong autocorrelation in measured SSC values.

The authors also found considerable evidence of within-channel-suspended sediment deposition and erosion between the terminus of Place Creek and a monitoring site located some distance downstream. In this case it was found that suspended sediment was predominantly lost to channel storage during low flows and re-suspended during high flows. Similar processes have been reported in other proglacial environments (e.g. Hodson *et al.*, 1998), indicating that some caution may be necessary in associating SSCs measured at proglacial stream gauging stations with those emerging from an upstream glacier.

Hodson, A., Gurnell, A., Tranter, M., Bogen, J., Hagen, J.O. and Clark, M. 1998. Suspended sediment yield and transfer processes in a small high-Arctic glacier basin, Svalbard. *Hydrological Processes*, 12(1), 73–86.

Richards, G. and Moore, R.D. 2003. Suspended sediment dynamics in a steep, glacier-fed mountain stream, Place Creek, Canada. *Hydrological Processes*, 17(9), 1733–1753.

relatively high SSC events are frequently associated with tapping new sediment sources during, for example, high discharge outburst floods and the recently recognized 'spring event', during which the subglacial drainage system opens up in a stepwise fashion. Both of these occurrences involve the penetration of effective subglacial drainage pathways into areas that were hitherto untapped.

4.4.2.1   Sampling for suspended sediment concentration

Suspended sediment concentration (SSC) is usually expressed as the weight of suspended sediment per unit volume of water ($g\,l^{-1}$, equivalent to $kg\,m^{-3}$). Importantly, rigorous guidelines need to be followed in sampling for SSC because the property varies markedly and systematically across the stream channel. If sampling is performed by hand, the instrument most commonly used is a USDH-48 sampler, consisting of a cast aluminium sampler attached to a rod (Figure 4.4). The sampler comprises a narrow nozzle that is pointed

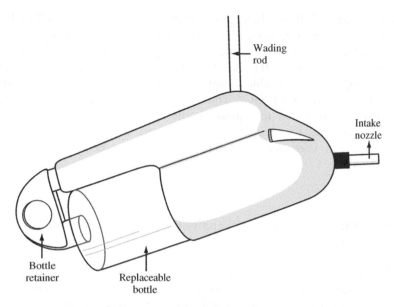

Figure 4.4 Drawing of a USDH-48 suspended sediment sampler

directly into the stream flow, and a rear section that houses a 450-ml bottle to receive the sampled water. When immersed in the stream, water enters the instrument through the front nozzle and the air displaced from the empty sample bottle escapes through a separate vent. A complete (450 ml) sample has been collected once air stops exiting the vent as bubbles. Sampling can be either at-a-point or depth-integrated. At-a-point samples are collected at specific depths by lowering the instrument into the stream with the nozzle pointing downstream until it reaches the depth at which the sample is required. The sampler is then rotated so that the nozzle points upstream and is held steady while a sample is collected. Once collected, the instrument is removed rapidly and steadily from the stream. In contrast, depth-integrated samples are collected continuously from along a vertical profile by moving the sampler slowly and steadily up and down the water column while it fills. It is important in collecting depth-integrated samples that the instrument is lowered and raised at a constant speed in order to avoid over-representation of any particular depth range. Ideally, sampling, whether at-a-point or depth-integrated, should be carried out at horizontal increments right the way across the stream, providing a complete 2D picture of SSC variation. Once a water sample has been collected, the sampler should be transferred to the bank for sample removal and storage or immediate filtration (section 4.4.2.2).

Not all proglacial stream sampling for SSC involves the use of a USDH sampler: A sampler may not be available or its use may be considered unsafe at night or at high flows. In such situations, researchers may collect hand samples following as reproducible a sampling regime as is possible in the circumstances. It may be possible to calibrate such hand samples against a smaller number of USDH-collected samples. One benefit of hand sampling, though, is that variable sample quantities can be collected according to a visual assessment of SSCs. Thus, sample size, and therefore filtration time, can be optimized to equate to the minimum acceptable sediment quantity required for accurate measurement: even under a powerful vacuum, it can take over an hour for a sediment-rich sample of only 200 ml to filter.

4.4.2.2   Sample treatment and analysis: Suspended sediment concentration

Analysis for SSC involves filtering a meltwater sample and measuring the volume of filtrate and the weight of sediment retained on the filter paper. It is important that all of the initial sediment and meltwater are filtered and measured accurately. Thus, samples that have been in storage for some time may need to be shaken vigorously to ensure that any settled sediment is re-suspended prior to decanting into a filtration chamber. Filter papers used should have a small pore space, since suspended sediment can be very fine (with a significant fraction in the clay-size range, i.e. <1 μm in diameter). Typically, cellulose nitrate filter papers are used, with a nominal pore size of 0.45 μm. Field filtration commonly involves the use of hand-pumped vacuum filtration units such as those manufactured under the Nalgene label by Nalge Nunc International, USA. These units comprise an upper chamber into which the raw sample is decanted and a lower chamber into which the filtrate is sucked under an imposed pressure gradient. The two chambers are separated by a filter bed, on which the filter paper rests, and the pressure is reduced in the lower chamber by a vacuum pump. Typical field instructions for using such an apparatus may be as follows:

- Assemble the filter apparatus with the filter paper in the filter holder. Ensure that the filter is trapped intact between the two chambers and has not been ruptured as they were screwed together.
- Agitate the meltwater sample to ensure all sediment is fully suspended, then rapidly and smoothly pour about 80% of the sample into the upper chamber of the filter apparatus. Use the remaining 20% to ensure that all sediment is suspended in the sample bottle, and then decant that suspension into the upper chamber.
- Place the lid over the upper chamber, ensuring one or more of the air vents in the lid is open to atmospheric pressure. If the upper chamber is sealed,

a vacuum will be generated in it as the meltwater flows into the lower chamber, reducing the gradient between the two chambers and reducing the effectiveness of the operation. Use the hand pump to create a vacuum in the lower chamber, and ensure that visually clear filtrate is being drawn through the filter paper into the lower chamber. If the filtrate is not clear, then a problem with the filter paper integrity or placement is likely, and the apparatus should be cleaned and the operation repeated.

- If the filtration rate is slow, the lower chamber should hold its vacuum for some minutes to tens of minutes. If it does not, then the seals around the lower chamber should be checked. It is worthwhile maintaining the vacuum intermittently to maximize the rate of filtration. A sensible vacuum magnitude in the lower chamber is between 40 and 50 cm of mercury.
- When filtration is complete, unscrew the upper chamber from the lower chamber and use tweezers to remove the filter paper from its bed. Fold the filter paper in half and place it in a clearly labelled storage container. Small lockable polythene bags are ideal for this purpose in the field. Return the paper to the laboratory for drying and weighing.
- Remove the filter bed from the lower chamber and decant the filtrate into a measuring cylinder to record its volume to the nearest millilitre. Note the volume measured – ideally both on the filter paper container and in a field notebook. Discard the filtrate, or store it if it is required for further chemical analysis (section 4.4.3.7).
- Back in the laboratory, the filter paper should be dried in an oven overnight at $\sim 40\,^\circ$C, and weighed. The weight of suspended sediment is then calculated by subtracting the weight of the clean filter paper (weighed before use) from the weight of the dirty filter paper, and the SSC of the sample is calculated as this weight (in grammes or kilogrammes) divided by the volume of sample filtrate (in litres or cubic metres respectively).

Once SSC has been measured, suspended sediment transport rate ($kg\,s^{-1}$) in the meltwater stream sampled can be calculated by multiplying the SSC ($kg\,m^{-3}$) by the stream's discharge ($m^3\,s^{-1}$).

4.4.3 Meltwater chemistry – background

Meltwater chemistry refers to the chemical properties of the meltwater itself rather than the character of the solid matter transported by the meltwater. In its broadest sense, therefore, any discussion of meltwater chemistry could include a large variety of individual physical properties. In this section we concentrate on those most frequently recorded and reported by glacier hydrochemists: ionic chemistry, $p(CO_2)$, pH and isotopic chemistry.

4.4.3.1 Ionic chemistry

An ion is an atom or group of atoms that has become electrically charged by losing or gaining electrons. Those that have lost electrons are positively charged and are termed cations, and those that have gained electrons are negatively charged and are termed anions. The sign and magnitude of the charge is indicated by a superscript following its chemical symbol. For example, the monovalent sodium ion is denoted Na^+ (lost one electron), and the divalent sulphate ion is SO_4^{2-} (gained two electrons). Many compounds are made up of combinations of positive and negative ions: sodium chloride (NaCl), for example, is formed from sodium ions (Na^+) and chloride ions (Cl^-). When these compounds dissolve in water they dissociate into their individual ions. The concentration of these ions in a liquid is normally expressed as ionic mass per unit volume of liquid, with units of $kg\,m^{-3} = g\,l^{-1}$ or $mg\,l^{-1}$. Concentrations may also occasionally be expressed as parts per million ($ppm = mg\,l^{-1}$), moles per litre ($mol\,l^{-1}$) or as equivalents per litre ($equiv\,l^{-1}$). The last of these expresses concentration in terms of the reaction potential of the ion concerned, and is defined as the ionic mass that would be required to react with 1 mole (1.008 g) of hydrogen (H^+) per litre of solution.

A practical conversion that is often required in solute concentration ranges typical of proglacial environments is between $mg\,l^{-1}$ (or ppm) and $\mu equiv\,l^{-1}$. To convert the former to the latter, (i) multiply the ppm concentration by the ion's charge (valence), (ii) divide the product by the ion's atomic mass, and (iii) multiply the product by 1000 (to convert from milli, or per thousand, to micro, or per million). For example, for Ca^{2+}, which has an atomic weight of 40.1, 1 ppm (or $mg\,l^{-1}$) $Ca^{2+} = (1 \times 2 \times 1000)/40.1 = 49.9\,\mu equiv\,l^{-1}$.

Meltwaters discharging from glaciers are ionically enriched relative to their initial precipitation inputs. While ionic species such as chloride (Cl^-) and nitrate (NO_3^-) are principally derived from atmospheric aerosols, others such as calcium (Ca^{2+}), magnesium (Mg^{2+}), sulphate (SO_4^{2-}) and silica (Si) are acquired from contact with reactive debris during water flow through the glacier drainage system. Since such contact is achieved most effectively at the glacier bed, temporal variations in meltwater ionic concentration may be interpreted in terms of the flow pathways followed by those waters. High bulk meltwater solute concentrations thereby indicate contact with reactive sediments, possibly for extended periods of time. Such a signature is generally interpreted in terms of flow through the subglacial drainage system, via hydraulically inefficient pathways with slow transit speeds and involving much contact with large amounts of freshly eroded debris. In contrast, low ionic concentrations are consistent with no, or only a short period of, flow at the glacier bed, via a hydraulically effective drainage pathway characterized by large water fluxes, rapid transit speeds and contact with low concentrations of debris (Wadham et al., 1997). These differences led to the application of EC-based mixing models to glacier hydrology (Collins, 1978), but these have been found to be somewhat too simplistic for useful application (Box 4.3).

Box 4.3   Meltwater discharge partitioning by the application of EC-based mixing models: Swiss Alps

Chemically based mixing models have been applied to glacier hydrology based on the assumption that the bulk meltwaters delivered from glaciers are made up of two components: an ionically rich, delayed-flow component and an ionically dilute, quick-flow component. This approach was pioneered by Collins (1978, 1979), who formalized this balance as:

$$Q_t C_t = (Q_s C_s) + (Q_e C_e)$$

where Q is discharge, C is solute concentration (EC), and the subscripts t, s and e refer to total (bulk) flow and its subglacial (delayed-flow) and englacial (quick-flow) components respectively. Rewriting this equation using the substitution ($Q_e = Q_t - Q_s$) gives:

$$Q_s = \left[\frac{(C_t - C_e)}{(C_s - C_e)} \right] Q_t$$

This allows the subglacial component of total glacier discharge (Q_s) to be calculated from the terms that comprise the right-hand side of this equation. Here, Q_t (bulk discharge) and C_t (bulk EC) are recorded in the proglacial outflow stream, C_e (quick-flow solute concentration) is given by the EC of supraglacial meltwaters (considered by Collins (1978) to be $\sim$2 μS cm^{-1}), and C_s (delayed-flow solute concentration) is given by the EC of bulk outflow when no supraglacial meltwater is being generated, such as following a snow storm (considered by Collins (1978) to be $\sim$44 μS cm^{-1}). Once Q_s is calculated, it can be subtracted from Q_t to yield Q_e.

Collins used this technique to reconstruct temporal variations in Q_s and Q_e at Gornergletscher and Findelengletscher, Switzerland, concluding, for example, that Q_s varied out-of-phase with Q_t at the former glacier, while they varied in-phase at the latter glacier. However, the applicability of this approach has been questioned, particularly in the light of spatially distributed measurements of subglacial water chemistry provided by arrays of boreholes. Sharp et al. (1995), for example, cited

---

Box 4.3   (Continued)

borehole measurements from Haut Glacier d'Arolla, Switzerland, and laboratory dissolution tests to argue that it may not be valid to assume the presence of two chemically uniform and discrete drainage components that do not react chemically once mixed at the glacier bed. This critique does not necessarily mean that chemically based mixing models have no place in studies of glacier hydrology, only that they need to be devised and interpreted more cautiously and in the light of a full knowledge of the operation of the drainage systems concerned. For example, focusing on a conservative indicator species such as SO_4^{2-} would overcome some of the problems inherent in an EC-based approach (e.g. Tranter and Raiswell, 1991).

Collins, D.N. 1978. Hydrology of an alpine glacier as indicated by the chemical composition of meltwater. *Zeitschrift für Gletscherkunde und Glazialgeologie*, **13**, 219–238.

Collins, D.N. 1979. Quantitative determination of the subglacial hydrology of two Alpine glaciers. *Journal of Glaciology*, **23**(89), 347–362.

Sharp, M., Brown, G.H., Tranter, M., Willis, I.C. and Hubbard, B. 1995. Comments on the use of chemically based mixing models in glacier hydrology. *Journal of Glaciology*, **41**(138), 241–246.

Tranter, M. and Raiswell, R. 1991. The composition of the englacial and subglacial components in bulk meltwaters draining the Gornergletscher. *Journal of Glaciology*, **37**, 59–66.

Glacier hydrochemists responded to the limitations of EC-based mixing models by investigating in more detail the chemistry of the reactions occurring in and around glaciers (Raiswell, 1984; Tranter *et al.*, 1993). Of these reactions, acid hydrolysis (the chemical reaction of a compound with acidic water), including ion exchange, surface exchange and lattice dissolution, is generally considered the most important subglacial chemical weathering process. The hydrolysis of carbonates is particularly rapid, although silicate and aluminosilicate minerals may also be weathered by hydrolysis. These reactions are described by equations of the form (here presented for carbonate hydrolysis):

$$CaCO_3(s) + H^+(aq) = Ca^{2+}(aq) + HCO_3^-(aq) \qquad (4.14)$$

where (aq) and (s) respectively denote aqueous and solid forms. The availability of free protons (H^+ – the concentration of which defines acidity) to drive these reactions represents one of the primary controls over solute acquisition by glacial meltwaters. In addition to transient snowpack leaching, protons may be derived from (i) the dissociation of H_2CO_3 (provided by atmospheric CO_2 coming into contact with water) and (ii) the oxidation of sulphide minerals such as pyrite, represented respectively by:

$$CO_2(aq) + H_2O(aq) = H^+(aq) + HCO_3^-(aq) \qquad (4.15)$$

and

$$4FeS_2(s) + 15O_2(aq) + 14H_2O(aq) = 16H^+(aq) + 4Fe(OH)_3(s)$$
$$+ 8SO_4^{2-}(aq) \qquad (4.16)$$

In reality, these reactions are probably coupled such that protons released by sulphide oxidation simultaneously fuel accompanying carbonate hydrolysis:

$$4FeS_2(s) + 15O_2(aq) + 14H_2O(aq) + 16CaCO_3(s)$$
$$-4Fe(OH)_3(s) + 8SO_4^{2-}(aq) + 16Ca^{2+}(aq) + 16HCO_3(aq) \qquad (4.17)$$

Reference to equations (4.13) to (4.15) indicates that the relative contributions of these proton sources may be reconstructed from the relative concentrations HCO_3^- and SO_4^{2-} in glacial meltwaters.

4.4.3.2   Partial pressure of carbon dioxide ($p(CO_2)$)

The relationship between the supply of protons to meltwaters and their consumption by weathering reactions may also be investigated on the basis of the partial pressure of the CO_2 in solution in those meltwaters, $p(CO_2)$. This approach forms the basis of characterizing the chemical composition of bulk meltwaters in terms of flow through *open* or *closed* systems (Raiswell, 1984).

In the absence of practicable devices for the direct field measurement of $p(CO_2)$, it is usually calculated from:

$$\log p(CO_2) = \log(HCO_3^-) - pH + 7.7 \qquad (4.18)$$

where concentrations are in $mol\,l^{-1}$ (Ford and Williams, 1989).

4.4.3.3   Potential of hydrogen (pH)

It is clear from the summary above that pH is a key chemical property of meltwaters. pH is a quantitative expression of the acidity or alkalinity of a solution, and is defined as the negative logarithm of the concentration (in moles per cubic decimetre; $mol\,dm^{-3}$) of hydrogen ions (H^+) in a solution:

$$pH = -\log_{10}(H^+) \qquad (4.19)$$

Thus, pH decreases as hydrogen ion concentration, and therefore acidity, increases, and vice versa. pH is measured on a scale of 0–14: a neutral solution such as pure water has a pH of 7, that is it has a hydrogen ion concentration of $10^{-7}\,mol\,dm^{-3}$. pH values above 7 indicate the degree of alkalinity and numbers below 7 indicate the degree of acidity. Most pH values in glacier systems lie in the range 5–8. As a result of the acidifying effect of atmospheric CO_2, rainfall has a pH of $\sim$5.8.

Because the reactions that govern pH can occur rapidly in response to temperature, pressure and gaseous conditions, pH is usually measured in the field as soon as possible after sample collection. Numerous pH probes providing digital readouts are available commercially.

4.4.3.4   Bicarbonate (HCO_3^-)

The concentration of HCO_3^- is strongly dependent on the physical properties of the meltwater sample (including pH and $p(CO_2)$) and should therefore also be measured in the field as soon as possible following sampling. This is commonly achieved by a dilute acid titration to pH 4.5, and the reader is advised to consult a standard geochemical text for procedural details. If field titration is not possible, then HCO_3^- concentration may be approximated by balancing the ionic charge of the solution. This calculation must be carried out in equivalence units, and is based on the assumption that the concentrations of all other major ions present in the sample have been accurately measured. Thus, since the net charge of the ions in solution must be zero, all of the 'missing' negative charge can be ascribed to HCO_3^-.

4.4.3.5   Stable isotopic chemistry

Most elements can exist in more than one atomic form, resulting from variations in the number of neutrons in their nuclei. These *isotopes* therefore have different atomic weights, calculated as the number of neutrons and protons in each atom. These atomic weights are normally denoted by a superscript preceding the atomic symbol. The two most common forms of oxygen are ^{16}O ('oxygen sixteen'), having 8 protons and 8 neutrons in the nucleus, and ^{18}O ('oxygen eighteen'), with 8 protons and 10 neutrons in the nucleus. The two most common isotopes of hydrogen are ^{1}H (one proton, no neutrons) and ^{2}H (one proton, one neutron). ^{2}H is also known as deuterium (D). Different isotopes of each element exist in fixed proportions globally but local departures from these ratios can occur because the higher vibrational frequency of light atoms within crystals or molecules means that they form bonds that are weaker than those of heavy atoms. This leads to isotopic fractionation during phase changes, the amount of which is given by the fractionation factor (α) where:

$$\alpha = \frac{R_a}{R_b} \tag{4.20}$$

Here, R_a and R_b are the ratios of heavy to light isotopes in phases a and b respectively. In the case of water the dominant isotopes are $H_2{}^{16}O$, $HD^{16}O$ and $HD^{18}O$. One critical ratio in the context of meltwater studies is the ratio of ^{18}O to ^{16}O. This is expressed using the δ notation, where the $^{18}O/^{16}O$ ratio in a sample is expressed relative to that of a standard concentration, normally in parts per thousand (‰). Thus, for ^{18}O:

$$\delta^{18}O = \frac{\left(\frac{^{18}O}{^{16}O}\right)_s - \left(\frac{^{18}O}{^{16}O}\right)_r}{\left(\frac{^{18}O}{^{16}O}\right)_r} \times 1000\,‰ \tag{4.21}$$

where s refers to sample and r to reference, usually SMOW (standard mean ocean water) in the case of water analysis. Hydrogen isotope ratios can similarly be expressed in terms of δD. Studies of meltwater isotope chemistry take advantage of the fact that at least two sets of physical processes can induce systematic variations in isotopic composition. First, the local-scale isotopic composition of snow is partially dependent on the temperature, and therefore the altitude, at which it formed. In general, snow composition gets lighter with altitude, although the precise rate of decrease depends on local factors (Box 4.4). Second, the freezing of water to ice is accompanied by an increase in $\delta^{18}O$ of up to 3‰ in that ice. In a closed system, this leaves the remaining liquid, and the ice formed from it, increasingly depleted in ^{18}O (Box 3.6).

4.4.3.6 Sampling for chemical composition

Because the chemical composition of meltwater does not depend on local flow hydraulics, stream channel cross-sections are probably characterized by far less spatial variability in meltwater chemistry than in suspended sediment concentration. Consequently, it is possible to sample for meltwater chemistry anywhere within the main channel flow. Moreover, this sampling can make use of a range of commercially available automated

Box 4.4   Reconstructing precipitation inputs and ice flow patterns from the distribution of oxygen isotopes over a glacier surface: Saskatchewan Glacier, Canada

As part of a broader oxygen-isotopic investigation conducted in the early 1950s, Epstein and Sharp (1959) sampled snow and ice from along one longitudinal and one transverse profile in the ablation area of Saskatchewan Glacier, Canada. The authors reported a general decrease in $\delta^{18}O$ in samples recovered from lower down the longitudinal transect (Figure B4.4a), suggesting a corresponding increase in the elevation at which those ice samples originally formed. This pattern was similar to trends measured elsewhere and was consistent

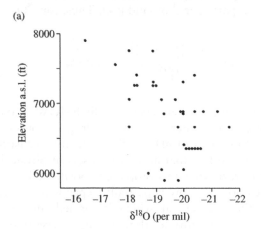

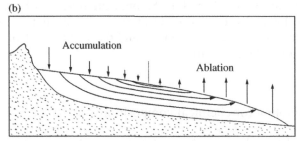

Figure B4.4   (a) Plot of $\delta^{18}O$ against elevation above sea level along a long profile at Saskatchewan Glacier; and (b) illustration of the centreline flow vectors in long section at a typical valley glacier (after Reid, 1896). Reproduced from Epstein and Sharp (1959) with the permission of University of Chicago Press

with Reid's (1896) model of ice redistribution by flow along a glacier (Figure B4.4b). Across glacier, Epstein and Sharp (1959) were able to identify the relatively light ice constituting a flow unit that had originated from a tributary glacier with a relatively high accumulation area. More generally, the authors found that ice located towards the sides of Saskatchewan Glacier tended to be isotopically lighter (i.e. characterized by lower $\delta^{18}O$ values) than that located towards the glacier's centreline. The authors interpreted this pattern in terms of (i) the particular topography and flow field of Saskatchewan Glacier and (ii) a more widespread flow pattern at valley glaciers involving the slower and deeper transfer of marginal ice relative to centreline ice, suggesting the former originates at a higher altitude in the accumulation area.

Epstein, S. and Sharp, R.P. 1959. Oxygen-isotope variations in the Malaspina and Saskatchewan Glaciers. *Journal of Geology*, 67, 88–102.

Reid, H.F. 1896. The mechanics of glaciers. *Journal of Geology*, 4, 912–928.

water samplers – such as those produced by Isco Inc., USA. These samplers are battery-powered, electronically programmed, and a full-sized model typically has the capacity to recover 24 samples, each of 1 litre. The samplers are generally robust and require little maintenance beyond routine battery charging and bottle replacement. However, despite the ready availability of such samplers, many glacier hydrochemistry studies also (or even solely) collect samples by hand: normally simply by dipping a plastic bottle (on the end of a pole if required) into the main flow of the stream. Although such hand sampling is more laborious than automated sampling, particularly where routine sampling is concerned, it also has the important advantage of flexibility, both in terms of the timing of sample collection and in allowing precise sample location to be chosen in the light of flow conditions at the time of sampling. Further, since samples may need to be treated soon after collection, a researcher may have to be present at the time of sampling anyway. It is easier in such situations to collect a hand sample than it is to retrieve one from an automated sampler.

Sampling proglacial meltwaters by hand involves little expertise beyond recognition of the requirement to minimize contamination. For most analyses, and in particular hydrochemical analyses, this requires, at each stage from sample collection to sample storage, that all surfaces contacting the sample should be cleaned first. This generally involves using small volumes of the sample itself to pre-rinse containers and sensor surfaces two or three times prior to coming into contact with the remainder of the sample. This rinsing can require a significant proportion of the initial sample collected and should be factored into deciding the volume of sample to be collected at the outset.

4.4.3.7   Sample treatment and analysis: Chemical composition

Great care must be taken during sampling for hydrochemical analysis to avoid sample contamination. The concentration of elements or ions in a small low-concentration sample can increase markedly upon incorporating only a small quantity of high-concentration liquid, or upon contacting a reactive solid. Similarly, sample bottles should not be rinsed out with deionized water between samples unless they are to be thoroughly cleaned and dried. This is because incorporating deionized water into a sample contaminates by decreasing the concentration of the impurity to be measured in the sample. It may also be necessary to wear clean rubber gloves to sample meltwaters if they are to be analysed for species that are anticipated to be present in very low concentrations. A typical meltwater sampling procedure for hydrochemical analysis would be as follows:

If an acid-washed, rinsed, dried and sealed 500-ml polythene bottle is not available, rinse a bottle out three times in the main stream flow.

- Collect 500 ml of meltwater in that bottle from the main stream flow.
- Pour ∼50 ml of the sample into the upper chamber of a filtration unit (section 4.4.2.2 above) and use it to rinse the sides and base of the chamber. Discard the sample and, by continued agitation, as much as possible of the sediment suspended in it. Repeat the procedure at least once more.
- Agitate the remainder of the meltwater sample to ensure all sediment is fully suspended and pour about 80% of the sample into the upper chamber of the filtration unit. Agitate the remaining 20% to ensure that all sediment is suspended in the sample bottle, and then decant the remaining suspension into the upper chamber.
- Place the lid over the upper chamber, ensuring one or more of the air vents in the lid is open to atmospheric pressure. Use the hand pump to create a vacuum in the lower chamber, and draw ∼30 ml of sample into it from the upper chamber. Seal the upper chamber and use the filtrate in the lower chamber to rinse its interior. Discard the water through a hole in the side of the lower chamber. Repeat once (remembering to release the seal on the upper chamber during filtration and to seal it for rinsing by turning the apparatus on its side and rotating it about its now-horizontal central axis).
- Draw the remainder of the sample through the apparatus into the bottom chamber.
- If an acid-washed, rinsed, dried and sealed 30–50 ml (depending on instrumental requirements) polythene bottle is not available, pour

~10 ml of the sample from the bottom chamber into a sample bottle, replace the top and rinse the interior of the bottle vigorously. Discard the water and repeat at least once.

- Pour the remainder of the filtrate into the sample bottle until the bottle is full and replace the top securely.
- Label and store the sample bottle, preferably out of direct sunlight and in refrigerated conditions. If there is any fear of micro-biological activity tainting the sample, it may be acidified by adding a small amount of acid to the sample bottle after it is rinsed out and before it is filled with the sample. Here, it is common to acidify the sample to ~pH 2 with HNO_3 (Brown *et al.*, 1970).

It may also be the case that samples are filtered for both the determination of suspended sediment concentration and the determination of the chemistry of the filtrate. In such cases, the procedure above must be introduced into that described in section 4.4.2.2 to ensure that the filtrate remains uncontaminated. Further, it is also necessary to record the total volume of filtrate that passed through the filter into the lower chamber. Since the lower chamber was pre-rinsed, that volume includes both the discarded pre-rinse and the filtrate that provided the final sample.

Sampling for stable isotope analysis alone need not be quite so rigorous because contamination from reacting with solids is negligible and contamination from the incorporation of a high-concentration liquid is unlikely. Thus, it is not necessary to pre-filter meltwater samples collected for isotopic analysis alone. However, laboratory sample preparation and mass spectrometry require samples to be free of particles of sand size and above. Further, isotopic differences between sample groups may be very small, and it is always desirable to minimize contamination by contacting another liquid. Pre-rinsing bottles with sample liquid is therefore always advisable for isotopic analysis. A number of other factors may also alter the isotopic composition of a sample:

1. *Partial freezing or evaporation.* It is vital that waters are not decanted for analysis from samples that have been partially frozen – in such cases it is highly likely that fractionation accompanied the freezing process, depleting the remaining liquid sample in heavy isotopes (section 4.4.3.5 above). Similarly, partial evaporation has the capacity to deplete the remaining liquid sample in light isotopes (that form the weaker bonds and therefore evaporate preferentially). Where such evaporation is feared, for example due to an anticipated long period of sample storage prior to analysis, bottles can be sealed to air with sealing film or tape placed over the bottle thread and under the cap.

2. *Micro-biological activity.* Biological photosynthesis and respiration can alter the isotopic composition of a sample; although the effect is likely

to be minor in glacigenic meltwaters unless samples are stored for long periods. This can be minimized by using amber-coloured bottles and storing samples out of sunlight at cold temperatures.

3. *Reaction with debris.* It may be possible over long time periods for isotopic exchange to occur between water and hydroxyl bearing minerals (Souchez *et al.*, 1990). Although it appears that high debris concentrations and long time periods are required for such exchange to be significant, the effect could be minimized by sample filtration.

4.5 AUTOMATED MEASUREMENTS BY SENSORS AND LOGGERS

Concentrations of many dissolved ionic species and gases can be measured directly in meltwaters by off-the-shelf sensors that will provide a digital readout on an associated meter. These systems carry their own operating instructions and their use requires little additional knowledge. Care must still be taken, however, to avoid sample contamination if readings are to be taken directly from the sample in the field. These portable sensors are also likely to provide data that are of lower accuracy than their larger and more stable laboratory counterparts, particularly when considering possible field conditions. For these reasons, direct field readings are often only taken when there is a risk that the value of the property being measured will change during storage. This is often the case, for example, with dissolved gases, pH and HCO_3, since these are partially dependent on site-specific conditions such as water temperature and pressure. Where on-site analysis is to be carried out, it is important to keep apparatus stable and free of dust. Although contamination by dust is less of a problem when working on ice or in snow (Figure 4.5), a shelter may be needed in snow-free proglacial environments.

Importantly, proxy data for some key water quality properties can be measured by data logger – with the advantage that time series of these properties can be generated automatically at very high resolution. The two most commonly measured properties are EC and turbidity, surrogates for total dissolved solids and suspended sediment concentration, respectively.

4.5.1   Automated measurement of electrical conductivity (EC)

Although decent hand-held EC sensors are available commercially, their use requires the presence of an investigation at the time of each reading. The use of fixed sensors, controlled by data loggers, is therefore generally preferable. The principles of the construction and measurement of such sensors are well laid out in relation to glacierized environments by Stone *et al.* (1993) and

Figure 4.5 Field filtration of a meltwater sample collected from the base of the snowpack for subsequent hydrochemical analysis

Smart and Ketterling (1997). The basic principle underlying these measurements is that the resistance between two electrodes immersed in a fluid will decrease as that fluid's EC rises. Field EC sensors therefore comprise a weighted housing containing two electrodes, a fixed area of each of which is exposed to the water to be measured. The wires from each electrode pass away from the housing to a nearby gauging station, where the resistance between them is recorded.

EC sensors are typically constructed of an electrically inert housing, for example a length of PVC pipe, perhaps 5 cm across and 30-cm long. Although the shape of the housing is not important, such pipe is also suitable for down-borehole use. Two electrodes are located in one end of the cylinder, set back some centimetres from its end for protection. The remainder of the cylinder is filled with resin to add weight and to protect the electrode wires and connections from shorting through water within the instrument. A typical electrode may be 5 mm in diameter (providing strength), and separated by ~10 mm over an exposed length of ~10 mm (Figure 4.6). They may be made from any conductive metal with a high resistance to corrosion and erosion: stainless steel is ideal. These electrodes may be protected by the cylinder, protruding from the end of the solid interior of the cylinder into its hollow end. This protective length of cylinder should be pierced with holes to ensure unrestricted water flow past the electrodes. A perfectly acceptable, though not terribly robust, EC sensor can be easily made from a standard 2-pin bulb socket. Here, the cable contacts are wired to the pins on the inside of the housing, which is then filled with resin ensuring that the contacts are covered.

Since EC sensors actually measure the resistance (R, in Ohms, Ω) of the water between electrodes of a certain geometry and separation, each sensor needs to be calibrated to allow that resistance to be converted into the water's resistivity (Ω m). This value then needs to be inverted to represent

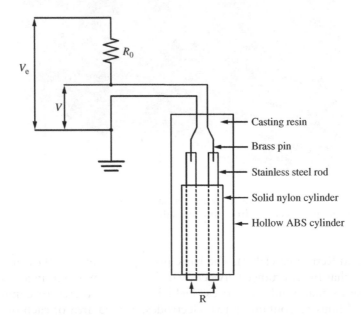

Figure 4.6 Illustration of a field EC probe, suitable for borehole and stream use. According to this design, stainless steel rods are pressed into pre-drilled holes in the nylon rod. The signal cable is soldered to these pins via brass pins attached to the interior end of these rods. Once attached, the entire assembly is cast in clear resin within a section of PVC pipe. The wiring for the AC half-bridge (illustrated) is located at the data logger end of the sensor cable. Reproduced from Stone *et al.* (1993) with the permission of the International Glaciological Society

conductivity (siemens per metre, $S\,m^{-1}$) and scaled to suitable units, often $\mu S\,cm^{-1}$ in glacial waters. Sensor calibration involves immersing the sensor into a solution of known EC and calculating its cell constant (K_c in m^{-1}), which describes the relationship between measured resistance (Ω) and the liquid's EC ($S\,m^{-1}$):

$$K_c = ECR \qquad (4.22)$$

Sensor measurement may be readily carried out by data logger, based on an AC half-bridge measurement, whereby sensor voltage is measured twice in rapid succession on the limbs of the AC cycle to avoid electrode polarization.

Once calibrated, the EC sensor is installed in a proglacial stream, preferably in a position that is protected from the erosional effects of bed-load and saltating particles while allowing free flow of water around the electrodes. Therefore, EC probes are commonly attached to the limbs of gauging station structures sufficiently far off the stream bed to minimize erosion yet low enough to ensure complete electrode immersion even at low flows. EC time series collected in this way can be used as a guide to general hydrochemical variations in the water.

4.5.2 Automated measurement of turbidity

Field turbidity sensor construction and measurement are described in relation to glacier borehole use by Stone *et al.* (1993). Turbidity expresses the reduction of transparency of glacial meltwaters due to the presence of suspended sediment. Turbidity sensors are based on the principle that the intensity of a beam of light transmitted across the fluid to be measured decreases as the turbidity of the fluid increases. Normally, the light source is a light-emitting diode (LED) and the intensity of the received signal is measured by a detector comprising a light-sensitive photodiode – although instruments based on lasers are now available. When measuring the intensity of the received signal it is necessary to compensate for variations in natural light that characterize proglacial streams. This can be achieved by using a signal with a wavelength of ~850 μm, in the infrared region of the spectrum (which minimizes, but does not eliminate the effect of natural light variations), and by using a second reference detector that is separated from the LED by an air space that is also open to variations in natural light. The use of a reference photodiode also allows the system to cater for possible variations in the luminosity of the LED.

Turbidity probes based on these principles can be purchased off the shelf from various companies – although such systems can be expensive and they may not be robust enough for use in proglacial streams. Fortunately, design and in-house turbidity sensor construction is relatively straightforward, requiring only a suitable housing, a LED, two photodiodes, and optionally a voltage regulator to ensure uniform voltage delivery to the LED. Of these components, the housing is the most tricky to construct. For glacier borehole use, Stone *et al.* (1993) used a resin-filled ping-pong ball with a V-shaped section cut through it for the sample to flow through (Figure 4.7). The advantage of filling a housing with resin is that it is clear, and can therefore be used to separate the LED and the reference photodiode, and waterproof, protecting the electronic components and connections from shorting.

Using such a system, the turbidity (T) is based on the ratio of the voltages recorded from the reference detector (V_0) and the sample detector (V), according to:

$$T = -\frac{1}{L}\ln\left(\frac{V}{aV_0}\right) \tag{4.23}$$

where L is the length of sample separating the source and the sample diodes. The correction factor a is specific to the precise geometry and construction

(a)

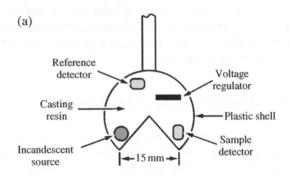

(b)

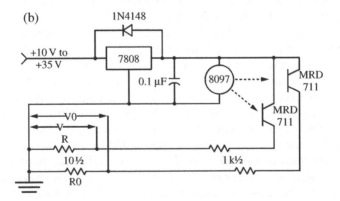

Figure 4.7 A field turbidity probe, suitable for borehole and stream use, illustrating: (a) the physical locations of the components, cast in clear resin within a plastic sphere (a ping-pong ball); (b) the instrument's circuit diagram. Reproduced from Stone *et al.* (1993) with the permission of the International Glaciological Society

of individual sensors, representing the correction that is required for V_0 to equal V at $T = 0$.

Although there is an absolute unit of turbidity, the nephelometric turbidity unit (NTU – based on standards containing a milk-like polymer called formazin), it is difficult to calibrate turbidity signals in terms of SSC because measured values are influenced by a number of additional properties. These include the grain-size distribution, particle shape and lithology of the suspended sediment. If the exact nature of the sediment that is expected to be in suspension is known, then it can be used as a basis for empirical calibration. However, this is unlikely in glacial settings and results are often reported as measured turbidity (m^{-1}) or relative turbidity, scaled to vary between 0 (clear) and 1 (opaque).

4.6   TRACER INVESTIGATIONS

4.6.1   Background and physical principles

Salt and dye tracers have been used for over a century to investigate glacier drainage systems. These tracers provide direct information on water flow characteristics between an input site, usually a freely draining, ice surface moulin or crevasse, and a detection site located on the proglacial stream close to the glacier terminus. The key requirements for an ideal tracer are that it should be readily soluble in cold water, it should be physically and chemically stable, it should not be easily adsorbed onto mineral surfaces, it should be detectable at low concentrations, it should have low background concentrations, and it should be non-toxic. Common salt (NaCl) has been used as a tracer in glacial meltwaters, which are typically of low salinity (commonly 10^0–10^1 $\mu S\,cm^{-1}$), mainly because salt is inexpensive, easily obtained, and can be readily detected by an electrical conductivity sensor (section 4.5.1). However, the use of salt as a tracer suffers from highly variable background concentrations and poor detection sensitivity, demanding the transport and use of large quantities. Consequently, the most commonly used tracers are fluorescent dyes (Smart and Laidlaw, 1976), particularly Rhodamine (termed 'Rhodamine WT' in its liquid form), which is a highly fluorescent red dye that absorbs green light and emits red light. Another well-used dye is Fluorescein, which emits a bright green fluorescence and is also detectable at very low concentrations. However, Fluorescein suffers from photochemical degradation (decreasing markedly over only some hours exposure to sunlight), is pH-sensitive and its fluorescence coincides with that of many natural materials. It is therefore generally used only in addition to Rhodamine, for example in dual tests based on synchronous measurement of both dyes.

At the simplest level, tracer experiments can be used at glaciers drained by several streams to delimit the spatial extent of individual drainage basins within that glacier (Box 4.5). However, tracer concentration–time curves (Figure 4.8) are capable of providing various quantitative parameters that may be related to the character of the glacier hydrological system. These are outlined in the following sections.

4.6.1.1   Tracer transit velocity

A minimum estimate of the mean water through flow velocity during a test can be obtained from the time elapsed between dye injection and peak concentration at the detection site and the distance between these two sites. If the straight-line distance is taken, then the calculated velocity is a minimum, since the actual path taken will be longer than that assumed. In general, the flow velocity through the subglacial drainage system depends on the character of the pathways through which the water has travelled. Rapid transit velocities of greater than $\sim 0.2\,\mathrm{m\,s^{-1}}$ are generally characteristic of flow through hydraulically efficient drainage systems. In contrast, lower velocities have usually been interpreted (e.g. Burkimsher, 1983) in terms of flow through less-efficient drainage pathways.

Box 4.5   Defining glacial drainage basins and their characteristics from proglacial stream records and tracer tests: South Cascade Glacier, USA

Fountain (1992) investigated the large-scale drainage characteristics of South Cascade Glacier, USA, by combining proglacial meltwater data with tracer experiments carried out both from the glacier surface (via moulins and crevasses) and from the base of boreholes drilled to the glacier bed. These tracer experiments indicated that the glacier was drained by three distinct drainage basins, each discharging through a separate proglacial stream (Figure B4.5). The discharge records of these streams allowed the author to draw conclusions about the responsiveness of each of the basins, which was found to be a function of the depletion of the surface snowpack over the basin and its principal flow pathways. These pathways were themselves reconstructed from records of the stream water's EC and turbidity, which were used to discriminate between waters that had primarily flowed through ice-walled conduits and those that had flowed in close contact with reactive debris at the ice–bed interface. In the latter case, Fountain (1992) compared the timing and discharge of the daily hydrograph from

stream 2 with the output from a model based on Darcian flow through a
1 m-thick confined subglacial aquifer. This indicated that the sediment
layer had a bulk hydraulic conductivity of $5 \times 10^{-3}\,\mathrm{m\,s^{-1}}$. This value is
higher than that reported elsewhere for subglacial sediments, and the
author considered it to reflect flow through the sediments supplemented
by that focused at the ice–sediment interface.

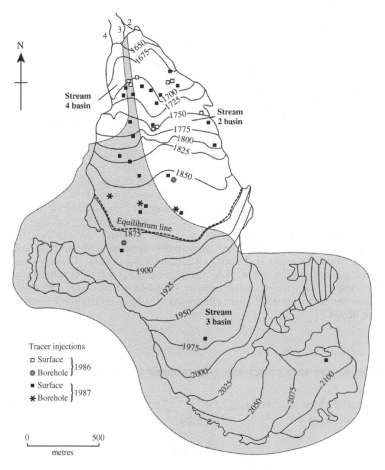

Figure B4.5 Three distinct drainage basins at South Cascade Glacier, USA,
identified by Fountain (1992) on the basis of proglacial stream water measurements and tracer experiments. Reproduced with the permission of the International Glaciological Society

Fountain, A.G. 1992. Subglacial water-flow inferred from stream measurements at
South Cascade Glacier, Washington, USA. *Journal of Glaciology*, 38(128), 51–64.

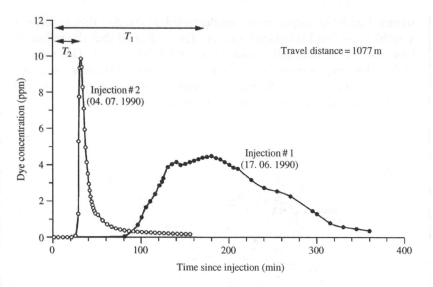

Figure 4.8 Two measured dye tracer breakthrough curves, expressed as dye concentration against time, conducted from a single moulin at Haut Glacier d'Arolla, Switzerland, in the summer of 1990. Injection #1, conducted in June, shows a delayed and dispersed concentration curve. The travel time to peak dye concentration (T_1) of ~180 minutes gives a mean transit velocity of $0.1\,\mathrm{m\,s^{-1}}$ ($= 1077/[180 \times 60]$), indicating flow principally through a distributed subglacial drainage network. In contrast, Injection #2, conducted in July, shows a much more rapid and peaked return curve. The travel time (T_2) of ~30 minutes to peak concentration yields a mean transit velocity of $0.54\,\mathrm{m\,s^{-1}}$ ($= 1077/[30 \times 60]$), indicating flow through a more efficient channellized subglacial drainage system by this time (see Box 4.7)

4.6.1.2   Tracer transit velocity – discharge relations

Channellized flow discharge (Q, $\mathrm{m^3\,s^{-1}}$), velocity (v, $\mathrm{m\,s^{-1}}$) and channel cross-section area (A, $\mathrm{m^2}$) are related by:

$$Q = vA \qquad (4.24)$$

Thus, researchers can use the relationship between Q and v as measured at the proglacial gauging station, to make inferences about the nature of A. Specifically, the relationship is used to identify whether subglacial channels linking the injection site to the measurement site are unfilled and flowing, like surface streams, at atmospheric pressure, or whether they are filled with meltwater and pressurized above atmospheric pressure. In the latter case, any change in Q can only be accommodated by an equivalent change in v, and a bivariate plot of Q against v will yield a positive linear relationship, the slope of which is a function of the net hydraulic geometry of the drainage system concerned. Further, this relation can be used to infer an averaged value of A, and it is possible to use turbulent flow theory to calculate the expected water flow velocity through a variety of alternative drainage configurations (Seaberg et al., 1988; Kohler, 1995). Alternatively, in the case of unpressurized or open channels, any change in Q will be accommodated by a change in A as well as in v, and a bivariate plot of v against Q will yield a curvilinear relation, the characteristics of which are again a function of the hydraulic geometry of the drainage system (Box 4.6).

Box 4.6 Reconstructing the nature of water flow through subglacial channels from tracer-based records of velocity and discharge: Storglaciären, Sweden

As part of their landmark dye-tracer study undertaken at Storglaciären, Sweden, Seaberg et al. (1988) pointed out that important information about the flow status of subglacial channels could be determined from plots of travel time against discharge for repeat experiments along the same flow pathway. This approach is based on the consideration that water discharge (Q) along a channel is a product of the velocity of that discharge (v) and the cross-sectional area of the flow (A):

$$v = \frac{Q}{A} \tag{1}$$

If that channel is subglacial and water-filled (i.e. pressurized) A cannot increase to accommodate an increase in Q, and the relationship between Q and v becomes linear:

$$v = kQ \tag{2}$$

where the value of the constant k (the slope of the straight-line bivariate plot of v against Q) reflects channel shape over the section length concerned. If, on the other hand, the channel section concerned is not water-filled (i.e. flowing at atmospheric pressure), changes in Q can be accommodated by adjustments in both A and v. In such open-channel flow cases, the relationship between Q and v becomes curved:

$$v = aQ^b \tag{3}$$

where the constants a and b reflect channel shape over the section concerned and the value of the exponent b is <1. Seaberg et al. (1988)

Box 4.6 (Continued)

plotted v against Q for a series of tracer tests undertaken at Storglaciären in 1984 and 1985, and compared these with tests undertaken in the glacier's open proglacial channel (Figure B4.6). While the latter experiments defined a relationship of the form of equation (3) (with $a = 0.69$ and $b = 0.27$), the subglacial tests indicated a relationship of the form of equation (2), with $k = 0.26$. The tests therefore demonstrated the utility of this approach and indicated that the subglacial flow tested at the glacier occurred through pressurized channels. These ideas weresubsequently applied in more sophisticated ways, principally by investigating the implications of the empirical constants k, a and b, by e.g. Kohler (1995) at Storglaciären, Fountain (1993) at South Cascade Glacier, USA, and Nienow *et al.* (1996) at Haut Glacier d'Arolla, Switzerland.

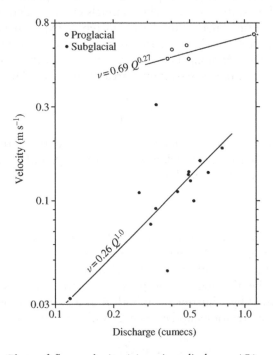

Figure B4.6 Plots of flow velocity (v) against discharge (Q) resulting from tracer tests undertaken on the glacier and in the proglacial area of Storglaciären, Sweden. Reproduced from Seaberg *et al.* (1988) with the permission of the International Glaciological Society

Fountain, A.G. 1993. Geometry and flow conditions of subglacial water at South Cascade Glacier, Washington State, USA – an analysis of tracer injections. *Journal of Glaciology*, 39(131), 143–156.

Kohler, J. 1995. Determining the extent of pressurized flow beneath Storglaciären, Sweden, using results of tracer experiments and measurements of input and output discharge. *Journal of Glaciology*, 41(138), 217–231.

Nienow, P.W., Sharp, M. and Willis, I.C. 1996. Velocity–discharge relationships derived from dye tracer experiments in glacial meltwaters: Implications for subglacial flow conditions. *Hydrological Processes*, 10(10), 1411–1426.

Seaberg, S.Z., Seaberg, J.Z., Hooke, R.L. and Wiberg, D.W. 1988. Character of the englacial and subglacial drainage system in the lower part of the ablation area of Storglaciären, Sweden, as revealed by dye-trace studies. *Journal of Glaciology*, 34(117), 217–227.

4.6.1.3   Percentage tracer recovery

The mass of tracer recovered during a test can be calculated by multiplying the area beneath the tracer breakthrough curve by the stream discharge. Similar to salt concentration–time curves measured for dilution gauging (section 4.3.2.2), this integral can be measured with specialist software, a compensating planimeter or by counting squares or trapeziums. Low-percentage recovery indicates tracer loss or storage within the glacier's drainage system. Such returns are more likely in a hydrologically complex drainage system, since open channels typically route water rapidly, as a concentrated tracer slug, with little potential for delay or storage. However, a number of non-hydrological variables may also influence tracer recovery. For example, Rhodamine B (but not WT) is susceptible to sorption onto particulate surfaces, and dye will therefore be preferentially lost from solution during periods when SSC is high. The presence of certain suspended sediments may also artificially raise the background fluorometer reading by fluorescing naturally at the same wavelength as the dye. These problems are most significant with green fluorescent dyes such as Fluorescein. Errors in either the proglacial discharge estimate, for example resulting from a poor rating curve, or in sensor calibration will also result in errors in the calculated amount of tracer recovered. Some caution should therefore be exercised when using tracer recovery to interpret the structure of the subglacial drainage system.

4.6.1.4   Tracer dispersion and breakthrough curve properties

The shape of a tracer breakthrough curve is determined by dispersive processes that spread water out during its passage through a flow system. Since these processes are controlled by flow hydraulics, the shape of a tracer return curve may be interpreted in terms of the structure of the flow pathways followed by that water. Following Seaberg *et al.* (1988) and Kohler (1995), the most commonly used parameter derived from the shape of breakthrough curves is *dispersivity* (d, m), which represents the rate of spreading of a tracer cloud (described by the *dispersion coefficient*, D, $m^2 s^{-1}$) relative to the rate of advection of the tracer during transit through a flow system (u, $m s^{-1}$):

$$d = \frac{D}{u} \tag{4.25}$$

The most commonly adopted form of D in glaciology is that derived by Brugman (1986) as:

$$D = \frac{\left(x^2(t_m - t_i)^2\right)}{\left(4t_m^2 t_i \ln\left(2\left(\frac{t_m}{t_i}\right)^{0.5}\right)\right)} \tag{4.26}$$

Here, x (m) is the straight-line distance from the point of injection to the point of measurement, t_m is the time (s) to peak concentration, and t_i is double-valued, representing the times (s) when the dye concentration reaches half its peak value on the rising (t_1) and falling (t_2) limbs of the breakthrough curve. Thus, equation (4.25) represents two equations ($i = 1$ or 2) and can be solved iteratively for t_m until both equations are satisfied, yielding a common value of D. In general, more dispersed tracer returns ($d > \sim10$) reflect inefficient drainage pathways, while highly peaked, less dispersed ($d < \sim10$) curves are interpreted in terms of flow through efficient channels (e.g. Burkimsher, 1983). Irregularities on the rising or falling limbs of breakthrough curves (including the occurrence of multiple tracer peaks) are interpreted in terms of flow divergence in the drainage pathways.

Tracer investigations therefore provide a simple, fairly inexpensive and valuable technique for investigating glacial drainage systems. Since dye can be injected at any point where water enters the englacial drainage system, extensive spatial coverage can be provided by tracer tests on any given -glacier. Tests can also be carried out over the course of a melt season to provide evidence for the temporal evolution of subglacial drainage systems (Box 4.7).

4.6.2   Dye injection

One of the major advantages of using a fluorescent tracer such as Rhodamine is that it can be detected by a field fluorometer at concentrations of ~0.1 ppb. This is equivalent to one drip of dye per $\sim500\,000$ litres of pure water. Thus, only small quantities of dye need to be transported and

Box 4.7 Dye tracer investigations of the seasonal evolution of the subglacial drainage system at a temperate glacier: Haut Glacier d'Arolla, Switzerland

Nienow *et al.* (1998) conducted repeated dye tracer tests from multiple moulins at Haut Glacier d'Arolla, Switzerland. The resulting dye returns allowed the authors to draw some important conclusions about spatial and temporal changes in the character of the subglacial drainage system of this temperate glacier through the summer. In general, dye returns became more rapid and less dispersed through the summer, indicating a general increase in the effectiveness of the flow pathways followed. However, by carrying out tests from many input locations the authors were also able to identify how these changes spread spatially through the glacier. In order to investigate this effect, the authors marked the net speed of each dye tracer experiment on a bivariate plot of distance up-glacier from the terminus of the injection site against the date of the test (Figure B4.7). Contouring these points by net flow

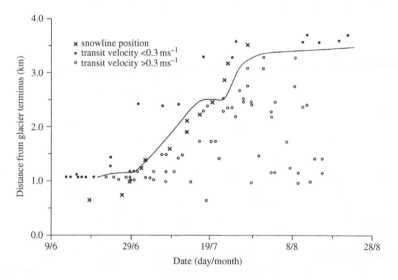

Figure B4.7 Plot of the distance of individual dye tracer tests from the terminus of Haut Glacier d'Arolla against the date on which the test was carried out. Tests characterized by a net transit velocity of faster than $0.3\,\mathrm{m\,s}^{-1}$ are plotted as closed dots and those with a net transit velocity of slower than $0.3\,\mathrm{m\,s}^{-1}$ are plotted as open circles. The boundary between the two speed classifications is marked by a solid line, and the seasonal retreat of the glacier's surface snowline is plotted as crosses. Reproduced from Nienow *et al.* (1998) with the permission of Wiley

injected in order to obtain useful returns. However, the precise quantity of dye required depends on glacier size and individual conditions, and a degree of trial-and-error is necessarily involved. We commonly use tens to hundreds of millilitre of Rhodamine WT on small valley glaciers, and useful guidelines are provided by Kilpatrick (1970). Dye injection should be designed to ensure instantaneous delivery of the full mass of dye to be used. In practice, this involves pouring dye directly into a supraglacial meltwater stream at a point a safe distance upflow of that stream's entry into the glacier via a crevasse or a moulin (Figure 4.9a). The dye receptacle will also need to be well flushed by the stream flow to maximize the amount of dye introduced into the stream. For this reason, dye should not be introduced into standing water. It is advisable when carrying out an injection to wear rubber gloves and protective clothing. It is also advisable to insert dye into a channel that is ice-walled and free of snow, since dye will permeate the latter, possibly causing its delivery to be delayed considerably. The precise time of injection should, of course, be noted immediately.

4.6.3   Dye measurement

Once injected, dye concentration needs to be recorded some distance downflow to infer hydraulic properties of the flow pathways separating the injection site from the measurement site. The measurement site is commonly located on the proglacial outflow stream as near as possible to the glacier terminus (section 4.3.1). Continuous dye concentration measurements can be made by using a pump or gravity to re-route stream water through a fluorometer (Figure 4.9b). Field fluorometers are available commercially from companies such as Turner Designs Inc., USA. Where a gravity-driven siphon is used, the system may need frequent checking, particularly if the

(a)

Figure 4.9 Rhodamine WT dye experiment: (a) injection into a supraglacial melt-water stream on the surface of Haut Glacier d'Arolla, Switzerland (Photograph courtesy: Peter Nienow); and (b) field fluorometer set up adjacent to the proglacial stream at Haut Glacier d'Arolla, Switzerland

(b)

Figure 4.9   (Continued)

local slope precludes the generation of a strong siphon. It is also important to ensure that sunlight does not enter the instrument's detection cell, best achieved by ensuring the plastic pipe used to route the water through the cell is opaque for several tens of centimetre either side of it and by covering the instrument with an opaque tarpaulin. Where a continuous-flow system cannot be established or where location-specific samples are required (for example, to investigate which glacier portal dye is emerging from where more portals exist than there are fluorometers available), discrete samples can be collected by hand or by automatic pump sampler (section 4.4.3.6) and run through the fluorometer manually. Where a fluorometer measures concentration continuously, tracer concentration may be recorded at frequent time intervals by outputting the signal to a data logger. One useful feature of many data logger programmes is their ability to allow the measurement interval to be changed depending on the magnitude of the value measured. Thus, a data logger can not only be left to sample each minute throughout a test, but also be instructed to decrease this interval to 10 seconds once the measured dye concentration exceeds a critical threshold placed well above the background level, perhaps 1 or 2 ppb.

4.7   STUDENT PROJECTS

Glacier hydrology and hydrochemistry are two major aspects of field glaciology, and there is a very broad range of student projects that can be carried out in these areas. Some of these are outlined in the following sections.

4.7.1 Meltwater discharge projects

- Measuring temporal variations in supraglacial meltwater discharge and analysing them by location over a glacier surface. Results could be compared with controlling factors such as location of the snow line, ice albedo, meteorological variables (via surface energy balance; section 7.2) or local topography. Results could also be compared with proglacial discharge variations to investigate lag-times in the discharge time series.

- Measuring temporal variations in proglacial discharge and relating them to the bulk surface energy balance of the glacier (section 7.2) and to supraglacial discharge variations to investigate delivery times from the glacier and through the proglacial area. In mixed basins the relative importance and timing of discharge components supplied by glacierized and glacier-free areas could be investigated by mixing models.

- Measuring diurnal proglacial discharge cycles over an extended period through the melt season to investigate changing patterns of surface meltwater generation, englacial meltwater transfer and meltwater flow at the glacier bed.

4.7.2 Meltwater composition projects

- Measuring spatial and temporal variations in supraglacial meltwater chemistry. Results can be compared with the chemistry of fresh snow and the existing snowpack, with meltwater source (e.g. whether the ice surface or the snowpack), with meltwater flow routing (e.g. whether flow has come into contact with medial moraine debris), and with time within the diurnal discharge cycle.

- Measuring the chemical and/or suspended-sediment composition of meltwater sampled from within and at the base of glacier boreholes (section 5.4.2). Results could be used to investigate the character of meltwaters flowing at the glacier bed (as an aid to reconstructing subglacial hydrology) and to borehole plumbing (section 5.4.1). If borehole sampling is not feasible, then EC profiling can provide a useful surrogate for the total ionic concentration of borehole waters.

- Measuring the chemical and/or suspended-sediment composition of meltwaters within proglacial meltwater streams as they emerge from the glacier terminus. The resulting data can be used to identify differences in the flow pathways followed by the waters through the glacier.

- Measuring the chemical and/or suspended-sediment composition of meltwaters within proglacial meltwater streams as they emerge from the glacier terminus and further downstream to investigate the nature of glacially derived meltwater composition and how that composition can change during flow through the proglacial area.

4.7.3   Tracer-based projects

- Although tracing with fluorescent dyes is the most effective technique, it requires specialist equipment (a powered field fluorometer and dyes) and operator knowledge. Thus, most student projects will be based on tracing using salt, measured by hand-held or automated EC sensor.
- Carrying out tracer experiments (salt is most suitable in the absence of a fluorometer and dye) between moulins on the glacier surface and proglacial meltwater streams to identify drainage basins beneath the glacier. The resulting data could also be used, through analyses of transit velocity and dispersion, to investigate the structure of the glacier's drainage system.
- Carrying out repeat tracer experiments between moulins on the glacier surface and proglacial meltwater streams over an extended period of the melt season to investigate changes in the structure of the subglacial drainage system.
- Carrying out tracer experiments along supraglacial meltwater streams to investigate spatial and temporal patterns in the rates of supraglacial meltwater transit. The results could be compared with possible controls including discharge in the streams concerned and with time in the diurnal discharge cycle.

5

Hot-water borehole drilling and borehole instrumentation

5.1   AIM

The aim of this chapter is to provide an overview of the techniques and instruments involved in (i) using hot, pressurized water to drill boreholes through glaciers, and (ii) deploying sensors to measure a variety of physical conditions within and at the base of those boreholes. Following a brief introduction, the chapter is divided into two main sections. The first considers the principles and design of hot-water drills, the supply of water to the drill, and the use of drills to supply hot, pressurized water to the melting front at the base of a borehole. The second section considers borehole instrumentation according to three categories: those that sense (i) borehole character and drainage, (ii) borehole water quality, and (iii) the glacier substrate at or close to the borehole base.

5.2   INTRODUCTION

Methods of hot-water drilling and borehole instrumentation have developed rapidly over the past 20 years. These developments reflect the efforts of many researchers working at numerous glaciers, including South Cascade Glacier, USA (Hodge, 1976, 1979; Fountain, 1994), Trapridge Glacier, Canada (Blake and Clarke, 1991, 1992; Blake *et al.*, 1994; Fischer and Clarke, 1994; Murray and Clarke, 1995; Waddington and Clarke, 1995;

Stone and Clarke, 1996; Fischer and Clarke, 1997; Stone *et al.*, 1997; Blake and Clarke, 1999; Fischer *et al.*, 1999), Storglaciären, Sweden (Hooke *et al.*, 1997; Fischer *et al.*, 1998; Hanson *et al.*, 1998), Small River Glacier, Canada (Smart, 1996), Bakaninbreen, Svalbard (Porter *et al.*, 1997), and Haut Glacier d'Arolla, Switzerland (Hubbard *et al.*, 1995; Sharp and Richards, 1996; Harbor *et al.*, 1997; Tranter *et al.*, 1997; Gordon *et al.*, 1998). While many of the more significant of the instrumental developments associated with these projects have been published as individual papers (many under the *Instruments and Methods* section of the *Journal of Glaciology*), they have not been brought together into a single, comprehensive review. While much of the material presented in this chapter is borrowed from these papers, the sections are also strongly directed by the authors' own experiences over several years as part of a collaborative drilling programme based at Haut Glacier d'Arolla, Switzerland. This programme, for which over 300 boreholes have been drilled, has involved a broad spectrum of borehole-based research, including studies of glacier hydrology, glacier hydrochemistry and glacier motion.

5.3   HOT-WATER DRILLING

Drilling boreholes involves melting a narrow, cylindrical hole through the ice with a pressurized jet of hot water. Three distinct aspects of this drilling may be identified: (i) the drill itself, (ii) the supply of water to the drill, and (iii) the delivery of water to the melting front.

5.3.1   Hot-water drill

Numerous hot-water drills have been constructed by combining water pump and heating units with individual specifications matched to achieve certain drilling rates over anticipated ranges of borehole depths, borehole widths and ice temperatures. Although it is common for these units to run on diesel, cleaner fuels such as propane may also be used, particularly for the heating unit. They also generally use an automated winch to lower the drill hose off a spool located at the ice surface. Such systems have been used, for example, at the Siple Coast Ice Streams, Antarctica (Engelhardt *et al.*, 1978; Tulaczyk *et al.*, 1998) (Figure 5.1), Unteraargletscher, Switzerland (Gudmundsson *et al.*, 1999), and Jakobshavn Isbrae, Greenland (Iken *et al.*, 1989) (Box 5.1). The construction and operation of bespoke hot-water drilling systems require a level of technical expertise that is beyond the remit of this text. However, the most commonly used hot-water drill,

Figure 5.1 The specially constructed hot-water drill used for many years by US researchers to drill to the base of Siple Coast Ice Streams, Antarctica (Photograph: H. Engelhardt)

Box 5.1 Deep hot-water drilling through cold ice streams: West Antarctica and West Greenland

Two notable research programmes have used bespoke, modular hot-water systems to drill boreholes to depths of greater than 1 km through cold ice. In West Antarctica, USA-based researchers developed a high-powered modular hot-water drill to melt deep boreholes to the bed of Whillans Ice Stream (formerly ice stream B) (Engelhardt *et al.*, 1990). The numerous resulting investigations of ice, meltwater and bed properties have revolutionized our understanding of ice stream hydrology and mechanics. For example, measurements of combined basal sliding and bed deformation at the base of one borehole indicated that these components account for between 70 and 100% of the total measured surface motion of $\sim 0.45 \, \text{km} \, \text{a}^{-1}$ (Engelhardt and Kamb, 1998). Mechanical tests of the fine-grained marine sediment that underlies the ice stream also reveal extreme weakness, better characterized by a plastic rheological model than a viscous one (Kamb, 1991; Tulaczyk *et al.*, 2000).

In West Greenland, Iken *et al.* (1993), Funk *et al.* (1994) and Lüthi *et al.* (2002) report on a series of boreholes drilled to a maximum depth

Box 5.1 (Continued)

of 1630 m through Jakobshavn Isbræ, Greenland. These studies revealed the presence of a thick basal layer of temperate Wisconsin-Pleistocene age ice, the enhanced deformation of which is responsible for almost all of the measured $7\,km\,a^{-1}$ surface motion of the ice stream. In this case, one of the fastest-recorded ice streams on Earth may have been moving largely by enhanced ice deformation rather than by basal motion.

Engelhardt, H., Humphrey, N., Kamb, B. and Fahnestock, M. 1990. Physical conditions at the base of a fast moving Antarctic ice stream. *Science*, 248(4951), 57–59.

Engelhardt, H. and Kamb, B. 1998. Basal sliding of ice stream B, West Antarctica. *Journal of Glaciology*, 44(147), 223–230.

Funk, M., Echelmeyer, K. and Iken, A. 1994. Mechanisms of fast flow in Jakobshavns Isbræ, Greenland: Part 2, Modeling of englacial temperature. *Journal of Glaciology*, 40(136), 569–585.

Iken, A., Echelmeyer, K., Harrison, W. and Funk, M. 1993. Mechanisms of fast flow in Jakobshavns Isbræ, Greenland: Part 1, Measurements of temperature and water level in deep boreholes. *Journal of Glaciology*, 39(131), 15–25.

Kamb, B. 1991. Rheological nonlinearity and flow instability in the deforming bed mechanism if ice stream motion. *Journal of Geophysical Research*, 96(B10), 16585–16595.

Lüthi, M., Funk, M., Iken, A., Gogineni, S. and Truffer, M. 2002. Mechanisms of fast flow in Jakobshavn Isbræ, West Greenland: Part III, Measurements of ice deformation, temperature and cross-borehole conductivity in boreholes to the bedrock. *Journal of Glaciology*, 48(162), 369–385.

Tulaczyk, S., Kamb, B. and Engelhardt, H. 2000. Basal mechanics of ice stream B: I. Till mechanics. *Journal of Geophysical Research*, 105, 463–481.

capable of melting boreholes up to ~150-m long, can be purchased commercially as a pressure car wash unit manufactured by Kärcher GMBH, Germany, and many of the general principles of hot-water drilling may be illustrated with reference to this widely available system.

The most commonly used system is the Kärcher HDS1000BE (Figure 5.2). The unit is fixed within a robust frame, measuring approximately 1.2 m × 0.8 m × 0.9 m, and weighs 185 kg without fuel. The unit is composed of a pump, powered by a petrol (HDS1000BE) or diesel (HDS1000DE) generator, and a diesel burner. The unit generally requires little maintenance, although helicopter lifting, transport over rough surface ice, and the use of old or contaminated fuels can create problems that are all, with a well-stocked tool and spares kit, surmountable.

Figure 5.2 The Kärcher HDS1000 pressure washer commonly used to drill holes up to 150-m long

5.3.2 Water supply to the drill

Running at a regulated outflow pressure of ~90 bar, the HDS1000 uses approximately 800 litres of water per hour of drilling (~13 litres per minute). Since it is important that the pump is supplied with debris- and bubble-free water, the drill is best fed directly from a 200-litre water butt via a short, ~10 m, length of hose incorporating both inlet and in-line filters (Figure 5.3). The pore size of these filter membranes should be large enough to allow sufficient water flow to the drill, but small enough to prevent damaging particles passing through the pump and into the water delivery system (described in section 5.3.3). We have found the optimal filter pore size to be ~250 μm. During snow-free summer drilling, the water butt is supplied by between one and three hoses, each ~100-m long, that siphon water by gravity from supraglacial streams located up-glacier of the drill. A similar system may also be used earlier in the year, as long as adequate water is available at the base of the supraglacial snowpack. However, water supply for drilling can be a problem during the winter or early spring, before the onset of sufficient surface melting. At these times sufficient liquid water is needed to start drilling a borehole, but, once started, water exiting the top of the borehole can be re-circulated by pump. Start-up water can be melted from surface snow using any available heating apparatus – including a tar melter fuelled by diesel and a commercially available water boiler run on a generator. In order to re-circulate borehole water, the water is captured by

Figure 5.3   Kärcher HDS1000 pressure washer (in the background) fed by a 4000 litre water bladder (foreground) and a 200-litre water butt (middle ground). Next to the water butt is a propane-powered tar melter, used here to melt snow and heat water which is poured, along with fresh snow, to the bladder

electrical pump as it emerges from the top of the borehole and fed back into the water butt that supplies the drill.

Drilling through the supraglacial snowpack is possible, but an effort should be made to avoid this where it is possible that supraglacial debris is present on the underlying ice surface because clasts falling into the borehole can cause major problems at the melting front (section 5.3.3). Boreholes should also be located away from actual, or incipient, supraglacial stream channels, since these can exploit a connected borehole, eventually transforming it into a moulin. For these reasons, it is advisable to dig pits to expose the ice surface before early-season drilling and to clean that ice surface prior to drilling holes on topographic highs in the local longitudinal foliation.

5.3.3   Water delivery to the melting front

Water delivery to the ice front is achieved through lengths of high-pressure hose attached to a straight, rigid drill stem (Figure 5.4). The optimum length for a stem is between 1.5 and 3.0 m: in general, the longer the stem the straighter the resulting borehole, but the more difficult it becomes to manage on the glacier surface. Drill stems can be constructed of any strong material, stainless steel probably being the most common, and are formed from a tube that may or may not be unscrewed into sections for ease of transport and storage. A drill tip is screwed onto the non-hose end of the drill stem, and the tip, in turn, houses the nozzle, bored with a ~1-mm-diameter aperture that is responsible for forming the water jet (Figure 5.5). The drill tip needs to be designed to allow the nozzle to be readily replaced, since these are susceptible to damage during drilling, particularly as a result of contact with englacial or subglacial debris. Degradation of the nozzle can result in a non-uniform or non-centred water jet, which in turn results in a corresponding degradation in drilling speed and borehole quality. Nozzles, which are available commercially, may therefore need to be replaced every three or four times the glacier bed or debris-rich basal ice is encountered.

Numerous problems can be encountered during borehole drilling. Maintaining a constant, clean water supply to the drill is crucial, since a poor water supply can cause variations in drill-water pressure and temperature, correspondingly degrading the rate and direction of borehole drilling. Sudden borehole water-pressure changes may occur when the borehole intersects englacial voids. Such events are commonly manifested as a pressure drop, resulting in the borehole water-level falling rapidly from its normal filled status to some new lower level. The weight of the drill hose and stem can increase dramatically during such events – as the buoyancy afforded by the water in the borehole is lost. The driller and a nearby assistant should be mindful of the possibility of such occurrences at all times. Continued drilling is possible following an englacial contact, particularly if the void is hydrologically isolated and it refills with drill water, or if a continuously draining void is located near enough to the ice surface that the extra weight can be borne (the anticipated hole length, and therefore the potential drilling time, should also be considered). Where these alternatives do not seem feasible

Figure 5.4 Drill stem and hose, producing a water jet, as used in association with the Kärcher HDS1000 pressure washer shown in Figures 5.2 and 5.3

(i.e. if the water level does not rise and the void is too low to continue drilling), the borehole should be abandoned and the hose removed. Although such 'blind' boreholes can frustrate a drilling programme, they may be informative in themselves and the location and nature of all such englacial connections should be noted in the drilling log for future reference. On the relatively rare occasions that a borehole encounters a pressurized englacial void, the borehole water-pressure could rise instantaneously with very dramatic consequences, effectively inducing a geyser at the glacier surface. Such events are

Figure 5.5   Photograph of the drill tip and nozzle which create the water jet at the end of the stem shown in Figure 5.4. The tip has an outside diameter of 30 mm

disruptive and can be dangerous, and the driller and assistant should again be mindful of the possibility of their occurrence at all times.

At Haut Glacier d'Arolla the water that leaves the drill is set at a pressure of ~90 bar and at a thermostat setting of ~80 °C. This combination results in a drilling speed that falls progressively from ~2 m per minute near the ice surface to ~0.5 m per minute at ice depths of greater than ~120 m. This decrease in speed primarily results from heat loss from the lengthening section of hose suspended within the water-filled borehole. However, drill speed can be maximized with a few simple measures. First, heat loss through the hose located at the ice surface can be minimized by adding insulation. Expanded foam cylinders, which can be purchased as 1-m long sections with one side split to allow mounting, provide a readily available, inexpensive and versatile hose lagging. Where speed is of paramount importance, a long borehole requiring more than one hose length can be drilled in stages. Here, additional hose sections may be added between the drill and the hose entering the borehole as required as opposed to attaching the full anticipated hose length to the drill from the outset. In such situations, it is useful to introduce a re-circulating loop to the water butt to enable the drill to be run continuously while additional lengths of hose are attached. During such procedures, the drill stem need not be removed from the borehole, but it should be raised some metres up the borehole to avoid a non-vertical water jet contacting the borehole base. Further, whilst disconnected, the hose within the borehole stretches and care should be taken to raise the stem sufficiently far away from the borehole base to avoid contact between the drill tip and the ice borehole base. Drilling rate can also be maximized by minimizing water leaks from hose and stem junctions with teflon thread sealing tape, and by maintaining the quality of the nozzle at the end of the drill tip and, hence, the water jet. At the drilling rates outlined above, typical fuel consumption is approximately 2 litres petrol and 4 litres diesel on the HDS1000BE, or 6 litres diesel on the HDS1000DE, per hour.

Problems with drilling rate and direction may also arise if the drill encounters debris prior to reaching the glacier bed. In such cases, the presence of debris at the base of the borehole serves to both reduce the rate of ice melt at the drilling face and deflect the melting jet laterally, causing the borehole to deviate from the vertical. The effects of the release of such debris into a borehole can vary from a temporary slowdown (until particles are blasted out of the borehole base – borehole video at Haut Glacier d'Arolla has revealed particles left stranded on ledges and in small englacial voids) to abandoning the hole in favour of starting anew. The presence of debris at the base of the borehole may become particularly acute if the borehole base intercepts debris-laden ice. Such ice is often encountered within some metres of the glacier bed, and some persistence may be required in such situations as the rate of borehole advance slows down markedly. Frequently, where such basal ice is encountered, and in the absence of a 'hard' connection (whereby the borehole drains rapidly), it may not be possible to discern whether or not the glacier bed has actually been reached. A degree of judgement, guided by experience, is required in such situations. However, the ultimate purpose of the borehole may also be considered: water sampling and water-pressure monitoring require only an effective hydraulic connection to the basal water system, whereas ploughmeter insertion, tilt-cell insertion, drag-spool insertion and sediment sampling require unimpeded physical access to the bed.

The hot-water drilling system described above is capable of drilling boreholes up to about 150-m long. The main constraints on drilling longer holes are imposed by the continuously decreasing drilling speed and the ability of the driller to lower the hose down the borehole at the optimal rate. This rate of lowering is defined by the requirement to keep the water jet as close to the drilling front at the borehole base as possible, but without resting any of the weight of the drill stem directly onto the ice. Developing a feel for the system is important and our records indicate that, other things being equal, the greater experience a driller has, the straighter are the resulting boreholes.

5.4   BOREHOLE INSTRUMENTATION

A wide variety of borehole instrumentation has been developed to measure properties of (i) the borehole itself, (ii) the water within the borehole, and

(iii) the glacier bed immediately beneath the borehole. These instruments operate mechanically or electrically, the latter requiring command and control from data loggers (the most commonly used being manufactured by Campbell Scientific Inc.) located at the glacier surface.

A number of general considerations should guide the design and construction of borehole instrumentation. First, instruments should be narrow enough to be raised and lowered along the borehole without catching on the borehole walls or on clasts that may protrude from those walls. In our experience this imposes a maximum diameter on instruments of ∼5 cm for boreholes shorter than 150 m. While all boreholes may be widened by the continuous passage of meltwater to diameters markedly greater than 5 cm towards the ice surface, it needs to be borne in mind that such boreholes narrow to approximately the diameter of the drill stem at the depth of the melting front. If wider boreholes are needed, then it is possible deliberately to drill a borehole slowly (allowing the borehole to widen more than would be the case at a normal rate of progress), although a more secure option is to use a larger diameter drill tip, the size of which imposes an absolute minimum diameter on the borehole. Second, electrical contacts need to be waterproofed to withstand the pressure gradients induced by the overlying water column. In practice, this necessitates casting most electrically-based instrumentation in resin. Ideally, the resin used should be clear, allowing visual inspection of the encased components and wiring. Depending on the purpose, such resin may also need to harden without generating potentially corrosive vapour, for example in the case of casting pressure sensors. It is possible to splice electrical cables near to borehole probes, allowing the cable lengths and probes to be transported separately and assembled as and when required on the glacier. Such splices should maximize the separation distance between the connections by folding each contact back in opposite directions along the parent cable. If each contact is folded 25 cm back, then the total distance between the contacts is 50 cm. This whole length should then be sealed meticulously with self-amalgamating tape. Third, instruments should be of sufficient density and weight to allow them to be lowered rapidly through the water column and for their contact with the glacier bed to be readily detected by hand from the surface.

5.4.1   Investigations of borehole character and drainage

5.4.1.1   Borehole video

Borehole TV-video developed from the earlier use of borehole photography (Engelhardt *et al.*, 1978), and has been used at many glaciers, with early studies pioneered at Storglaciären, Sweden (Box 5.2) (Pohjola, 1993, 1994). The borehole camera used at Haut Glacier d'Arolla is a GeoVision Micro,

Box 5.2   Use of borehole TV-video to reveal the character of glacier ice and the glacier bed: Storglaciären, Sweden

Pohjola (1993) built on earlier research by Engelhardt *et al.* (1978) by using borehole-based TV-video to investigate the character of the basal interface and rates of basal sliding at Storglaciären, Sweden. The video revealed a mixed basal interface, with one borehole terminating in soft sediments and the other contacting bedrock. Analysis of the movement of particles across the field of view at the base of these boreholes indicated background basal motion speeds of $\sim$40 mm d^{-1} in one borehole and $\sim$100 mm d^{-1} in the other. Both records, however, are characterized by velocity spikes which the author interprets in terms of local basal stress release.

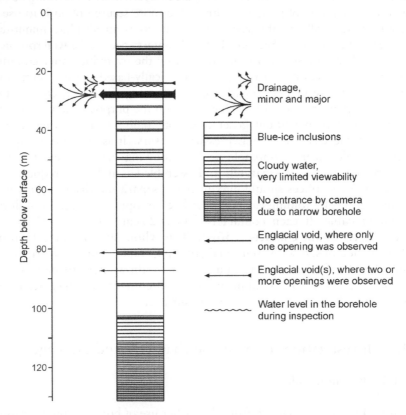

Figure B5.2   Stratigraphic log of borehole 89-3, Storglaciären, Sweden, as revealed by TV-video footage showing the visual nature of the englacial ice and the hydraulic connections intersected. Reproduced from Pohjola (1994) with the permission of the International Glaciological Society

In a later paper, Pohjola (1994) reported on the use of video to investigate the englacial structure of Storglaciären. In this case, the video revealed numerous englacial hydraulic connections and voids which together accounted for ~1.3% of the glacier's total thickness (Figure B5.2). Video-based logging also revealed the presence of bubble-rich and bubble-poor layers, the latter of which appeared to be particularly associated with the englacial voids. The author interpreted this association in terms of a common origin as crevasses in the accumulation area of the glacier.

Pohjola, V.A. 1993. TV-video observations of bed and basal sliding on Storglaciären, Sweden. *Journal of Glaciology*, 39(131), 111–118.

Pohjola, V.A. 1994. TV-video observations of englacial voids in Storglaciären, Sweden. *Journal of Glaciology*, 40(135), 231–240.

Engelhardt, H.F., Harrison, W.D. and Kamb, B. 1978. Basal sliding and conditions at the glacier bed as revealed by bore-hole photography. *Journal of Glaciology*, 20(84), 469–508.

manufactured by Marks Products Inc. and supplied by Colog Inc., USA, and is described by Copland *et al.* (1997a). The video camera is lowered down a water- or air-filled borehole, and real-time narration is recorded by an operator at the ice surface. Resulting video footage may be used to identify the nature of the substrate at the base of the borehole, glacier sliding speed (where a fixed reference can be identified on the substrate surface), the nature and size of englacial hydraulic connections, and the character of the ice forming the borehole walls.

5.4.1.2   Initial borehole inclinometry

The most commonly used borehole inclinometer is available commercially from Icefield Tools Corp., Canada (Figure 5.6) and is described in some detail by Blake and Clarke (1992). The inclinometer's tilt cells and magnetometer respectively record dip and strike in real time as the instrument is lowered down or raised up the borehole in regular steps, normally of 0.5 or 1.0 m. This information is continually updated and displayed graphically on a laptop computer operated at the ice surface. Repeat borehole inclinometry results from Haut Glacier d'Arolla reveal a mean borehole deviation from the vertical of less than 1% of the borehole depth (Copland *et al.*, 1997b).

Figure 5.6   A field inclinometer (MI-3) manufactured and supplied by Icefield Tools Corp., Canada, being prepared for use at Tsanfleuron Glacier, Switzerland

5.4.1.3   Cable-following and repeat borehole inclinometry

Repeat borehole inclinometry has the capacity to reveal 3D patterns of net ice deformation integrated over the time period separating the surveys (Box 5.3). It is important during such resurveys to ensure that the inclinometer follows the (now closed and deformed) path of the initial borehole as closely as possible. This is generally achieved by following a cable inserted into the original borehole prior to its closure, under the assumption that the cable has deformed passively with the enclosing ice since that time. Since the purpose of such borehole re-drilling is to follow a cable as closely as possible and not to drill a vertical hole (as is the case during drilling for any other purpose), the drill stem, tip and nozzle are slightly adapted from those normally used and described above. Thus, for re-drilling (i) a relatively short (~0.5 m), heavy (4–5 cm diameter) stem is used, which allows large lateral deviations in the borehole direction to be followed; (ii) a cable-following tip is used, through which the wire being followed is allowed to pass; and (iii) a reaming nozzle is used, which produces a diffuse (conical), low-pressure (typically 40–50 bar) water jet relative to that of the standard nozzle (again available commercially, requiring no alteration). While cable-following in general tends to proceed more slowly than initial drilling, rates of cable following can vary markedly depending on the integrity of the initial borehole, now forming the ice or void around the cable. Where the ice is solid, rates can be very slow indeed (some centimetres per minute), but sections of the borehole that may not have closed completely can be drilled very rapidly: Finally, care should be taken to ensure that the cable being followed is kept taut to allow the stem to pass over the cable smoothly and to avoid any loops forming and snagging the stem at the melting front.

5.4.1.4 Englacial tilt cells

Tilt cells may be left in place within a borehole to provide continual time series of local dip. In a similar manner, 3D ice deformation may be reconstructed from many cells placed at different depths within an array of boreholes. The most robust of these are electrolytic tilt cells, initially described by Blake *et al.* (1992) for use at Trapridge Glacier, Canada, that operate on the

---

Box 5.3 Repeat borehole inclinometry as an indicator of 3D patterns of ice deformation: Worthington Glacier, USA and Haut Glacier d'Arolla, Switzerland

Repeat inclinometry readings from arrays of boreholes can provide important information about spatial patterns in driving stress and the rheology of the ice responding to those stresses. For example, results from a dense array of 31 boreholes drilled through temperate Worthington Glacier, USA, revealed that the ice may be characterized by two layers: an upper layer of 115 m thickness best characterized by a linear viscous flow law and a layer beneath this to the glacier bed, at 198 m, best characterized by a non-linear flow law with a power exponent of ~3 (Marshall *et al.*, 2002).

Repeat borehole inclinometry at Haut Glacier d'Arolla, Switzerland, has also revealed interesting patterns of ice deformation, in this case related to spatial variations in the drag exerted by the bed on the overlying ice. Initially, Harbor *et al.* (1997) noted that an inter-annually persistent subglacial meltwater channel acted as a longitudinal axis of locally enhanced sliding or *slippery zone*. More recently, Willis *et al.* (2003) presented 3D velocity data based on borehole repeat inclinometry, which indicate that this sliding generates extrusion flow in the ice located

Box 5.3   (Continued)

immediately above the slippery zone (Figure B5.3), a phenomenon that has also been reported above a subglacial riegel at Storglaciären, Sweden (Hooke *et al.*, 1987).

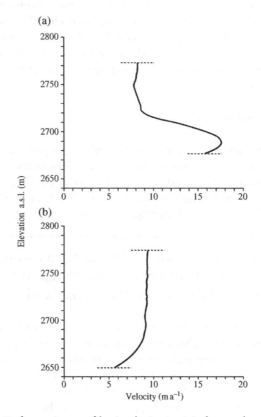

Figure B5.3   Deformation profiles in glacier ice (a) above a basal zone of low traction corresponding to the location of a major melt season basal channel; and (b) above adjacent ice characterized by higher basal traction at Haut Glacier d'Arolla, Switzerland. Reproduced from Willis *et al.* (2003) with the permission of the International Glaciological Society.

Harbor, J., Sharp, M., Copland, L., Hubbard, B., Nienow, P. and Mair, D. 1997. Influence of subglacial drainage conditions on the velocity distribution within a glacier cross section. *Geology*, 25(8), 739–742.

Hooke, R.L., Holmlund, P. and Iverson, N.R. 1987. Extrusion flow demonstrated by bore-hole deformation measurements over a riegel, Storglaciären, Sweden. *Journal of Glaciology*, 33(113), 72–78.

Marshall, H.P., Harper, J.T., Pfeffer, W.T. and Humphrey, N.F. 2002. Depth-varying constitutive properties observed in an isothermal glacier. *Geophysical Research Letters*, 29(23), art. no. 2146.

Willis, I.C., Mair, D., Hubbard, B., Nienow, P.W., Fischer, U.H. and Hubbard, A. 2003. Seasonal variations in ice deformation and basal motion across the tongue of Haut Glacier d'Arolla, Switzerland. *Annals of Glaciology*, 36, 157–167.

principle that the conductivity between two electrodes partly immersed in an electrolyte is proportional to the total wetted surface area of those electrodes. A dual-axis tilt cell, for example, has five electrodes arranged in a cross pattern with the central electrode common to both circuits. As such a cell tilts, its outer electrodes are immersed to different depths and the resistances between them change accordingly. To make a tilt measurement, the tilt cell is connected to an AC bridge containing two reference resistors. The magnitude of the output voltage of the cell is proportional to the tangent of the tilt angle, and the sign determines the tilt direction. Individual cells are commonly housed within rigid cylinders between 50- and 100-cm long in order to increase contact with the borehole wall and to smooth out centimetre-scale deformation, which may be induced by factors such as ice crystal growth patterns as the borehole closes around the cells.

Prior to use in the field, tilt cells are calibrated on a jig that rotates the cell around its long axis at various axial tilts. The method by which calibration functions may be derived and an inversion scheme used to interpret field data is described by Blake and Clarke (1992). The azimuth of the tilt cell can change due to: (i) rotation of the tilt cell about the vertical axis; (ii) rotation of a tilted cell around its long axis (possibly due to rotation of the cable to which it is attached); and (iii) tilting of a cell through the vertical plane (at which point an immediate 180° change in the azimuth is recorded). The last of these types of behaviour can be easily identified by observation of the azimuth record through time. The first two causes of azimuth change cannot be differentiated. However, if it is assumed that ice deformation is unlikely to cause the tilt cell or cable to twist, then change in azimuth must be attributed to the rotation of the tilt cell about the vertical axis.

Once tilt cells have bedded into their host material (such that changes in tilt reflect changes in local deformation), tilt values must be resolved to the flow direction in order to estimate longitudinal strain rates and velocities. In contrast to the borehole inclinometer, however, tilt cells do not currently incorporate magnetometers, and tilt direction cannot therefore be recorded directly. The orientation of tilt is therefore commonly assumed to be in the same direction as the motion of the top of the borehole as recorded by ice-surface surveying or from detailed scrutiny of the tilt-cell records. For example, if tilt increases steadily on a constant azimuth over a prolonged period of time (i.e. weeks to months), then the average of the azimuth values throughout this period may be assumed to equate to the ice flow direction. Output azimuth values may then be corrected such that the azimuth value taken as the flow direction is assigned a value of zero. Tilt rates in, and away from, this plane can then be calculated and transformed geometrically into rates of change in the 3D co-ordinate location of the cell.

5.4.2 Investigations of borehole water quality

5.4.2.1 Water sampler

Water may be sampled directly from the borehole water column using a variety of devices, the most commonly used of which is an adapted *Niskin* design (Blake and Clarke, 1991) (Figure 5.7). This cylindrical sampler is open at both ends as it is lowered down the borehole, allowing the borehole water to pass freely through it. Once the sampler has been lowered to the required sample depth it is sprung shut by a messenger weight dropped from the ice surface, trapping local water within it. The

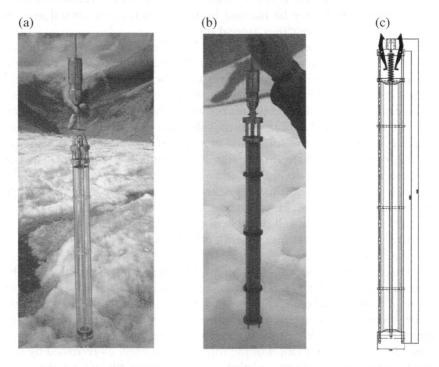

Figure 5.7 An adapted Niskin water sampler: (a) empty before use in a borehole; and (b) full of turbid meltwater following sampling from the base of a borehole, and (c) in CAD

sealed sampler is then raised to the ice surface containing the trapped water sample. Water may then be treated according to standard procedures (section 4.4.3.6) and analysed for a variety of quality characteristics, including suspended sediment properties, ionic composition, trace element composition and dye concentration.

5.4.2.2 Water pressure

Subglacial water pressure exerts a key control on ice-mass hydrology and dynamics and its measurement has accordingly formed a key part of almost all borehole-based research programmes (Box 5.4). Most water-pressure

---

Box 5.4 Relationship between water pressures measured in boreholes and glacier motion: Findelengletscher, Switzerland, and Storglaciären, Sweden

Iken and Bindschadler (1986) reported simultaneous measurements of horizontal ice surface velocity, measured several times a day over four rows of stakes, and subglacial water pressure, measured as water level within 11 boreholes, over a five-week period early in the summer melt season at Findelengletscher, Switzerland. This pioneering study revealed synchronous and systematic fluctuations in borehole water levels across the array, and a positive relationship between these levels and ice surface velocities. The nature of the relationship between borehole water level and surface velocity was positive, but non-linear (Figure B5.4). In a similar subsequent study, Jansson (1995) plotted subglacial effective pressure (borehole-measured water pressure minus local ice overburden pressure) against measured surface velocity at Storglaciären, Sweden, and argued that this relationship and that earlier identified by Iken and Bindschadler (1986) could both be described by:

$$U = aN^{-b}$$

where U is ice surface velocity, N is effective pressure and a and b are empirical constants. The best-fit values of these constants were found to be $a = 30$ and $b = 0.40$ at Storglaciären, and $a = 371$ and $b = 0.40$ at Findelengletscher.

More recently, detailed analysis of time series of the relationships between subglacial water pressure and velocity has revealed out-of-phase behaviour and phase lags. This unsurprisingly suggests that sliding

Box 5.4 (Continued)

velocity is probably controlled in detail by a combination of factors, including basal water pressure, rate of (positive) change of basal water pressure, and perhaps total bed separation.

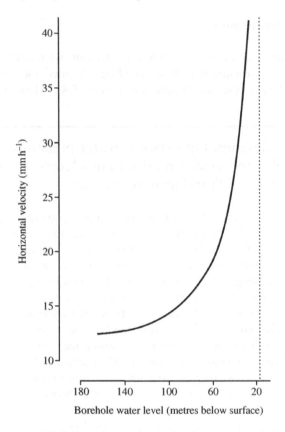

Figure B5.4 Simplified plot of the relationship between borehole water level (a surrogate for subglacial water pressure) and horizontal surface velocity measured early in the summer melt season at Findelengletscher, Switzerland. Reproduced from Iken and Bindschadler (1986) with the permission of the International Glaciological Society

Iken, A. and Bindschadler, R.A. 1986. Combined measurements of subglacial water-pressure and surface velocity of Findelengletscher, Switzerland: Conclusions about drainage system and sliding mechanism. *Journal of Glaciology*, 32(110), 101–119.
Jansson, P. 1995. Water-pressure and basal sliding on Storglaciären, northern Sweden. *Journal of Glaciology*, 41(138), 232–240.

transducers record the resistance of a metal membrane as it flexes in response to the pressure differential across it. This resistance is recorded using the excite-delay differential voltage programme on a Campbell Scientific data logger. The inner cavity of such transducers is therefore usually sealed at a fixed pressure (often termed *absolute range*, in contrast to *relative range*, in which a gland in the supply cable maintains the cavity at atmospheric pressure). Water-pressure transducers, which are typically 40–50-mm long and 20–30 mm in diameter, may be purchased as sealed units with wire attached (expensive but effective) or as open units that may be cast in clear, non-exothermic resin with minimal vapour emission. The most common problems encountered using such sensors are shorting by water due to imperfect sealing, and damage to the conductor-membrane contacts within the body of the transducer. These membrane contacts may be particularly fragile and damage may be caused by water ingress, build-up and discharge of static electricity, and rough handling.

Most water-pressure transducers are supplied pre-calibrated, but it is also advisable (particularly for use at high altitude and following transportation) that they also be (re)calibrated on site prior to installation. This is achieved by lowering the transducer down a water-filled borehole and reading the voltage registered at incremental steps, and by repeating the process back up the borehole. The depth increments adopted should be chosen to provide >30 calibration data points over the full pressure range anticipated for the transducer.

5.4.2.3 Electrical conductivity (EC)

The EC of glacigenic meltwater acts as a direct surrogate for its total ionic concentration (section 4.4.3.1). Measured EC can therefore be used as an important tool to inform investigators of that water's flow history through a glacier's drainage system. The construction and operation of borehole EC probes have been described by Stone *et al.* (1993) and Smart and Ketterling (1997). The basic unit comprises two electrode contacts separated by the water whose EC is to be measured (section 4.5.1). The EC of this water is given as the inverse of the resistivity recorded between the electrodes (the recorded resistance scaled to account for electrode geometry). Resistance is best measured across a switched AC half-bridge, which minimizes electrode polarization. Such EC sensors may be either installed at a fixed location within the borehole to provide a time series of the EC of waters at that location, or moved along the borehole, normally at steps of 0.5 or 1 m, to record vertical EC profiles. Repeating such profiles through time may be used as a powerful tool in the identification of along-borehole water sources and sinks, or *borehole plumbing* (Gordon *et al.*, 2001; Box 5.5).

Box 5.5   Along-borehole EC measurements as an indicator of temperate glacier drainage: Haut Glacier d'Arolla, Switzerland

Gordon *et al.* (2001) used repeated along-borehole EC profiling to investigate patterns of water flow into, along and out of boreholes at Haut Glacier d'Arolla, Switzerland. In this study, the authors raised and lowered a single sensor along the borehole water column to record EC profiles. Since basally-sourced water can be distinguished by its relatively high EC from englacially sourced water, the rate of passage of the boundary between different water types within the water column may be compared with the rate of change of the overall borehole water level to identify the locations of water sources and sinks. The authors also injected saline solution to the base of some boreholes to provide a traceable marker. Although water only entered and left through the base of many boreholes, some flow patterns were surprisingly

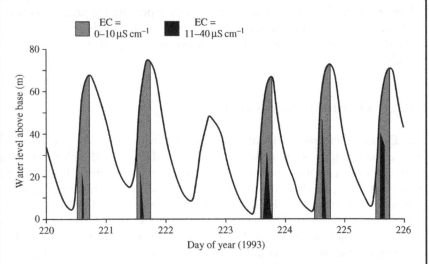

Figure B5.5   Plot of EC and borehole water level against time over a six-day period in the summer melt season at Haut Glacier d'Arolla, Switzerland. The EC profile indicates that, over each diurnal cycle, the borehole was initially filled with dilute, englacial or supraglacial water. After some time of rising borehole water level, high-EC water entered the base of the borehole and that high-EC water then drained from the base of the borehole while the overall water level continued to rise. This pattern indicates that englacial and/or supraglacial inputs pressurized the borehole relative to the subglacial system, thereby forcing the high-EC water back out of the base of the borehole. Reproduced from Gordon *et al.* (2001) with the permission of Wiley

complex, indicating multiple englacial and basal hydraulic connections (Figure B5.5). As a result of the study the authors proposed a classification of four borehole drainage types. (i) *Unconnected* boreholes do not connect to an effective drainage system. Those that terminate englacially were labelled as *blind unconnected* and those that terminated at the glacier base were labelled *apparently unconnected*. (ii) *Englacially connected* boreholes intersect only englacial effective drainage channels. (iii) *Subglacially connected* boreholes intersect only a basal effective drainage system. (iv) *Multiply connected* boreholes intersect both englacial and subglacial effective drainage systems. Finally any of the categories above may also be labelled as *sporadically connected*, indicating that they have been recorded to change category over time.

Gordon, S., Sharp, M., Hubbard, B., Willis, I., Smart, C., Copland, L., Harbor, J. and Ketterling, B. 2001. Borehole drainage and its implications for the investigation of glacier hydrology: Experiences from Haut Glacier d'Arolla, Switzerland. *Hydrological Processes*, 15(5), 797–813.

5.4.2.4 Turbidity

Borehole meltwater turbidity can provide a useful indication of the origin of that water, as well as a direct indication of its erosive capacity. For example, Hubbard *et al.* (1995) cited turbidity as evidence of the basal transport of fine suspended sediment during periods of strong lateral hydraulic gradients across the bed of Haut Glacier d'Arolla, Switzerland. Turbidity is usually measured in terms of the voltage returned by a light-sensitive diode separated from a LED by the water being characterized. A control, light-sensitive diode (separated from the LED by a material of constant light-transmitting properties, such as oil) may also be used to account for possible variations in LED luminosity. Such devices may be constructed in-house by casting the diodes into a housing made from a waterproof and non-corroding material (e.g. PTFE or resin), as discussed in section 4.5.2.

5.4.3 Investigations of the glacier substrate

5.4.3.1 Subglacial hammer

The subglacial hammer supplies the impulse to instruments that are designed to be inserted by force into the glacier substrate. Hammer design should minimize the chances of either damaging or becoming entangled with the cable of the instrument being inserted. The hammer should also be robust, and easy to use and repair. It is also desirable, but not necessary, for the hammer to provide a consistent impulse.

Subglacial hammers are controlled from the glacier surface via a single cable that is used to raise and lower a weight (the hammer) on to an anvil, onto the base of which the subglacial instrument to be inserted is attached. The hammer must be free-running relative to the anvil, allowing an impulse to be transferred at the base of both the downstroke (to insert an instrument) and the upstroke (to detach a hammer from an inserted instrument or to remove an inserted instrument such as a ploughmeter) (section 5.4.3.4). The hammer must also be attached to the anvil, allowing the entire assemblage to be lowered down and raised up the borehole on a single cable. Most hammers currently in use are adapted versions of that designed for use at Trapridge Glacier, Canada. According to this design, a cylindrical outer hammer slides up and down an inner rod of approximately twice the hammer's length, being held in place by stoppers at the ends of the rod.

An adaptation of this design involves the use of two retaining pins protruding from the inner rod (attached to the anvil) to guide the stroke of a grooved outer hammer (Figure 5.8). This enables simpler wiring attachments (just one attachment point at the top of the instrument) and a reduction in the instrument length for transporting of ~50%, since the rod and hammer are now of similar length. The lower tip of the rod houses the anvil, the base of which has a male thread for the attachment of subglacial tools. The hammer slides freely along the rod, but cannot detach

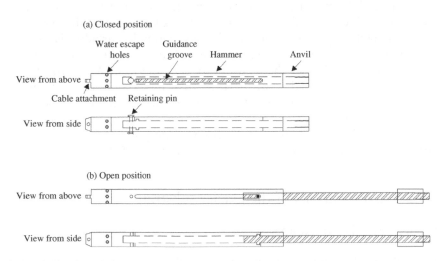

Figure 5.8 CAD image of a compact subglacial hammer design whereby the outer hammer slides over an inner rod of similar length: (a) closed position; and (b) open position

completely from it, being restricted at the base of the stroke by the anvil and at the top of the stroke by retaining pins. The length of the grooves in the hammer dictates the maximum length of the hammer stroke, ~50 cm should suffice. Hammering, and raising and lowering the hammer along the borehole, requires only a single length of wire: multi-strand stainless steel wire of ~2.0-mm diameter serves this purpose well. The wire is attached via a loop to the top of the hammer. For transport and storage, the hammer is collapsed to its minimum length and held in place by a retaining pin. The screw attachment point for subglacial instruments may also be protected from damage during transport by screwing the anvil closed over its length (Figure 5.8).

The depth to which an instrument has been inserted into the subglacial material can be measured by reference to a mark on the hammer wire (electrical tape suffices for this purpose). The position of the edge of the tape is then noted before and after insertion relative to a fixed point, normally the rim of the borehole at the ice surface, and the depth of insertion is given by the difference between the two readings (Figure 5.9). These measurements form the basis of penetrometry, whereby a solid spike is driven into the glacier bed and the depth of penetration recorded as an indication of the nature of the glacier bed.

5.4.3.2 Subglacial sediment sampler

The ideal down-borehole sediment sampler should be capable of recovering large, representative sediment samples from the base of water- or air-filled boreholes. One of the simplest and most successful designs is formed from a hollow inner barrel with a window and an outer sleeve that can slide up and down the length of the barrel to reveal or cover the window (Figure 5.10). The outside diameter and length of the barrel are ~50 mm and ~200 mm respectively, and a conical tip is attached to its lower end to ease insertion into the subglacial material to be sampled. The sampler works on the principle that friction with the sediment rising up the sampler during insertion raises the window relative to the barrel, exposing the aperture into which sediment falls. Upon removing the instrument the window falls back into place, sealing the aperture and trapping the sediment inside the barrel as it is raised back up the borehole.

5.4.3.3 Subglacial tilt cells

Subglacial tilt cells are similar in purpose and design to englacial tilt cells, but they are inserted into the subglacial substrate, often as a series of two to four cells. Subglacial tilt cells are also housed in cylindrical containers that are shorter than their englacial counterparts, since higher shear strain rates

Figure 5.9   Electrical tape (white) fixed to the wire attached to the subglacial hammer. Here, the change in height of the marker relative to the ice surface provides a quantitative indication of the minimum insertion depth of the subglacial tool being hammered

are expected in the subglacial sediment layer. Where inserted as strings of cells, the lowermost cell is attached to the sediment by an anchor and to other cells by a 20–30-mm length of flexible wire (Figure 5.11). In this way, the cells are positioned at different depths within the sediment, providing information relating to variations in strain rate with depth. Cells are inserted via a stainless steel tube within which they sit, allowing the lip of the anchor to protrude from one end and the cables to exit from the top of the groove.

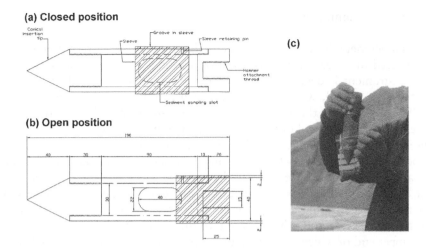

Figure 5.10 Subglacial sediment sampler illustrated: (a) as CAD in open position; (b) as CAD in closed position; and (c) photographed in use yielding subglacial sediment at Haut Glacier d'Arolla, Switzerland

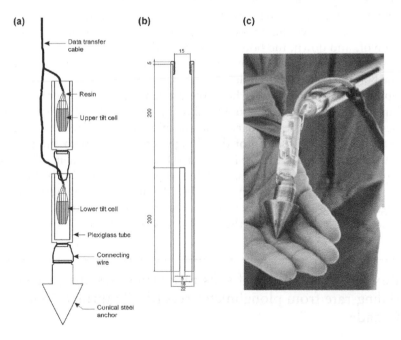

Figure 5.11 A chain of subglacial tilt cells illustrated as: (a) a line drawing of the cells; (b) a line drawing of the insertion tool that is attached to the end of the subglacial hammer and retrieved with the hammer after the cells are emplaced; and (c) photograph of cells and insertion tool prior to emplacement at the base of a borehole at Haut Glacier d'Arolla, Switzerland

5.4.3.4   Ploughmeter

The ploughmeter is designed to measure the bending force exerted on it as it is dragged in a near-vertical orientation through the subglacial sediment. The instrument is described in its current form in some detail by Fischer and Clarke (1994), who developed an earlier concept by Humphrey *et al.* (1993). The basis of the instrument is a ~1.5-m-long and ~20-mm-diameter steel rod, the lower end of which is sharpened into a conical tip for ease of insertion into the subglacial sediment. The rod is sheathed in a clear vinyl tube (internal diameter ~25 mm), beneath which strain gauges are bonded by adhesive to polished sections of the rod. The space between the rod and the sheath is filled with clear resin. Strain gauges are arranged as two sets of four, each set forming a ring around the rod with individual gauges separated by 90°. The rod is hammered into the sediment via an insertion tool, the upper end of which is screwed to the base of the hammer and the lower end of which fits snugly into a cylindrical hole (~100-mm long and ~10-mm wide) bored into the top of the ploughmeter rod. The ploughmeter and hammer are lowered down the borehole together, with the total weight being taken by the ploughmeter cable to prevent the two instruments from becoming detached during lowering. Experience indicates that the most effective way of lowering involves one person to direct the instruments into the top of, and down, the borehole, and another to walk the cables, which are extended down-glacier of the borehole to just over the anticipated length of the borehole, to the borehole.

Upon contacting the glacier base, the ploughmeter is hammered into the sediment to a depth that should submerge both sets of strain gauges. A bending force is then imparted on the ploughmeter, and registered by the strain gauges, as the top of the ploughmeter moves with the borehole into which it protrudes and the base is held back by the sediment. The forces recorded by the strain gauges, which are pre-calibrated by bending the ploughmeter in the laboratory, can then be interpreted in terms of the strength (Fischer and Clarke, 1994; Fischer *et al.*, 1998) and sedimentology (Fischer and Clarke, 1997) of the subglacial sediment layer (Box 5.6).

Box 5.6   Estimation of substrate properties and glacier sliding rate from ploughmeter records: Trapridge Glacier, Canada

Fischer and Clarke (1994) developed a theoretical basis for approximating the character of subglacial sediments from ploughmeter records under assumed viscous or plastic deformation models. Applying this theory to data from Trapridge Glacier, Canada, resulted in an effective sediment viscosity of 3.0×10^9 to 3.1×10^{10} Pa s according to the viscous model, and a yield strength of 48–57 kPa according to the plastic model. Fischer and Clarke (1997) extended this theory to incorporate the effects of collisions with individual clasts to allow the basal sliding rate of the glacier to be estimated from ploughmeter records. In this case, the authors matched spectral power density functions of ploughmeter force time series to effective sediment viscosities and ploughing velocities (Figure B5.6). Results of the analysis yielded a value of sediment viscosity of $\sim2.0 \times 10^{10}$ Pa s, which agreed closely with the authors' earlier results, and sliding velocities of ~45 mm d^{-1}, which agreed closely with those measured directly by drag spools at the base of the glacier by Blake *et al.* (1994).

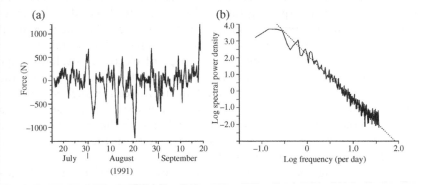

Figure B5.6 Ploughmeter force time-series records from a borehole drilled through Trapridge Glacier, Canada: (a) raw cross-flow force; and (b) the same record expressed as a spectral power density function, the best-fit line with which allowed Fischer and Clarke (1997) approximated local sediment viscosity and sliding speed. Reproduced from Fischer and Clarke (1997) with the permission of the International Glaciological Society

Fischer, U.H. and Clarke, G.K.C. 1994. Plowing of subglacial sediment. *Journal of Glaciology*, **40**(134), 97–106.

Fischer, U.H. and Clarke, G.K.C. 1997. Clast collision frequency as an indicator of glacier sliding rate. *Journal of Glaciology*, **43**(145), 460–466.

Blake, E.W., Fischer, U.H. and Clarke, G.K.C. 1994. Direct measurement of sliding at the glacier bed. *Journal of Glaciology*, **40**(136), 595–599.

5.4.3.5 Drag spool

The drag spool has been designed to measure the motion of the base of the glacier relative to its underlying substrate. The instrument was initially described by Blake *et al.* (1994), and a larger and more sophisticated variant has been described by Engelhardt and Kamb (1998). Drag spool operation relies on measuring the distance from the base of a moving borehole to an anchor inserted into the underlying substrate. Thus, the instrument comprises an anchor, which is hammered to a known depth into the substrate, a multi-turn potentiometer or metering spool that is housed within a protective case near the base of the borehole, and a tether line linking the two (Figure 5.12). This design is based on the assumption that the anchor remains in place within the sediment and, as the base of the borehole slides over that sediment layer, the line is spooled out, turning the potentiometer. The resistance of the potentiometer is then recorded by a data logger located at the glacier surface, providing high-resolution time series of the length of line paid out. However, this potential performance is tempered by some ambiguity in interpreting the resulting records. This difficulty arises principally from the precise geometries of (i) initial emplacement, and (ii) subsequent motion being unknown at the glacier base. The initial insertion of drag spools is difficult, with greased loops of electrical insulating tape attached to the outside of the protective housing designed to slide off the insertion tool (a steel rod of ~7.5-mm diameter) once the anchor has been hammered into the glacier bed. The end of the insertion tool inserts into a cylindrical hole bored into the top of the anchor such that the two are held together by friction at this join and tension (~1 N) in the line linking the anchor to the potentiometer housing.

In practice, drag spool insertion is frequently not straightforward, particularly through ice thicknesses greater than ~100 m or where sediment is not uniformly soft. Problems can also arise from the housing not sliding off the insertion tool easily enough, in which case the housing will come back up the borehole, paying out its line, as the hammer and insertion tool are retrieved. The success of insertion, however, can be monitored closely from the glacier surface by measuring the resistance of the potentiometer with a hand-held multi-meter. Once a successful insertion is suspected and the hammer has been removed from the borehole, the potentiometer housing can be gently raised away from the bed until the line tightens and the potentiometer begins to spool out, registering as a change in resistance on the multi-meter. At this point the instrument is ready for use, and the cable is tied off at the glacier surface (a series of hitches over a dowel resting on the borehole top is sufficient) and wired into a data logger.

Interpreting drag spool data with confidence may also be problematic since the line spooling rate depends on the path that the line takes through the sediment between the base of the borehole and the anchor, which is unknown. The possibility of anchor movement also cannot be discounted because the instrument requires a soft, and therefore potentially deformable, substrate for successful emplacement. This problem cannot be overcome with the current instrument design.

(a)

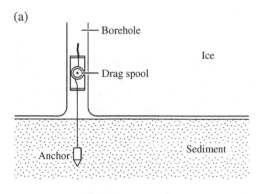

(b)

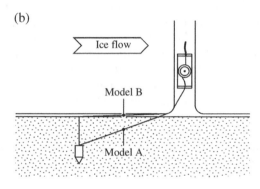

(c)

Figure 5.12 Drag spool emplacement and operation illustrated as: (a) line drawing of instrument as initially emplaced at the base of a borehole; and (b) line drawing of instrument some time after emplacement; and (c) a photograph of the instrument attached to a subglacial hammer before emplacement. Note that analysis of spool records depends on the path the line follows between the spool in the borehole and the anchor inserted within the subglacial sediment layer; in (b) two assumed paths are given (labelled Model A and Model B). (a) and (b) reproduced from Blake *et al.* (1994) with the permission of the International Glaciological Society

5.5 SUMMARY

Hot-water drilling has been used at all scales of glaciological investigation over the past 30 years, and the technique will most probably be of equal or greater importance in at least the medium-term future. The technique allows researchers close access to both the interior and the basal interface of ice masses at numerous, predetermined locations. In providing direct access to the ice–bed interface, hot-water drilling has allowed researchers to make important discoveries about basal conditions and processes at some of the Earth's largest ice masses, including the Antarctic. For example, US-supported hot-water drilling (principally based at the California Institute of Technology; Figure 5.1) at the Siple Coast ice streams, Antarctica, has led the way in our understanding of ice stream dynamics for some 20 years and a current UK-based project aims to gain access to the 2.2-km-deep base of the Rutford Ice Stream by hot-water drilling. The latter project will use sophisticated hot-water drilling technology and procedures developed at the British Antarctic Survey (Makinson, 1994) including, for example, storing water drilling for the deeper sections of each hole in a large englacial cavity initially formed by drilling with priming water melted at the ice surface. Other teams are developing hot-water drilling technology that will allow the recovery of ice core (e.g. Engelhardt et al., 2000), and an even more ambitious project may be realized in the next decade – to use hot water to drill through almost 3.5 km of ice into subglacial Vostok Lake.

5.6 STUDENT PROJECTS

Hot-water borehole drilling requires specialist equipment, field logistical support and operator expertise. However, once drilled, boreholes are available for multiple, repeated research use until they become inaccessible by deformation, or more commonly, refreezing, or are required for research purposes to be left undisturbed. Thus, most students who have the opportunity to relate a project to boreholes in glaciers do so through affiliation with an ongoing research programme, using the boreholes as part of, or supplementary to, the main focus of that programme. If boreholes are available for student use, then almost any of the techniques presented in this chapter may be used. However, where flexibility of choice is available, some research themes would be particularly well suited to student investigation. These are summarized below.

- Plotting the incidence and nature of englacial voids intersected during drilling by investigating drilling logs. These resulting data could be related to potential controls such as crevasse patterns and theories of the location and development of englacial drainage networks.

- Investigating spatial and temporal patterns of subglacial and englacial water drainage by EC profiling (to investigate borehole plumbing) and time-series survey.
- Investigating the nature of the ice–bed interface and sediment thickness where sediments occur by penetrometry. Results could be plotted spatially and related to glacier topography and perhaps basal drainage patterns where these are known or can be approximated.
- Investigating spatial patterns in borehole water levels by use of a water-level probe (an EC probe would suffice). Results might be used to characterize and define certain modes of water-level behaviour and these could be investigated in terms of possible controls, principally spatial patterns in meltwater delivery to, and removal from, areas of the glacier bed.
- Investigating glacier hydrochemistry (and particularly subglacial meltwater chemistry) by sampling water from the borehole water column and analysing it in the laboratory. Such a study could easily be combined with or used in association with borehole EC profiling (above).

6

Ice radar

6.1 AIM

Ice radar (Radio Detection and Ranging) or radio-echo sounding (RES) uses the transmission and detection of electromagnetic radiation at frequencies of between 5 and 1000 MHz to investigate ice-mass properties. This chapter provides an overview of the physical principles that underpin ice radar, including summaries of the main controls over radar wave velocity, power loss and detectability. The chapter then considers the hardware used in radar investigations, providing, for example, instructions for assembling home-constructed antennae. The presentation of radar data is then considered and illustrated, followed by an overview of the most commonly used field radar survey techniques. These include common offset survey, wide-angle reflection and refraction and common mid-point surveys, borehole-based transmission survey and time-series survey. The chapter then considers radar signal processing, providing a brief overview of signal filtering, amplification and the migration of common offset profiles. Finally, some of the most frequent glaciological applications of ice radar are summarized and illustrated.

6.2 BACKGROUND AND PHYSICAL PRINCIPLES

Radar wave propagation through ice is principally controlled by two material properties: (i) permittivity, and (ii) conductivity (Table 6.1). Electrical permittivity (ε, m^{-1}) describes the capacity of ice to store an electrical charge, effectively impeding the flow of an applied electrical current. Permittivity is normally described relative to that in free space (8.854×10^{-12} F m^{-1}), and should therefore accurately be referred to as

Field Techniques in Glaciology and Glacial Geomorphology   Bryn Hubbard and Neil Glasser
© 2005 John Wiley & Sons, Ltd

Table 6.1 Typical electrical properties of a variety of common earth surface materials. Modified from Annan (1999)

Material	Relative electrical permittivity (ε_r)	Electrical conductivity (σ) (mS m^{-1})	Velocity (V) ($\times 10^8$ m s^{-1})	Attenuation (α) (dB m^{-1})
Air	1	0	3.0	0
Distilled water	80	0.01	0.33	0.002
Fresh water	80	0.5	0.33	0.1
Salt water	80	3000	0.1	1000
Dry sand	3–5	0.01	1.5	0.01
Saturated sand	20–30	0.1–1.0	0.6	0.03–0.3
Silt	5–30	1–100	0.7	1–100
Clay	5–40	2–1000	0.6	1–300
Granite	4–6	0.01–1	1.3	0.01–1
Ice	3–4	0.01	1.67	0.01

relative permittivity (ε_r), commonly called the *dielectric constant*. In the following text, the term 'permittivity' will refer to 'relative permittivity' unless otherwise stated. The permittivity of ice is ~3 and that of pure water is ~80. The latter value rises further with the presence of impurities such as acids and salts. The permittivity of ice is also sensitive to material properties, including crystal orientation and, to a lesser degree, density and temperature. Electrical conductivity (EC in S m^{-1}; section 4.4.3.1) describes the ability of a material to conduct an electrical current. The EC of ice ($\sim 10^{-9}$ S m^{-1}) is principally controlled by its ionic, or impurity, content. Natural polar ice therefore tends to be more conductive than natural temperate ice, from which impurities are commonly flushed out.

Three sets of properties are important to the use and interpretation of ice RES: (i) radar wave velocity, (ii) radar wave power loss, and (iii) radar wave resolution and detectability.

6.2.1 Radar wave velocity

The propagation velocity of radar waves through a material (V) is given by:

$$V = \frac{c}{\sqrt{(\varepsilon/2)[(1 + P^2) + 1]}} \tag{6.1}$$

where c is the speed of the radar wave in free space (i.e. the speed of light $= 3.0 \times 10^8$ m s^{-1}), ε is permittivity, and P is a loss factor, defined as $P = \sigma\omega\varepsilon$ where σ is EC, and ω is angular frequency ($= 2\pi f$, where f is the frequency of the radar wave). Radar wave velocity in ice is therefore mainly a function of permittivity (and the material properties that control it) and, to a lesser extent, EC and radar wave frequency. Since the EC of ice is low, equation (6.1) simplifies to:

$$V = \frac{c}{\sqrt{\varepsilon}} \qquad (6.2)$$

Field and laboratory measurements of the velocity of radar waves in ice indicate an approximate velocity of $1.67 \times 10^8 \, \mathrm{m \, s^{-1}}$ (i.e. $167 \, \mathrm{m \, \mu s^{-1}}$); slightly higher in cold, relatively dry ice and slightly lower in temperate, relatively wet ice. However, natural ice is heterogeneous and often anisotropic, exhibiting spatial variations in a variety of dielectric properties, especially water content and composition, density and crystal structure. Since each of these variations induces a related variation in permittivity, the velocity of radar waves through natural ice masses also varies with location, scale and orientation. Further, the seasonality of certain glaciological processes, particularly drainage, introduces additional temporal variations that may need to be considered.

A key consequence of knowing the velocity of radar waves in ice is that it allows the distance d to a reflector to be calculated from the time t it takes for an emitted wave to travel to that reflector and back (i.e. a total distance of $2d$). This quantity t is known as the two-way travel time. Since distance equals velocity V multiplied by time t, equation (6.2) can be expressed in terms of d as:

$$d = \frac{tc}{2\sqrt{\varepsilon}} \qquad (6.3)$$

It should be remembered that using this equation to determine the distance to a reflector is only as accurate as the value adopted for the mean permittivity. In the case of ice this value can change at many different scales.

6.2.2   Radar wave power loss

Loss of radar signal strength or wave amplitude (in dB) may conveniently be expressed in terms of the skin depth (commonly denoted δ), which is the distance (in m) from the source over which the strength has decreased to $1/e$ (i.e. $\sim$37%) of its initial value. In ice, this loss can occur as a result of a number of processes, summarized in Figure 6.1, including: (i) geometrical spreading, (ii) attenuation, (iii) scattering, and (iv) absorption.

6.2.2.1   Geometrical spreading

Radar signals are generally transmitted orthogonal to the antennae as conical beams with a cone angle of $\sim$90°. This spreading results in a loss of signal energy at a rate of $1/d^2$, where d is distance from source.

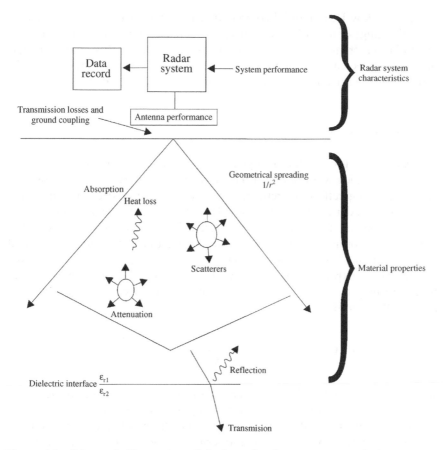

Figure 6.1 Schematic illustration of the loss of radar wave energy during passage through ice. Reproduced from Reynolds (1997) with the permission of Wiley

6.2.2.2 Attenuation

Attenuation represents a fundamental cause of energy loss as a radar wave passes through ice. This loss depends on the frequency of the radar signal as well as the properties of the ice itself (electrical conductivity and permittivity). The loss of radar wave energy per unit distance travelled is given (in $dB\,m^{-1}$) by the attenuation factor or attenuation coefficient (α), where:

$$\alpha = \sqrt{\omega\left\{\left(\frac{\varepsilon_r}{2}\right)\left[\left(\frac{1+\sigma^2}{\omega^2\varepsilon_r^2}\right)^{\frac{1}{2}} - 1\right]\right\}} \tag{6.4}$$

The attenuation coefficient in ice is therefore inversely related to the skin depth (δ) and typically has a value of $0.01\,dB\,m^{-1}$. This makes pure ice a relatively low-loss material compared with other Earth surface materials. Over the range of ice radar frequencies, α increases with ice permittivity, ice EC and radar frequency. One important consequence of this relation is that penetration decreases at higher radar frequencies. This has important implications for the selection of radar frequency in field investigations.

6.2.2.3   Scattering

In practice, a great deal of signal strength loss occurs as a result of 'scattering', which is an umbrella term covering a variety of energy-loss processes including reflection, refraction and diffraction. Indeed, much of the appeal of ice radar derives from desirable scatter, or signal, which is produced by wave reflection from the target of interest. In contrast, unwanted scatter is termed clutter or noise. Scattering losses are a function of the number, size and type of scattering bodies in the ice and their dielectric and geometrical contrasts.

Reflectivity expresses the quantity of radar energy returned after reflection, and may be described quantitatively by the 'power reflection coefficient' (PRC), in dB:

$$PRC = A - G + 20 \, \log\left(\frac{\lambda_0}{8\pi(r_0 = (d/n))}\right) + D \qquad (6.5)$$

where A is the amplitude of the received radar wave relative to the transmitted power, G is the two-way antenna gain (or amplification), D is the dielectric absorption (all in units of dB), and the bracketed term accounts for the geometric spreading and refraction of the waves, where λ_0 is the signal wavelength in air, d is the ice depth of the internal reflecting horizon (IRH) under investigation, r_0 is the terrain clearance of the radar where, for example, the system is aircraft-mounted, and n is the refractive index of ice $(=\sqrt{\varepsilon})$.

For normally incident waves, the strength of radar energy reflection, as measured by the PRC, at a dielectric boundary is a positive function of the difference between the refractive indices of the two materials forming that boundary. Thus, the reflected energy E_r at such a boundary is related to the incident energy E by:

$$E_r = E\left(\frac{n_2 - n_1}{n_2 + n_1}\right) = E\left(\frac{\sqrt{\varepsilon_2} - \sqrt{\varepsilon_1}}{\sqrt{\varepsilon_2} + \sqrt{\varepsilon_1}}\right) \qquad (6.6)$$

where the properties of the host material are indicated by subscript 1 and the reflecting material by subscript 2. The ratio of reflected energy to incident energy, E_r/E, may also be referred to as the amplitude reflection coefficient (R). One important product of this relation is that, since E and E_r represent the amplitudes of the electrical field vectors of the radar waves, a change in the sign of E_r relative to E indicates a change in the polarity of the reflected wave relative to that of the incident wave. Such a change occurs for all cases of $n_1 > n_2$ (and thus for $\varepsilon_1 > \varepsilon_2$). Consequently, analysis of the phase polarity sequence of a reflected radar wave relative to the incident or generated radar wave provides a means of distinguishing the relative permittivities of the boundary materials producing the reflection (Box 6.1).

Box 6.1   Phase analysis of reflected GPR waves to determine material properties within ice: Matanuska Glacier, USA

The phase polarity sequence of reflected radar signals (i.e. whether the reflected waveform is positive-negative-positive $(+-+)$ or negative-positive-negative $(-+-)$ indicates the sign of the permittivity contrast between the host material and the reflector. In glaciology, reflections from rock and water, with a higher permittivity than ice, thereby produce a signal sequence of the form $+-+$, the opposite of that

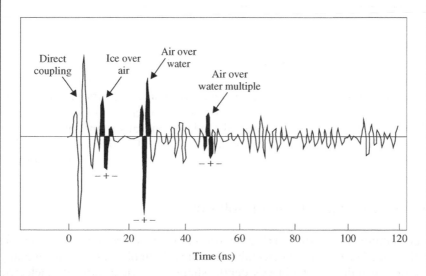

Figure B6.1 Radar scan from within an ice mound at Matanuska Glacier, illustrating different phase polarity sequences and their interpretations. Reproduced from Arcone *et al.* (1995) with the permission of the International Glaciological Society

Box 6.1   (Continued)

produced by a reflection from air $(-+-)$, with a lower permittivity than ice (e.g. Walford *et al.*, 1986). Arcone *et al.* (1995) applied the technique to deduce the material properties of ice sequences and voids within Matanuska Glacier, Alaska, USA. These authors analysed the phase polarity of the first three major half-cycles of reflected waveforms at 50 and 400 MHz following signal filtering and profile migration (Figure B6.1). The authors calibrated their up-glacier interpretations by reference to observed phase sequences at the glacier margins, for example between the englacial ice and the upper surface of the basal ice layer. One of the main glaciological results of the study was that englacial voids near the glacier terminus were interpreted as being air-filled, whereas many deeper voids located further up-glacier were interpreted as being debris-filled or water-filled.

Arcone, S.A., Lawson, D.E. and Delaney, A.J. 1995. Short-pulse radar wavelet recovery and resolution of dielectric contrasts within englacial and basal ice of Matanuska Glacier, Alaska, USA. *Journal of Glaciology*, 41(137), 68–86.

Walford, M.E.R., Kennett, M.I. and Holmlund, P. 1986. Interpretation of radio echoes from Storglaciären, Northern Sweden. *Journal of Glaciology*, 32(110), 39–49.

6.2.2.4   Dielectric absorption

Radar waves are absorbed in ice by *conduction* and *relaxation*. Conduction causes electrons to shift slightly relative to their nuclei, losing energy from the electric field into the host material. Relaxation causes energy to be lost through the oscillation of water molecules. Such absorption is increased in impure ice (high permittivity and EC) and is positively related to temperature.

6.2.3   Detectability and resolution

Detectability refers to whether an object or layer produces a detectable reflection. The minimum detectable size of such an object depends on many factors, including that object's depth, shape and orientation, its dielectric contrast with its host medium, signal loss, and noise and radar system performance. In contrast, resolution may be defined as the ability of a radar system to reproduce the size and shape of reflectors accurately. In practice this may be considered as the smallest scale at which adjacent reflectors can be differentiated from each other. In this case vertical resolution differs slightly from horizontal resolution.

Vertical resolution is a temporal variable, and is theoretically equal to one quarter of the signal's centre wavelength (i.e. $\lambda/4$). In reality, actual resolution may be reduced to up to $\lambda/3$ to $\lambda/2$ by, for example, complexities in the form of the penetrating wave-train and the progressive filtering of higher-frequency elements from it (according to equation (6.4)).

Horizontal resolution is controlled by the illumination area of a radar system's antennae, or its footprint, which defines the first Fresnel zone around the object of interest. If an object is larger than the first Fresnel zone, then that object's outline will be faithfully reproduced in the resulting radargram. However, if an object is smaller than the first Fresnel zone, then its reflection will contain diffraction patterns and the resulting radargram will not be represented to its fullest extent faithfully. Similarly, several objects located within a single Fresnel zone cannot be distinguished from each other. The radius of the first Fresnel zone in ice (F_r) may be calculated from the centre wavelength of the radar used (λ) and the depth of the object being imaged (h) according to (e.g. Robin et al., 1969):

$$F_r = \sqrt{\frac{\lambda h}{2} + \frac{\lambda^2}{16}} \qquad (6.7)$$

Welch et al. (1998) investigated this effect empirically in a simulated study of the radargram representation of hummocks of different sizes located beneath 300 m of electrically transparent ice at 5 MHz. These authors found that, following profile migration (section 6.6.3), hummocks longer than $\lambda/2$ were reproduced accurately while those smaller than $\lambda/2$ contained false reflections. Welch et al. (1998) therefore proposed that a maximum horizontal resolution may be approximated as $\lambda/2$ in migrated profiles (Figure 6.2). This and some other ways of improving the resolution of radar investigations are outlined in section 6.6.

Since radar signal resolution is improved at high frequencies (e.g. equation (6.7)) while radar signal penetration is improved at low frequencies (equation (6.5)), the frequency selected for field radar investigations often represents a trade-off between penetration and resolution.

6.3   ICE RADAR EQUIPMENT

Radar systems include a transmitter and a receiver, each of which is equipped with a similar antenna. The transmitter generates a pulse of radar waves that penetrates into the ice body, where they are eventually reflected or absorbed. The receiver detects both the direct signal travelling directly from the

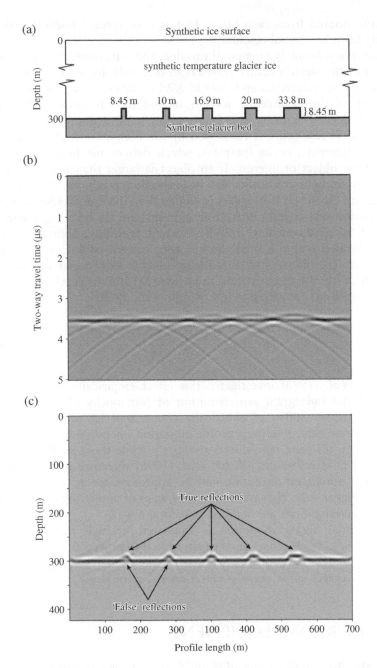

Figure 6.2 Synthetic analysis of the role of migration in resolving bed hummocks of different sizes using 5-MHz radar: (a) bed hummock sizes; (b) unmigrated hyperbolic reflections; and (c) migrated radargram. Reproduced from Welch *et al.* (1998) with permission from the International Glaciological Society

transmitter (the 'airwave') and the components of the transmitted signal that are reflected within the ice body. The most commonly used antenna, a dipole, consists of two identical wire arms. The transmitter causes electrical currents to oscillate along the wire, such that one complete round trip of the electrons (i.e. to each end of the antenna) represents one oscillation cycle. Thus, the length of each dipole antenna arm represents one quarter of a wavelength. Since the radar wave moves at a fairly constant velocity V through the ice, the frequency of the emitted and received signal is inversely related to the length of each individual antenna arm (L, in m) according to:

$$L \approx \frac{\lambda}{4} \approx \frac{V}{4f} \qquad (6.8)$$

where f = frequency (in MHz). If the approximation is made that the mean velocity of the radar wave in ice (V) is $167\,\mathrm{m\,\mu s^{-1}}$, then

$$L \approx \frac{42}{f} \qquad (6.9)$$

Although readily purchased in association with ground-penetrating radar (GPR) systems, dipole antennae can also be easily constructed, requiring only conductive wire (rubber-coated test wire is ideal) and resistors. The antennae are resistively loaded to prevent ringing, and the number of resistors n stepped along each arm depends on the application; $n = 5$ is suitable for general use, for example for detecting the glacier bed, although ten or more may be preferred where a high-quality signal is important. Although not critical, resistors should be spaced equidistant along each arm, separated by $(1/n)L$ m from each other, and with the first and the last resistors being placed $(1/2n)L$ m from each end. The resistance of each resistor (R) depends on n and the chosen damping constant (φ, Ω) such that:

$$R = \frac{\varphi}{\left[n\left(1 - \frac{l}{L}\right) \right]} \qquad (6.10)$$

where l is the distance along each arm of the antennae from the central feed point. For most uses, a damping constant of $300\,\Omega$ is suitable. Thus, a typical resistively loaded dipole antenna for use at $10\,\mathrm{MHz}$ on ice is illustrated in Figure 6.3. Once constructed, it is advisable to enclose each wire–resistor–wire connection in at least one layer of heat-shrink sleeve to strengthen the connections. Clear heat shrink allows ready inspection of each connection should a problem arise.

Dipole antennae are highly directional, transmitting and receiving most of their energy orthogonal to the length of the wire, and little energy parallel to it. Thus, careful consideration of antennae orientation can be used both to

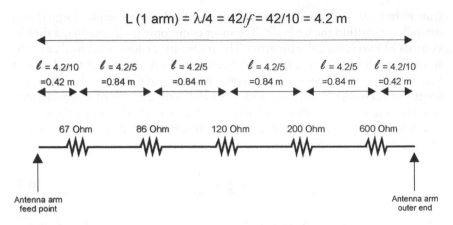

Figure 6.3 Schematic illustration of the resistor sizes and spacing along one arm of a resistively loaded dipole antenna, using a centre frequency, f, of 10 MHz as an example. Each transmitter antenna and receiver antenna usually comprises two such arms

reduce spurious reflections from valley walls and to focus radar signals in a particular direction.

6.3.1   Radar systems

Current impulse radar systems represent relatively minor adaptations of those first successfully applied to temperate ice in the 1960s. Transmitters used in these systems are based on avalanche transistors, capable of generating a continuous high-voltage signal over a wide range of frequencies, linked to resistively loaded dipole antennae. Such systems have been applied with centre frequencies of between 5 MHz (e.g. Watts and England, 1976) and 840 MHz (e.g. Narod and Clarke, 1980). The basis of such systems is straightforward. Signal output from a transmitter is fed to the transmitting antenna such that one arm transmits the positive part of the waveform and the other the negative part of the waveform. This is achieved by connecting one antenna arm to the positive output terminal of the transmitter and the other to the negative terminal of the transmitter. At the receiving end, the antenna feeds directly into an oscilloscope that samples the received waveforms and averages or stacks them typically 64 or 128 times (thereby improving the signal-to-noise ratio). Hand-held digital oscilloscopes can be controlled and interrogated digitally by palmtop computers, allowing instrument control and data acquisition to be almost wholly automated. Suitable digital oscilloscopes are manufactured by companies such as Fluke and Tektronix Inc. and are available from standard distributors.

The most commonly used impulse transmitter is available from Icefield Instruments Inc., Canada (Narod and Clarke, 1994).

In recent years, the widespread commercial availability of complete radar systems incorporating dedicated control and analysis software has led to a massive increase in the scale and breadth of field-based ice radar applications. These high-specification GPR systems, provided by companies such as Malå GeoScience, Sweden, and Sensors and Software Inc., USA, are back-pack-portable, provide user-friendly operation and produce high-quality data. The transmitters on these systems emit a strong ($\sim$120–160 dB) and relatively short radar pulse via antennae that are housed within a protective housing, often mounted to a frame or sledge. Transmitters and receivers are also usually linked by an optical cable, automating the co-ordination of signal timing.

6.4   RADAR DATA PRESENTATION

Field radar surveys typically involve investigating signal variations as the antennae are moved through space, usually across the ice surface but occasionally along a borehole (section 6.5). However, it is also possible to collect radar returns continually from stationary antennae in order to investigate temporal variations in dielectrically sensitive processes such as glacier hydrology (Box 6.2). In either case, received signals are usually presented as one of three forms. First, individual radar traces may be plotted as returned signal strength (i.e. amplitude) on the x-axis against time on the y-axis, sometimes termed *A-scope* (Figure 6.4a). Second, several such traces may be plotted side-by-side in their true spatial relationship to each other (i.e. with distance along a profile as a primary x-axis) to produce a *radargram*, or wiggle plot, that can be evaluated visually (Figure 6.4b). Finally, a *Z-scope* radargram is similar to a wiggle plot but signal intensity is plotted as a colour or grayscale shade (Figure 6.4c).

Most of the resulting field radargrams have certain features in common. The strongest and earliest event is almost always the direct wave or *air wave*. This represents the signal transmitted directly from the transmitter to the receiver through the air, travelling at the speed of light. Next, the ground-coupled wave is slightly less dominant than the air wave, having passed directly between the antennae through snow or ice in the near-surface of the glacier. Since the permittivity of ice is greater than 1, this ground-coupled wave or *ground wave*, travels slower than the air wave, and arrives slightly later, and is correspondingly lower in the trace. All waves received subsequent to the ground wave represent reflections from the ice body being surveyed. A spatially coherent and often dominant basal reflector may be notable amongst these where waves penetrate to the ice-mass base.

Box 6.2   Interpretation of fixed-location RES time series in terms of glacier hydrological changes: Variegated Glacier, USA

Jacobel and Anderson (1987) reported ice radar results as part of an extensive research programme investigating the build-up and surge of Variegated Glacier, Alaska, in 1982–1983 (Kamb *et al.*, 1985). These authors focused on the use of RES to reconstruct certain hydrological properties of the glacier. As part of these investigations the authors repeated daily soundings from stationary 4- and 8-MHz ice surface radar over a ~20-day period. Returns from these soundings (Figure B6.2) indicated the presence of two pronounced echoes, located at depths of ~50 m and ~190 m, and a more minor reflection at ~120 m. All three echoes changed in intensity and phase over the period of investigation, while the location of the deepest echo also moved deeper into the glacier. These echoes were considered by the authors to be englacial cavities and the time series of echoes from them was interpreted in terms of changes in their position, size and water content. Analysis of the returned waveforms at the two frequencies used indicated that these cavities had a perpendicular dimension of

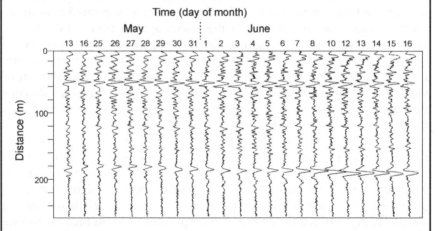

Figure B6.2   Time series of radar wave-forms between 13th May and 16th June, 1983, recovered from a fixed position on the surface of Variegated Glacier, Alaska, USA. Note the prominent echoes assumed to be returned from water bodies at ~50-m and 180-m depth beneath the ice surface. Reproduced from Jacobel and Anderson (1987) with the permission of the International Glaciological Society

10^{-1}–10^{0} m, and correlation with other data, including water levels in nearby boreholes, indicated that the englacial drainage identified responded to surface water inputs rather than to basal water pressures.

Jacobel, R.W. and Anderson, S.K. 1987. Interpretation of radio-echo returns from internal water bodies in Variegated Glacier, Alaska, USA. *Journal of Glaciology*, 33(115), 319–323.

Kamb, B., Raymond, C.F., Harrison, W.D., Engelhardt, H., Echelmeyer, K.A., Humphrey, N., Brugman, M.M. and Pfeffer, T. 1985. Glacier Surge Mechanism: 1982–1983 Surge of Variegated Glacier, Alaska. *Science*, 227(4686), 469–479.

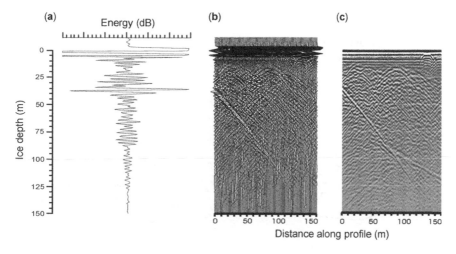

Figure 6.4   The most commonly used radio-echo sounding signal depiction formats: (a) single A-scope trace; (b) wiggle plot radargram; and (c) Z-scope radargram. Note that the single trace illustrated in (a) originates from the start of the profiles illustrated in (b) and (c)

6.5   FIELD RADAR SURVEYS

Once the fundamentals of RES are understood, there is no limitation to the type or style of field survey that can be devised in the light of specific glaciological problems. However, the overwhelming majority of radar investigations utilize only a small number of field survey techniques. GPR survey design and operation are covered in detail in Annan and Cosway

(1992) and Annan (2001). However, the most common of these surveys are summarized in the following sections.

6.5.1   Common offset radar survey

Common offset (CO) profiling represents the standard mode of spatial radar survey and the only mode of airborne survey. During a CO survey, the transmitter and receiver are separated by a fixed distance and moved together along a profile (Figure 6.5). Typically, the separation distance between the two antennae is equal to the antenna length (or $\lambda/2$) and for maximum resolution to be achieved the step distance for high-resolution survey should not be greater than the theoretical resolution of the system, or $\lambda/4$ (Table 6.2).

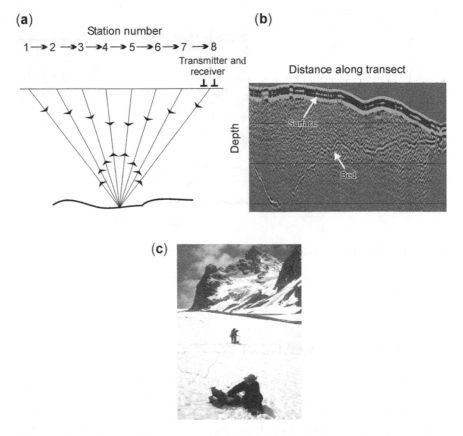

Figure 6.5   Common offset radar profiling: (a) schematic illustration of transmitter and received positions; (b) unmigrated common offset profile radargram; and (c) photograph of common offset profiling using GPR

Table 6.2 Summary of typical radar system frequencies used in ice surface radar and corresponding field survey parameters

Centre frequency (MHz)	Wavelength in ice (m)	Antenna length* (m)	Optimum antenna separation (m)	Maximum step size (m)	Vertical (m)	Horizontal (m)
200	0.8	0.4	0.5	0.25	0.21	0.42
100	1.7	0.8	1	0.5	0.42	0.84
50	3.3	1.7	2	1	0.84	1.67
20	8.4	4.2	5	2.5	2.09	4.18
10	16.7	8.4	10	5	4.18	8.35
5	33.4	16.7	20	10	8.35	16.70

Column groups: "System parameters" (Centre frequency, Wavelength in ice, Antenna length); "Common offset profiling" (Optimum antenna separation, Maximum step size); "Approximate RES resolution[#]" (Vertical, Horizontal).

* Each antenna arm is one-half this length.
[#] Assuming a clean signal with little interference or noise.

Thus, for a 100-MHz survey, the transmitter antenna and receiver antenna are placed 1 m apart and the entire apparatus moved in individual steps of ≤ 0.5 m, while for a 5-MHz survey, the transmitter and receiver are placed 20 m apart, and moved in steps of ≤ 10 m. Resulting CO profiles are normally presented as wiggle plots or Z-scope radargrams with two-way travel time (t) on the y-axis and the horizontal position of the system on the x-axis (Figure 6.5c). Commonly, the vertical axis is presented in terms of distance under a measured or assumed ice velocity ($167 \, \mathrm{m \, \mu s^{-1}}$ is commonly used). The technique thereby often produces sections that are readily interpreted by eye, particularly once hyperbolae are collapsed by migration (section 6.6.3).

6.5.2 Wide-angle reflection and refraction and common mid-point radar surveys

Wide-angle reflection and refraction (WARR) and common mid-point (CMP) surveys are used to investigate the distribution of radar signal velocity with depth. The principle of both surveys is that the antennae spacing is altered above a particular internal reflector, yielding variations in the two-way travel time to that reflector. Since total travel distance is constrained for each antennae offset, changes in two-way travel time for each offset can be used to calculate the averaged radar wave velocity through the ice above the reflector. The difference between WARR and CMP is that the former involves moving either the receiver or the transmitter over a mid-point, which therefore moves itself, while the latter involves moving both the receiver and the transmitter above a static mid-point (Figure 6.6a). Consideration of the

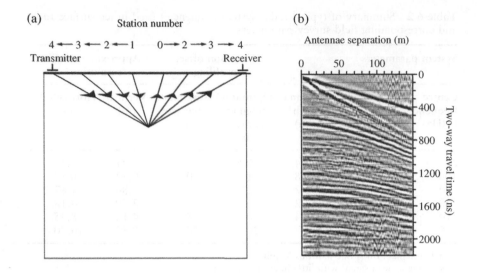

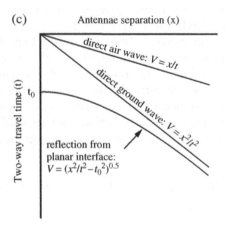

Figure 6.6 Common mid-point radar survey: (a) schematic illustration of transmitter and receiver positions; (b) wiggle plot radargram of survey results; and (c) its interpretation

geometries of these two schemes indicates that the former only works perfectly in the case of a horizontal planar reflector. Since such a reflector does not exist in nature, CMP studies are preferable to WARR studies.

The radargrams from such schemes, normally presented as antennae separation on the x-axis and two-way travel time on the y-axis (Figure 6.6b), are typically characterized by a gently dipping linear air wave, a slightly steeper dipping linear ground wave, and a curvilinear reflected wave (Figure 6.6c). Information yielded by CMP radargrams includes the following:

(i) The velocity of the air wave V_A, which is given by the slope of the arrival time of the air wave t_A:

$$V_A = \frac{S}{t_A} \qquad (6.11)$$

(ii) The velocity of the near-surface ground (snow, ice or firn) wave V_G, which is given by the slope of the arrival time of the air wave t_G:

$$V_G = \frac{S}{t_G} \qquad (6.12)$$

(iii) The permittivity of the near-surface ground (snow, ice or firn) ε_G, which is given by the relative velocities (and therefore time for the same distance) of the arrival of the air wave and the ground-coupled wave:

$$\varepsilon_G = \left(\frac{t_A}{t_G}\right)^2 \qquad (6.13)$$

(iv) The average velocity of the radar signal between the surface and the internal reflector. In this case the slope of the arrival time of the reflected wave t_R is not linear but hyperbolic where, from geometrical considerations:

$$t_R V_R = L_G = \sqrt{(S^2 + 4z^2)} \qquad (6.14)$$

Thus, V_R is most readily calculated by transforming this slope into a straight line, which is achieved by squaring both axes on the plot of two-way travel time (t) against offset (S). On such a plot, the inverse gradient of the arrival time slope of the reflected wave t_G is equal to the squared velocity of the reflected wave V_R. Indeed, on such a plot all three arrival time slopes are straight lines with slopes that are equal to their respective velocities squared.

Where multiple internal reflectors, located at different depths, can be imaged by CMP survey, the resulting data can be used to reconstruct the wave velocity through each of the layers separating those reflectors (Box 6.3). In this case, the interval velocity of any individual layer must be calculated by considering the net velocity through all layers located between that layer and the radar apparatus. Similar to seismic processing, this may be achieved via the Dix formula, where the interval velocity in a confined layer V_{int} over the nth interval is given by:

$$V_{int} = \sqrt{\left[\frac{V_n^2 t_n - V_{n-1}^2 t_{n-1}}{t_n - t_{n-1}}\right]} \qquad (6.15)$$

where V_n, t_n and V_{n-1}, t_{n-1} are the average velocities and reflected wave two-way travel times to the nth and, preceding, $n-1$th reflectors, respectively.

6.5.3   Borehole-based transmission radar survey

Much of the discussion in this chapter refers explicitly to radio-echo sounding, which involves analysing a signal reflected from the surface of, or within, an ice mass. Borehole transmission radar differs fundamentally from echo sounding in that it involves analysing a one-way signal that passes through the material of interest directly from the transmitter to the receiver. While one antenna is located down the borehole, the other may be located at the ice surface or down a second borehole. Although the principles of borehole radar are similar to ice surface radar, borehole antennae are housed within narrow-diameter waterproof tubes and can either have on-board

---

Box 6.3   Use of surface and borehole ground penetrating radar to determine temperate glacier water content: Falljökull, Iceland

Radar waves propagate through glacier ice at a velocity that is inversely proportional to the bulk permittivity of that ice. Since most glacier ice may be considered a mixture of (dry) ice and water, each with a different characteristic permittivity, the bulk water content of an ice mass can be calculated from the measured velocity of radar waves through it. The calculation, however, can only be made on the basis of a mixing model that describes the geometrical relationship between the ice and the water held within it. Two such mixing models are currently used, following Paren (1970) and Looyenga (1965). Murray *et al.* (2000) applied both of these models to radar velocity data collected by surface-based common mid-point survey and borehole-based vertical radar profiling in a study of the internal structure of temperate Falljökull, Iceland. As a result of these analyses the authors identified several zones within the glacier, each with characteristic permittivity, and therefore water-content, patterns (Figure B6.3). At the ice surface, the measured radar wave velocity of $166 \, m \, \mu s^{-1}$ equates to a water content of 0.23–0.34%. This velocity falls to $149 \, m \, \mu s^{-1}$ (3.0–4.1% water) at a depth of 28 m, which the authors interpret as the glacier's internal piezometric water level. Below this, the velocity rises gradually to $152 \, m \, \mu s^{-1}$ (2.4–3.3%), until a depth of $\sim$102 m where it rises dramatically to $167 \, m \, \mu s^{-1}$ (0.09–0.14%), remaining roughly constant to the glacier bed at a depth of $\sim$112 m. Although some of these bulk water contents are a little high, the study clearly indicates the potential of borehole-based radar to investigate the internal structure of temperate glaciers in a non-invasive and quantitative manner.

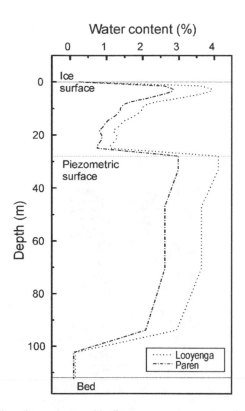

Figure B6.3 Plot of reconstructed bulk water content against depth in temperate Falljökull, Iceland, based on surface and borehole radar wave velocities according to the mixture models of Paren (1970) and Looyenga (1965). Reproduced from Murray *et al.* (2000) with permission from the International Glaciological Society

Looyenga, M. 1965. Dielectric constants of heterogeneous mixture. *Physica*, 31(3), 401–406.

Murray, T., Stuart, G.W., Fry, M., Gamble, N.H. and Crabtree, M.D. 2000. Englacial water distribution in a temperate glacier from surface and borehole radar velocity analysis. *Journal of Glaciology*, 46(154), 389–398.

Paren, J.G. 1970. *Dielectric Properties of Ice*. Cambridge, University of Cambridge.

power or be powered from the glacier surface (Figure 6.7). The technique is often used to generate a tomographic image, or tomogram, of the velocity or attenuation through the plane of ice separating two boreholes or a single borehole and the ice surface (Box 6.4). To create such a tomogram, it is

Figure 6.7   100-MHz borehole antenna (Malå GeoScience) ready to be lowered into a borehole at Tsanfleuron Glacier, Switzerland (Photograph: T. Murray)

Box 6.4   Use of inter-borehole transmission radar to investigate the near-surface ice properties of a temperate valley glacier: Haut Glacier d'Arolla, Switzerland

Although applications of inter-borehole radar are widespread in ground geophysics, few glaciological applications have been reported. One study, however, has recently been carried out by Middleton (2000) as part of wider borehole-based investigations of the hydrology and dynamics of Haut Glacier d'Arolla, Switzerland. Here, the authors

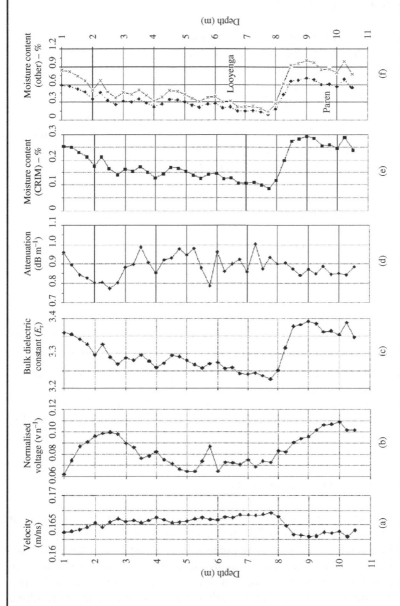

Figure B6.4 Zero-offset inter-borehole transmission radar results from a single borehole plane at Haut Glacier d'Arolla, Switzerland. Reproduced from Middleton (2000) with the permission of the author

---

Box 6.4   (Continued)

used zero-offset inter-borehole transmission radar and vertical radar
profiling between the borehole and the ice surface, both at 100 MHz,
to record spatial variations in signal travel time and peak-to-peak
voltage. The resulting wave velocity and attenuation data allowed
spatial variations in controlling physical properties to be reconstructed
in the top ~20 m of the glacier (Figure B6.4). As a result of this study,
supplemented by inter-borehole electrical resistivity tomography,
Middleton (2000) identified systematic electrical heterogeneity and
anisotropy in the near-surface zone of the glacier. For example, the
authors interpreted a zone of relatively rapid transmission velocity
(and correspondingly low bulk water content) between 9- and 15-m
depth as a remnant of the preceding winter's cold wave, while
signal attenuation was found to be greater across-foliation (i.e.
across-glacier) than along-foliation (i.e. along-glacier).

Middleton, R.T. 2000. Hydrogeological characterisation using high-resolution
    electrical resistivity and radar tomographic imaging. Unpublished PhD Thesis,
    Lancaster University.

---

necessary to acquire numerous alternate wave paths through the ice
separating the antennae transects. Specialist processing software then
associates the resulting velocity or wave power data with the specific
geometries investigated to produce a tomogram – in two or, potentially,
three dimensions.

6.5.4   Time-series radar survey

It may be desirable to track some time-varying changes by recording radar
records continually from a fixed location. Such time-series surveys are under-
utilized in glaciology, despite their potential for investigating temporal variations in, for example, glacier hydrology, where major variations in
water–air–ice boundaries may occur at a time scale of hours to days. In such
surveys, the station interval is a time rather than a distance, and the resulting
radargram plots signal strength on axes of two-way travel time against time
of survey rather than distance along profile (Box 6.2). These diagrams may
be interrogated, similar to CO profile radargrams, although the fundamental
difference in the x-axis property cannot be overlooked; for example, migration is not required as point reflectors will not generate hyperbolae.

6.6 PROCESSING

6.6.1 Frequency filtering

The effect of applying filters during data acquisition is to remove or reduce stipulated frequency ranges. Commonly, an appropriate high-pass (high frequency retained) or band-pass (a frequency range retained) filter is determined by trial-and-error inspection of the amplitude spectrum. In the case of a high-pass filter, a low-frequency cut-off is defined, usually in order to remove artificial instrument noise (often referred to as *dewow*). In applying any filter though, it is important to remember that once an initial filter has been applied, filtered data are lost from the saved data set. Thus, it is better not to apply any filter during data acquisition (or at least to apply a relatively broad band filter) and to re-filter the data subsequently than to realize, following fieldwork, that too narrow a band-pass filter has been applied during acquisition.

6.6.2 Amplification or gain

Due to power loss as a radar signal propagates through ice (section 6.2.2), most radar applications use some form of signal amplification, or *gain*. Most processing software packages offer various user-defined and automatic gain options. Of the latter, the most commonly used gains include deterministic corrections, for example, for attenuation by geometrical spreading, and semi-empirical methods such as automatic gain control (AGC). Since most processing programmes allow a visual preview of the effect of applying a gain, the precise choice of gain used may be dictated by trial and error as much as by preconceived theory.

6.6.3 Migration

Since radar waves are transmitted conically, reflections from point sources traversed by CO profiles appear in traces located either side of the reflector as well as those located directly above it. In such situations, the more oblique the radar location the longer the wave-path to the reflector, causing it to appear lower in the radargram. The consequence of this effect is that a single point reflector is expressed on a radargram as an upward-pointing hyperbola with that reflector located at its apex. Migration is the process of correcting geometrical irregularities such as this, including the dip of sloping planar reflectors and hyperbolae. This is achieved by associating spatial variations in wave-path length with wave velocity to collapse hyperbola limbs back to their apex or source. Hyperbolae on unmigrated radargrams therefore appear as point reflectors in their true positions on migrated radargrams (Figure 6.2). Indeed, if sufficient data are recorded it is possible to migrate radar profiles in two orientations, with a second migration pass orientated orthogonal to the initial radar profile additionally reducing the effect of off-transect reflectors (Box 6.5).

Most commonly, the steps outlined above are combined into a specific processing scheme that is developed for, and applied universally to, each GPR data set. The details of each such protocol depend on the nature of the field data and on the purpose of the study. A typical protocol, for example, was used by Murray *et al.* (2000a) in their 50-MHz and 100-MHz investigation of the character of englacial ice at Falljökull, Iceland. For their common offset profiles:

- a high-pass filter, with a cut-off of 2 MHz, was applied to remove low-frequency, instrument-generated noise.
- the first arrival waves (the air wave) were set to define zero on the time scale.
- automatic gain control was applied to amplify deeper reflections.
- the elevation of each radar trace was adjusted to account for variations in elevation along each profile.

Box 6.5   Two-pass migration of radio-echo sounding data of the internal structure of a temperate valley glacier: Worthington Glacier, USA

Radio-echo sounding surveys of valley glaciers are complicated by complex topography and oblique valley-side reflections. Profiles are therefore commonly migrated to correct for geometrical artifacts, most notably to collapse the limbs of profile hyperbola to their point source. Welch *et al.* (1998) extended this approach to migrate high-resolution orthogonal 5-MHz profiles in three dimensions in a study of the internal structure of temperate Worthington Glacier, Alaska, USA. The authors adopted this approach to address two breaches in the assumptions implicit in standard 2D migration: (i) that no reflections are recorded from objects or surfaces oblique to the profile being migrated, and (ii) that no migrated slopes are oblique to the direction of the profile being migrated. To achieve pseudo-3D migration, the authors first migrated along their profiles, which were collected in the cross-glacier direction, then migrated adjacent sequences of traces (including interpolated traces to decrease point-to-point spacing) in the down-glacier direction. The resulting radargrams depict notably better spatial resolution following 3D migration relative to 2D migration, which is in turn better than unmigrated profiles. Indeed, the three-dimensionally migrated elevation model of the glacier bed has a resolution that approaches the maximum obtainable horizontal resolution of the system ($\lambda/2 = 16.9$ m at 5 MHz).

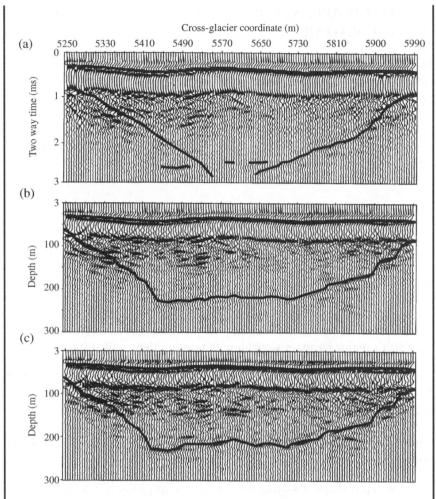

Figure B6.5 Radargrams of a cross-glacier profile from Worthington Glacier, Alaska, USA: (a) raw, unmigrated profile; (b) two-dimensionally (cross-glacier) migrated profile; and (c) pseudo-three-dimensionally (cross-glacier and along-glacier) migrated profile. Note the improved resolution of the bed echo generated by each successive processing stage. Reproduced from Welch *et al.* (1998) with the permission of the International Glaciological Society

Welch, B.C., Pfeffer, W.T., Harper, J.T. and Humphrey, N.F. 1998. Mapping subglacial surfaces of temperate valley glaciers by two-pass migration of a radio-echo sounding survey. *Journal of Glaciology*, **44**(146), 164–170.

6.7   FIELD APPLICATION AND INTERPRETATION OF ICE RADAR

The relative transparency of pure ice to radar wave penetration and the dielectric contrasts between that ice and water or rock (Table 6.1) allow ice radar to be used to investigate a variety of physical properties, many of which are summarized by Plewes and Hubbard (2001). These studies may be broadly classified in terms of the interpretation of reflected radar waves (generally ice surface radio-echo sounding) and the interpretation of direct radar waves (generally borehole-based transmission radar).

6.7.1   Radio-echo sounding

Radar wave reflections can be produced from within the interior of ice masses as well as from their upper (ice–air) interface and their basal (ice–bed) interface. Where radar wave penetration is sufficient, a strong and spatially coherent basal reflection is often recorded, allowing ice-mass thickness to be calculated. This technique has been widely adopted, to provide baseline information relating to the thickness and volume of all of the Earth's major ice masses, yielding maximum depths of ~4000 m in cold ice and ~1500 m in temperate ice. For example, the BEDMAP consortium was established with the remit of rationalizing existing data, collecting new data and producing and making available a new topographic model of the bed of the Antarctic Ice Sheet. These data are now largely available for research purposes from the BEDMAP Consortium and via the Internet at http://www.antarctica.ac.uk/aedc/bedmap/. More sophisticated analysis of bed reflections may be used to characterize the nature of the basal interface; for example, analysis of phase indicates the presence or absence of water and thus basal thermal regime (Box 6.6). These techniques have, over the past 30 years, also been fundamental to the identification and demarcation of numerous sub-ice lakes located beneath the East Antarctic Ice Sheet (Siegert *et al.*, 1996). Indeed, partnering RES surveys with other geophysical data surveys provides a powerful investigative tool for reconstructing both ice and substrate properties. Such studies have been carried out, amongst other places, on the Siple Coast ice streams, Antarctica, where RES was combined with seismic data (e.g. Blankenship *et al.*, 1986) and more recently above Lake Vostok, Antarctica, where RES was combined with gravity data (Studinger *et al.*, 2004). The latter study indicates that the sub-ice lake has two distinct basins and that the water is up to 800-m deep in the deeper, southern basin.

In addition to bed reflections, laterally extensive IRHs are routinely recorded in all of the Earth's major ice masses. Often the uppermost IRH represents the snow–ice or a snow–firn boundary, allowing snow depth to be mapped by high-frequency surface radar at a glacier-wide scale (e.g. Kohler *et al.*, 1997). Deeper IRHs are normally passive marker horizons caused by variations in the permittivity of ice layers formed from snow originally deposited in the ice mass' accumulation area. These permittivity contrasts are most probably caused by impurity variations, often associated with ash deposits from volcanic eruptions, supplemented by contrasts in ice density near the surface and ice fabric at depth. IRHs therefore act as chrono-stratigraphic markers, able to provide information relating to variations in mass balance (e.g. Siegert *et al.*, 2000) and ice flow. It is also possible to track IRHs across and between radargrams in order to transpose a known chrono-stratigraphy, for example, from an existing ice core site to distant locations. This can serve to enhance our understanding of spatial variations in ice flow dynamics or mass balance, or even to inform expectations relating to future ice coring programmes (see Box 3.5).

One particular boundary commonly identified by ice radar in polythermal ice masses is that between cold surface ice and warm interior ice. This boundary is most commonly identified by an IRH at the interface between

---

Box 6.6   Use of GPR to map the internal structure of a polythermal glacier: Stagnation Glacier, Bylot Island, Canada

Moorman and Michel (2000) applied GPR to map the internal structure of (unofficially named) Stagnation Glacier and a nearby proglacial icing at (unofficially named) Fountain Glacier on Bylot Island in the Canadian High Arctic. The authors used various combinations of GPR transmitter power and centre frequency and found that a high-power (1000 V) transmitter combined with 50-MHz antennae provided best results over the $\leq$200-m-thick glacier, while a lower-power (400 V) transmitter combined with 200-MHz antennae provided best results through the $\leq$20-m-thick icing. The authors used centre mid-point surveys to investigate the velocity structure of these ice bodies, finding, for example, that the propagation velocity decreases from $\sim$190 m µs^{-1} in packed snow and firn in the accumulation area of the glacier to $\sim$160 m µs^{-1} in drained slush to bubbly ice in the perennial proglacial icing. The authors also interpreted filtered and migrated common offset profiles in terms of the internal architecture of the glacier and the icing (Figure B6.6), using phase sequence analysis and power reflection coefficient analysis to identify the direction and magnitude of the permittivity contrasts identified.

Box 6.1 (Continued)

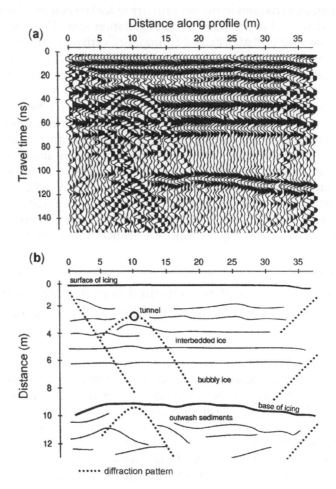

Figure B6.6 GPR map of the seasonal zone of an icing located in front of (unofficially named) Fountain Glacier, Bylot Island, Canada: (a) unmigrated GPR profile; and (b) its physical interpretation. Reproduced from Moorman and Michel (2000) with the permission of Wiley

Moorman, B.J. and Michel, F.A. 2000. Glacial hydrological system characterization using ground-penetrating radar. *Hydrological Processes*, **14**(15), 2645–2667.

the relatively low-permittivity surface cold ice and the relatively high-permittivity underlying wet ice. This boundary, for example, was found in over 50% of glaciers included in an extensive airborne radar survey of Svalbard's ice masses (Macheret and Zhuravlev, 1982; Dowdeswell *et al.*, 1984; Bamber, 1987). Alternatively, warm ice may be identified by signal loss, essentially due to attenuation below the cold ice–warm ice boundary, producing a zone of no reflections within the ice mass concerned. These effects are also felt at the glacier-wide scale such that IRHs are far less frequently identified in temperate glaciers than in partly or wholly cold glaciers. Although this is partially due to poor signal return, percolating meltwater also serves to homogenize any initial impurity-defined stratification at temperate glaciers. Scattering within temperate glaciers, however, can provide useful information relating to englacial drainage system, with discrete englacial reflections often being interpreted as water-filled channels (e.g. Goodman, 1975; Fountain and Jacobel, 1997).

6.7.2 Borehole transmission radar

To date there have been relatively few published applications of borehole-based radar, despite the technique's potential. In one such application, Murray *et al.* (2000a) used borehole-to-surface radar as part of a wider investigation of the water content of temperate ice at Falljökull, Iceland (Box 6.3). Borehole-to-borehole radar was also reported by Macheret *et al.* (1993) and its application has more recently been carried out by Middleton (2000), at Haut Glacier d'Arolla, Switzerland, to investigate the dielectric properties of temperate near-surface ice at the glacier (Box 6.4).

6.8 STUDENT PROJECTS

Clearly, any ice-radar-based student project requires the use of an ice radar and a skilled operator. While home-assembled impulse radars are small and very easy to move across an ice surface, GPR systems, although larger and often fairly unwieldy, generally provide better-quality data and have the advantage of bespoke processing software packages. Assuming a system and operator are available, many different investigations are possible, some of which are summarized below.

- Quantifying spatial and temporal variability in the velocity of radar wave propagation through snow and ice through the use of centre mid-point surveying.

- Locating the glacier bed by low-frequency (≤ 50 MHz), common offset profiling and migrating the profiles to collapse artificial hyperbolae to their reflective sources. This could be supplemented by using phase analysis to characterize spatial and/or temporal patterns in the direction of the dielectric contrast at the ice–bed interface.
- Identifying and characterizing discrete reflectors, linear reflectors and reflecting horizons located within the ice body by common offset profiling at intermediate frequency (~ 25–100 MHz). Discrete and linear reflectors may be related to englacial drainage where, for example, they can be traced down-glacier from large surface moulins into which supraglacial meltwater is delivered.
- Investigating spatial variations in surface snow thickness or the presence of the winter cold wave close to the surface of temperate glaciers by common offset profiling at high frequency (≥ 200 MHz).
- Investigating temporal changes within and beneath a glacier, most likely related to hydrological changes, by time-series radar survey.

7

Glacier mass balance and motion

7.1 AIM

The aim of this chapter is to provide the reader with an overview of the techniques used to investigate glacier mass balance and glacier velocity. First, the principles governing the surface energy budget of glaciers are introduced and the main field-based techniques used to measure the individual energy fluxes involved are summarized. Second, concepts underlying glacier mass balance are introduced and the main field-based techniques used to measure mass balance are presented. The techniques are divided into those concerned principally with measuring ablation and those concerned principally with measuring accumulation. Finally, glacier motion and ice velocity are introduced by way of theoretical principles of glacier motion and an introduction to the models used to mimic it. Next, the most common techniques used to measure ice emotion components are summarized. The techniques are classified into those addressing surface motion, those addressing englacial motion and those addressing basal motion.

Field Techniques in Glaciology and Glacial Geomorphology Bryn Hubbard and Neil Glasser
© 2005 John Wiley & Sons, Ltd

7.2 SURFACE ENERGY BUDGET

7.2.1 Principles

Although friction and geothermal heat can result in a small amount ($\sim 1\,\mathrm{cm\,a^{-1}}$) of melting at the base of temperate glaciers, most melting occurs at the glacier surface. Surface melting is strongly controlled by the weather, such that rates typically vary between nil in the winter months (while basal melting may well continue) and up to 10–$20\,\mathrm{cm\,d^{-1}}$ at alpine glaciers in the summer. However, at any given time the energy available to heat ice up and to melt it once it is at the melting temperature (Q_m, J) varies with altitude (increasing at lower elevations) and latitude (increasing at lower latitudes) according to a glacier's surface energy budget. This budget comprises four main energy inputs: (i) net radiation (Q_{nr}), (ii) sensible heat (Q_s), (iii) latent heat (Q_l), and (iv) heat provided by the cooling and freezing of precipitation (Q_p). Thus

$$Q_m = Q_{nr} + Q_s + Q_l + Q_p \tag{7.1}$$

Where this energy is used to melt ice already at the melting temperature, the mass of ice melted (M, g) is given by the energy available to melt that ice divided by the latent heat of fusion of the ice (L, $\mathrm{J\,g^{-1}}$). Thus

$$M = \frac{Q_m}{L} \tag{7.2}$$

Heat fluxes into subfreezing materials such as soils can be measured with a flux plate or calculated from temperature profiles where thermal conductivity is known. These methods, however, have rarely been applied on ice or snow. In order to derive Q_m accurately, field glaciologists must measure, or approximate, the four energy components as closely as possible. These are outlined below.

1. *Net radiation* (Q_{nr}). All matter emits radiation, the magnitude and wavelength of which varies with the temperature of the emitting matter. The wavelength of the peak emitted energy is linearly related to the inverse of the emitter's temperature, and the total radiation flux varies with the emitter's temperature above $0\,\mathrm{K}$ ($-273\,^{\circ}\mathrm{C}$) raised to the fourth power. This explains why the Sun emits more radiation that is of a shorter wavelength than does the Earth: the Sun emits short-wave radiation (of wavelength between ~ 0.2 and $4.0\,\mu\mathrm{m}$, which overlaps with the visible spectrum), while the Earth and its atmosphere emit long-wave radiation in the (invisible) infrared spectrum. Importantly, short-wave radiation is far more easily reflected and scattered than long-wave radiation. This complicates calculations

of radiation used for melting ice at the glacier surface because much of the incident short-wave radiation may be re-radiated within the atmosphere (in particular by clouds) and back into it from the ground. The latter effect may be particularly important in mountainous areas characterized by high relief, where local topography may play an important role in local radiation receipt. Consequently, much solar radiation arrives at ice surfaces as scattered, diffuse radiation. Indeed, one of the most important characteristics of ice is its high reflectivity (termed *albedo*), which can vary from between 0.1 and 0.2 for debris-covered and old (dirty) glacier ice to >0.95 for fresh clean snow. The globally averaged albedo value for the Earth is between 0.3 and 0.4.

2. *Sensible* (Q_s) *and latent* (Q_t) *heat fluxes.* These two fluxes are frequently combined since they are both strongly dependent on heat transfer between the air and an ice surface in contact with it. Sensible heat transfer relates to the warming or cooling effect of air, caused by a temperature difference between that air and the ice surface. The sensible heat flux is therefore principally controlled by the air temperature gradient directly above the ice surface. Latent heat transfer relates to the energy that is consumed or released during changes of state (or phase) between ice, water and water vapour. The latent heat flux is therefore principally controlled by the vapour pressure gradient above an ice surface. These gradients in temperature and vapour pressure dictate the direction of energy flow such that:

- Sensible heat is transferred to the ice if the air is warmer than the ice, and *vice versa*. The magnitude of this flux is primarily governed by the temperature difference between the air and the ice surface, the thermal conductivity of the ice surface and near-surface air turbulence.
- Latent heat is removed from the ice if melting occurs, and latent heat is released to the ice if water freezes onto its surface. The magnitude of this energy transfer is governed by the latent heat of fusion of ice ($334\,\mathrm{J\,g^{-1}}$), such that the freezing of 1 g of water releases 334 J of energy. Similarly, the melting of 1 g of ice consumes 334 J of energy.
- Latent heat is removed from the ice if ice sublimation occurs, and latent heat is released to the ice if water vapour condenses and freezes onto its surface. This transfer is governed by the vapour pressure difference between the ice surface and the overlying air. Thus, if the vapour pressure in the air is greater than at the ice surface, water vapour may condense onto the ice, releasing latent heat and warming the ice up. Conversely, if the vapour pressure in the air is less than at the ice surface then ice may sublimate, consuming energy and cooling the ice surface down. For an ice surface at $0\,°\mathrm{C}$, the water vapour pressure can be calculated to be $\sim\!610\,\mathrm{Pa}$ (Paterson, 1994). The magnitude of these energy fluxes is governed by the latent heat of vaporization

of water to water vapour ($2834\,\mathrm{J\,g^{-1}}$), such that the condensation of 1 g of water vapour releases $2834\,\mathrm{J}$ of energy. Similarly, the sublimation of 1 g of ice consumes $2834\,\mathrm{J}$ of energy.

In practice, since the rate of energy transfer is dependent on near-surface gradients in temperature and vapour pressure, the turbulence of that air becomes an important controlling factor. In this context, greater turbulence promotes faster energy transfer by bringing non-equilibrated air into closer contact with the ice surface. Thus, for example, for wind speeds above $\sim 3\,\mathrm{m\,s^{-1}}$, the latent heat flux may double for a doubling of wind velocity. These energy transfers are therefore frequently referred to as the *turbulent heat fluxes*. Wind turbulence, in turn, varies as a function of a number of factors, the most important of which are wind speed, ice surface roughness and air-temperature stratification.

3. *Heat provided by cooling and freezing precipitation* (Q_p).   Rain falling onto a snow or ice surface can release energy to that ice by cooling and, where the surface is cold enough, freezing. In general, the latter provides substantially more energy than the former since freezing releases $\sim 334\,\mathrm{J\,g^{-1}}$ of water frozen (governed by the latent heat of fusion – section 7.2.1) while cooling releases $\sim 4.2\,\mathrm{J\,g^{-1}}$ of water per $1\,°\mathrm{C}$ cooling (governed by the *specific heat capacity of water*). Thus, almost 100 times more heat is released by the freezing of a unit mass of water than is released by its cooling by $1\,°\mathrm{C}$. The consequence of this is that even heavy warm rain is of minor importance in melting a snowpack. In contrast, rainfall on a cold snowpack is very effective at raising the temperature of that snowpack to $0\,°\mathrm{C}$: Freezing 1 g of rain (releasing $334\,\mathrm{J\,g^{-1}} \times 1\,\mathrm{g} = 334\,\mathrm{J}$) raises the temperature of $\sim 80\,\mathrm{g}$ snow by $1\,°\mathrm{C}$ (consuming $4.2\,\mathrm{J\,g^{-1}\,°C^{-1}} \times 80\,\mathrm{g} \times 1\,°\mathrm{C} = 336\,\mathrm{J}$).

7.2.2   Measurement

Although all of the energy fluxes outlined above can be measured directly in the field, such a complete suite of measurements is rarely possible or desirable, and many studies therefore approximate the energy budget components from a mixture of primary and proxy data. Net radiation, for example, is composed of four individual fluxes: (i) incoming short-wave radiation, (ii) outgoing short-wave radiation, (iii) incoming long-wave radiation, and (iv) outgoing long-wave radiation. Each of these components can be measured by specialist sensors based on thermopiles which convert the radiation into a thermal response and a voltage signal that can be read by data logger. There are a variety of such instruments, depending on their precise use, including *radiometer* (an instrument that measures radiation), *pyranometer* (short-wave radiation on a plane surface), *net pyranometer* (net short-wave radiation), *pyrgeometer* (long-wave radiation on a horizontal surface) and *net pyrradiometer* (net all-wave radiation from above and below). These instruments are discussed at length in Appendix A2 of Oke (1987). Thus, while it is possible to use a net pyrradiometer to measure all of the radiation fluxes directly, such instruments yield only point-specific data and may require substantial field support. Thus, many glaciological studies (which are essentially designed to predict ice melting from readily-determined meteorological conditions) adopt some degree of substitution of proxy data for primary measurements. Long-wave radiation, for example, may be approximated from ice surface temperature and air temperature, which can be measured in the field, modelled or assumed. Although, such proxy records are less accurate than direct measurements, they are normally acquired more easily and provide greater spatial coverage. Indeed, such substitutions may be made at increasingly inaccurate approximations depending on the purpose of the study and the logistical support available for field-based measurements. For example, in the absence of any direct measurements, all four radiative fluxes may be estimated either theoretically or from calibrated empirical formulae relating the radiation term to surrogate atmospheric variables. Thus, long-wave radiation emitted by a cloudless atmosphere ($L_0, \mathrm{W\,m^{-2}}$) can be described by:

$$L_0 = \varepsilon_{a(0)}\sigma T_a^{\,4} \tag{7.3}$$

Here, T_a is atmospheric temperature (K), σ is the Stefan–Boltzmann constant ($\mathrm{W\,m^{-2}\,K^{-4}}$), and $\varepsilon_{a(0)}$ is the atmospheric emissivity of a cloudless sky, which can itself be related empirically to controlling factors such as geographic location and atmospheric temperature. Similarly, short-wave radiation may be approximated from solar radiation and number of sunshine hours at the surface while albedo may be approximated from the day-of-melt-season, since its value falls as snow is replaced by ice and ice becomes progressively dirtier through the melt season. Importantly, many of these variables may be derived with good spatial coverage without having to carry out exhaustive field measurements at the glacier concerned.

Equally, while the turbulent heat fluxes can be measured directly in the field by eddy fluctuation instruments (most commonly 3D devices such as differential-pressure sensors and acoustic anemometers) located at various heights above the ice surface, the approach is rarely used due to excessive logistical demands and instrument fragility (cf. Smeets *et al.*, 1998). As a more common alternative, aerodynamic theory may be applied, via the so-called bulk or profile methods, to allow turbulent fluxes to be approximated from fast records of wind speed and air temperature (for the sensible flux) or vapour pressure (for the latent flux) at two near-surface levels (e.g. Denby and Greuell, 2000) (Box 7.1). Fast temperature measurements can be made with fine-wire thermocouples (although their survival under field conditions can be short) or sonic anemometers (from variations in the speed of sound), and fast vapour pressure measurements can be made with a fast hygrometer. This method rests on the assumption of a constant flux layer and neutral stratification (which can be relaxed for all but very small fluxes), allowing gradients to be approximated by logarithmic functions. Such studies may be supplemented where possible by measures of ice surface roughness, which may itself be represented by a proxy such as day-of-melt-season, since the surface of a glacier tends to get rougher as the melt season progresses.

Thus, many of the instruments needed to measure energy fluxes directly provide only site-specific data and are costly, restricting the spatial coverage afforded by monitoring multiple sites. Further, such direct measurements may demand the continual presence of field researchers. These factors may place unacceptable constraints on field-based data collection, particularly in cases where spatial variations in energy fluxes may be as important as at-a-site variations through time. Consequently, much of this field-based research aims to develop and tune proxy records of energy fluxes that can be measured or approximated remotely. Ultimately, use of these proxy

---

Box 7.1   Use of data from an automatic weather station to calculate surface energy balance: Morteratschgletscher, Switzerland

Extending earlier research, Oerlemans (2000), based on micrometeorological data from an automatic weather station (AWS) located on the surface of Morteratschgletscher, Switzerland, and Oerlemans and Klok (2002) focused on a comprehensive data set collected over the full year 2000. The weather station, mounted on a tripod in the ablation area of the glacier, measured air pressure, windspeed, wind direction, air temperature and humidity, incoming and reflected solar radiation, incoming and outgoing longwave radiation, snow temperature and change in surface height. The authors used these data to calculate all of the principal components of the surface energy flux at the glacier. Since the turbulent fluxes were not measured directly they were approximated by the so-called bulk approach. This relates a single measurement of windspeed and measurements of temperature and humidity at the melting surface and in the near-surface atmosphere to the turbulent fluxes via a turbulent exchange coefficient.

This was empirically tuned by matching the calculated energy balance with measured surface melting, yielding a value of 0.00153 (with a reference height of 3.5 m). Using this constant exchange coefficient value for the entire year resulted in an excellent fit between measured surface lowering and that predicted by the energy balance model (Figure B7.1). Over the entire period about 75% of the melt energy was supplied by short-wave and long-wave radiation and 25% by the turbulent fluxes.

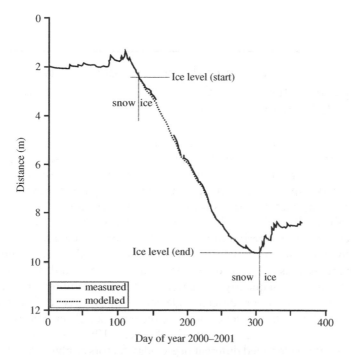

Figure B7.1 Ice surface lowering measured at Morteratschgletscher, Switzerland (solid line), compared with that predicted by a full surface energy balance model (dashed line) run using data measured by an automatic weather station mounted on the glacier surface. Reproduced from Oerlemans and Klok (2002) with the permission of the Regents of the University of Colorado

Oerlemans, J. 2000. Analysis of a three-year meteorological record from the ablation zone of the Morteratschgletscher, Switzerland: Energy and mass balance. *Journal of Glaciology*, **46**(155), 571–579.
Oerlemans, J. and Klok, E.J. 2002. Energy balance of a glacier surface: Analysis of automatic weather station data from the Morteratschgletscher, Switzerland. *Arctic Antarctic and Alpine Research*, **34**(4), 477–485.

data should allow energy fluxes to be calculated accurately with good spatial coverage and with a minimum of site-specific tuning – thereby reducing the need for logistically demanding measurements on the ground. However, achieving this aim is still some way off and some fairly coarse approximations are still used at large-scale and/or remote ice masses. For example, total energy receipt, and therefore melting, may be related to a single index of mean temperature such as the degree-day method (Box 7.2).

Box 7.2   Application of the positive degree-day temperature method to determine ice surface ablation: Iceland, Norway and Greenland

Where multi-sensor meteorological stations cannot be located and interrogated on-site, alternative, proxy data may be used to approximate surface ablation rates. One of the simplest and most effective of these proxy approaches is the positive degree-day method, which uses a degree-day factor to transform the cumulative sum of all temperatures above the melting point into an actual ablation rate at a point of interest and over a period of interest. However, the degree-day factor, usually expressed as $mm\,d^{-1}\,^{\circ}C^{-1}$, is neither spatially nor temporally constant as it is influenced by changeable surface and atmospheric conditions. Consequently, researchers have attempted to define degree-day factors for different glacial settings over different time periods, with particular focus on the Greenland Ice Sheet. For example, in refining earlier large-scale work of Reeh (1991), Braithwaite and Olesen (1993) identified different degree-day factors applying to different seasons on Qamanârssûp sermia, West Greenland. In this case, while the annually averaged factor for ice was found to be $7.9\,mm\,d^{-1}\,^{\circ}C^{-1}$, in good agreement with the generalized value of $8\,mm\,d^{-1}\,^{\circ}C^{-1}$ adopted by Huybrechts *et al.* (1991) for modelling purposes, more specific seasonal factors were found to be $9.4\,mm\,d^{-1}\,^{\circ}C^{-1}$ for the period September–May and $7.5\,mm\,d^{-1}\,^{\circ}C^{-1}$ for the period June–August.

Extending these studies to other ice masses, Jóhannesson *et al.* (1995) compared annually averaged and seasonal degree-day factors for ice melting from Qamanârssûp sermia with those derived from Sátujökull, Iceland, and Nigardsbreen, Norway. In this study, the annually averaged factors were found to be $7.3\,mm\,d^{-1}\,^{\circ}C^{-1}$ for the Greenland site, $7.7\,mm\,d^{-1}\,^{\circ}C^{-1}$ for the Icelandic site, and $6.4\,mm\,d^{-1}\,^{\circ}C^{-1}$ for the Norwegian site.

Braithwaite, R.J. and Olesen, O.B. 1993. Seasonal variation of ice ablation at the margin of the Greenland ice sheet and its sensitivity to climate change, Qamanârssûp sermia, West Greenland. *Journal of Glaciology*, 39(132), 267–274.

Huybrechts, P., Letreguilly, A. and Reeh, N. 1991. The Greenland ice sheet and greenhouse warming. *Global and Planetary Change*, 89(4), 399–412.

Jóhannesson, T., Sigurðsson, O., Laumann, T. and Kennett, M. 1995. Degree-day glacier mass-balance modeling with applications to glaciers in Iceland, Norway and Greenland. *Journal of Glaciology*, 41(138), 345–358.

Reeh, N. 1991. Parameterization of melt rate and surface temperature on the Greenland ice sheet. *Polarforschung*, 59(3), 113–128.

7.3   MASS BALANCE

7.3.1   Principles

A glacier's *mass balance* (or *mass budget*) describes its mass inputs and outputs over various spatial and temporal scales, providing a quantitative expression of volumetric change through time. Mass balance investigations thereby provide an objective means of relating the 'health' of a glacier to climate and climate change. Mass balance is generally expressed in terms of water equivalent depth (in units of metre of water).

Mass inputs are usually dominated by snowfall, but may also include avalanched and blowing snow. These may be lumped together as *accumulation*, which generally increases with altitude and latitude. Mass output is often equated with surface melting, although many of the world's ice masses lose much of their mass by avalanching or calving into water. Calving is particularly significant in ice sheets, whose margins are largely bounded by ocean and whose upper surface falls predominantly within the dry snow zone. Together, these mass losses are termed *ablation*, which generally decreases with altitude and latitude.

A glacier's *net balance* quantifies the net difference between accumulation and ablation over a whole ice mass (unless stated otherwise) per unit time. This balance year normally spans the period from minimum total mass in 1 year to minimum total mass in the next (i.e. end of ablation season to end of ablation season). This generalized index, however, masks systematic temporal variations through the year and spatial variations over the surface of a glacier. The latter may be addressed by expressing point-specific net balance values for different locations on the glacier surface (usually as a bivariate plot of local mass balance against altitude). Since accumulation increases with altitude and ablation decreases with altitude, one may define an area of upper glacier characterized by net mass gain (the *accumulation area*) and an area of lower glacier characterized by net mass loss (the *ablation area*). These two areas are separated by the *equilibrium line*, at which there is no net mass gain or loss over the year. The *balance gradient* is defined as the rate of change of net balance with altitude (i.e. the rate of decrease of ablation from the glacier terminus to the equilibrium line and the rate of increase of accumulation from the equilibrium line to the head of the glacier). Relatively warm and wet *maritime-type* glaciers (such as those located on the west coast of the South Island of New Zealand) are characterized by high accumulation rates and high ablation rates and therefore have high *balance gradients*. These contrast with dry and cold *continental-type* glaciers (such as those located in the high Arctic), which have low accumulation and ablation rates and correspondingly low balance gradients. A glacier's balance gradient exerts a strong control over its dynamics since, in a stable state, all of the mass gained in the accumulation area must be removed by ice flow and all of the mass lost in the ablation area (the same volume) must be replaced by ice flow. For this reason the balance gradient is sometimes referred to as the *activity index* of a glacier.

The information provided in the following sections represents only a very brief overview of mass balance measurement techniques and methods. Full accounts of glacier mass balance principles and methods are provided by, amongst others, Østrem and Brugman (1991) and Kaser *et al.* (2002).

7.3.2   Measurement

Several different methods exist for determining the mass balance of ice masses. These range from remote geodetic methods to direct field-based glaciological measurements. Some of the more common of these methods are outlined in the following sections.

7.3.2.1   Geodetic methods

Geodetic methods are based on calculating an ice mass's volumetric change through time from repeated topographic surveys of surface elevation or extent. These snapshots may be derived from different map editions, repeated aerial photographs, or repeated satellite images. The last of these may be costly at the resolution required, although some satellites, such as the recently launched ICESat, now provide accurate, high-resolution altimetry information. All of these methods can be used to investigate spatial patterns of change responsible for mass balance variations, indicating, for example, where on the ice mass of interest ice may be thickening and/or where it may be thinning. It is, however, essential for the success of such an approach that full coverage of the ice mass concerned is achieved and that the errors associated with each elevation data set are known and appreciated. The method also allows net change to be investigated only between the dates of specific surveys. The method also commonly assumes that elevation change can be equated directly with mass change. This only holds if changes in density do not occur – an aspect of the Earth's ice masses that remains virtually unknown.

Where repeated surface topography models are not available, mass balance may be approximated by identifying the proportion of the glacier's surface falling within the accumulation area. The lowermost limit of the accumulation area is given approximately by the location of the snow line at the end of melt season. This also equates to the equilibrium line altitude (ELA), and may be identified on aerial photographs or satellite imagery (e.g. Advanced Spaceborne Thermal Emission and Reflection Radiometer (ASTER) imagery from terra or enhanced thematic mapper plus (ETM+) imagery from Landsat 7). It is important in this process to identify a consistent snow line (which will have been retreating to higher elevations on the ground) and not to confuse it with, for example, the lower limit of a summer snow-fall event. In general, an unusually low ELA may be equated with a positive annual mass balance and an unusually high ELA with a negative mass balance. This relationship may be quantified in terms of the *accumulation area ratio* (AAR), which is the accumulation area divided by total glacier area. As a general rule of thumb, zero mass balance at a valley glacier equates with an AAR of ~0.7 (e.g. Kuhn *et al.*, 1999). This method, however, has several limitations, including the inaccuracies involved in equating a specific AAR with a glacier's medium-term equilibrium and with the failure to account for accumulation by superimposed ice located below the snow line.

7.3.2.2 Hydrological balance method

The hydrological balance method involves calculating the mass accumulated on an ice mass by subtracting water output from the glacierized basin from water and snow inputs to it. Mass balance is thereby calculated as total precipitation less: (a) runoff from the basin, (b) evaporation from the basin, and (c) storage that does not contribute to the glacier's mass, such as groundwater storage or storage at the glacier bed. These balance components can be measured, directly or indirectly, by field instrumentation. However, errors associated with the method can be large depending on the complexity of the basin concerned (for example, a partly glacierized drainage basin involves additional, unconstrained balance terms) and the quantity of instrumentation available to the project. For example, water storage at the glacier bed is difficult to isolate while spatial variations in snowfall may be only poorly measured (or not at all), as may be the contribution of wind-blown snow.

7.3.2.3   Glaciological methods

Glaciological methods of measuring mass balance involve repeated point measurements at the glacier surface to yield rates of ablation and accumulation. Such measurements may be acquired as frequently as required, either manually or automatically by data logger. It is common in routine monitoring programmes to acquire measurements once or twice per year at the end of the principal mass balance season(s). The locations of such measurements will clearly be constrained by logistical demands, but the ideal sampling regime would be designed to include as much variability as possible. This means accounting for variations in mass balance with factors including elevation, predominant weather (specifically wind) patterns, valley-side shading and slope. In practice, a typical sampling regime will involve a primary longitudinal transect of measurement sites along a glacier's centreline (extending along its full length if possible), supplemented by a smaller number of transverse transects designed to characterize lateral variations. Although there are no hard and fast rules on the number of sample stakes required to measure the mass balance of a glacier accurately, Fountain and Vecchia (1999) analysed empirical data from two glaciers to conclude that ten is sufficient for a small valley glacier.

The techniques used to make these measurements may best be classified into those designed to measure ablation and those designed to measure accumulation.

Ablation measurements   Ablation is generally measured by reference to stakes inserted to a depth of some metres into the glacier surface and fixed at that datum by packing and freezing-in. The distance from the ablating ice surface to a fixed point on the protruding stake, often its top surface, is then measured repeatedly using a standard tape measure or by reference to gradated marks on the stake (Figure 7.1). Stakes can be various lengths and constructed from a variety of materials. Often, they are wooden dowels between 2- and 4-m long and of diameter 25 ± 10 mm, although flexible plastic poles and wires may be inserted to greater depths where longer holes are available. Where cost is a consideration, lengths of bamboo may be suitable since they are cheap, strong and have low thermal conductivity. The latter of these properties is important as conductive materials can transfer heat to the ice, causing melting and enlargement of the holes into which the stake is inserted. Bamboo poles are particularly suited to short-term studies on snow, into which they can be inserted easily by hand due to their small diameter. Where longer lengths of material such as wires are used, multi-annual measurements may be obtained from the same sample site, and it may be necessary to trim the material periodically and remove excess. In such situations, it should be borne in mind that the sample site moves

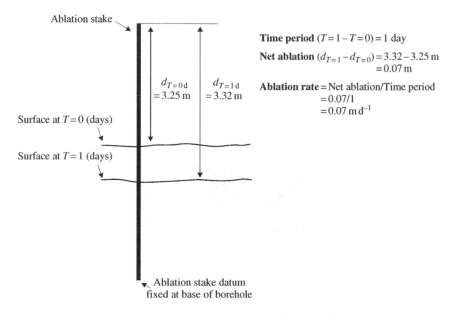

Figure 7.1   Illustration of the use of ablation stakes to measure surface lowering of ice or snow

continuously down-glacier with the motion of the glacier surface. Sample site locations should therefore be fixed by optical or GPS survey (section 7.4.2.1) as often as necessary. It is also necessary to redrill holes for ablation stakes periodically to prevent poles from melting out. Usually a new hole is drilled nearby, but it is also possible to extend an existing hole if the existing hole is not too shallow and degraded. In the event of such redrilling it is important to remember to record the height of the pole above the surface both immediately before and immediately after redrilling.

Numerous methods are available to drill holes for ablation stakes. In the simplest case, if the surface to be measured is snow then holes may not need to be drilled as stakes can normally be pushed manually into the surface. However, if the surface is ice then a hole needs to be pre-drilled to accommodate the stake. Such holes should be of a diameter slightly larger than that of the ablation pole to be used – remembering that a wooden stake will expand when wet and will bend with age. Ablation-pole holes may be drilled by hot water (Chapter 5), if such a drill is available on site, but more lightweight devices are usually used. These include hand augers (e.g. similar to that manufactured and sold by Kovacs Enterprises Inc., USA) (Figure 7.2), ice-fishing augers, of which there are various brands available including Jiffy, Strikemaster and Tanaka in USA, or even more widely available post-hole borers adapted to drive an ice auger. It is also possible to use in-house-constructed steam drills (see Gillet, 1975; Heucke, 1999). Narrow-diameter

Figure 7.2   Hand auger (Kovacs-type) in use on the surface of Tsanfleuron Glacier, Switzerland. Note the hole is being drilled to provide a new replacement hole for the velocity stake in the background

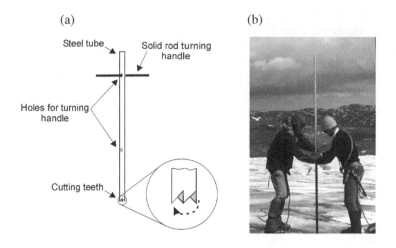

Figure 7.3   A simple, homemade ice borer suited to drilling holes of up to 2-m depth for wooden ablation stakes: (a) design illustration; and (b) a borer in use at Nigardsbreen, Norway. The drill can be of any practical length and have an external diameter a few millimetre larger than that of the ablation poles being used. The wall of the drill need be no more than ~1–2-mm thick in order to keep the instrument weight down. Four vertical cutting edges, bevelled inwards, are sufficient, allowing easy field re-cutting by hacksaw and sharpening by flat-faced file

(~10 mm) holes can be bored very easily to ~1-m depth through the use of a length of pipe welded onto the upper end of a standard climbers' ice screw. Finally, where specialist drill flights and a cutting head are not available, a hole borer can be constructed fairly simply by cutting four teeth into the end of a steel pipe (Figure 7.3).

Accumulation measurements   Short-term accumulation events may be recorded by standard ablation stakes, which will register a reduction in the distance measured from the fixed point on the stake to the newly deposited snow surface. Similarly, short-term or seasonal accumulation may be recorded automatically by a commercially supplied ultrasonic depth sensor located 2–3 m above the initial snow or ice surface or by a snow pillow emplaced immediately on top of the initial snow or ice surface. The thickness of transient snow located in the ablation area (i.e. where seasonal snow lies directly above ice) can be measured by inserting a long, thin graduated probe until it meets the (unyielding) ice surface. Commercially available avalanche probes are ideal for such snow-depth probing. Some care is needed with this technique to ensure that the probe comes to rest on the underlying ice surface rather than on ice layers or lenses located within the snowpack. However, for longer-term studies or in situations where several metres of accumulation are anticipated between measurements, other methods need to be used. In general, these utilize the presence of datable reference horizons within the accumulated snow.

For multi-year accumulation measurements, a snow pit may be dug and individual annual layers counted down from the surface. In this case seasonal snow layers may be identified by laterally extensive variations in snow structure. The contrasts between these layers may be subtle or highly distinctive, mainly depending on seasonal weather variability. For example, at cold ice masses, summer snow may be more variable in character than winter snow, giving it a more transparent appearance, while at warm ice masses, summer warming leads to the partial firnification and preferential densification of surface snow. These contrasts may be identified by GPR (e.g. Kohler *et al.*, 1997), viewed by eye, recorded as density variations in the field, or sampled as a snow core and analysed later in the laboratory. To view such layering by eye, Alley (2000) recommends digging two adjacent snow pits separated by a narrow (~30 cm) vertical wall of snow. One pit is covered and darkened, and the other is left open to the sunlight. The structure of the snow wall separating the two pits can then be viewed by an observer from within the darkened pit.

Snow density is generally straightforward to measure in the field. In this case, the researcher needs only to sample a known volume of snow (m^3) and to measure its weight (kg). The latter is divided by the former to yield density ($kg\,m^{-3}$). To measure a known volume of snow a solid-walled sampler is normally used, the size of which is dictated by the scale of sample required for the study in question. A standard snow-density sampler comprises a rectangular box cut in half across a diagonal and a closely fitting lid that slides over the open surface of the box to seal it (Figure 7.4). The box is then inserted horizontally into a clean, flat snow surface until the box-end is flush with the snow surface. The lid is then slid into place to seal the half-box while it is still emplaced in the snow. In this way a snow sample is sealed into the box without the snow having been compressed and with minimum deformation. If such a sampler is not available or if a sample of smaller footprint is needed then a cylindrical tube may be used. In such cases

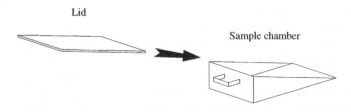

Figure 7.4   Illustration of a box snow-density sampler for horizontal insertion into the wall of a snow pit

it may be necessary to add retaining lugs (similar to core dogs on an ice corer; section 3.3.4) to the end of the cylinder to ensure a clean edge, or, if this is not possible, the recovered sample may need to be trimmed to a smaller, but known, volume. The volume (V, m^3) of such a sample is given by:

$$V = \pi r^2 L \tag{7.4}$$

where r (m) is the internal radius of the sampling cylinder and L (m) is the length of the sample.

Recently, a novel and logistically straightforward technique has been developed by researchers at the Byrd Polar Research Center, University of Ohio, USA, to measure local mass balance at a very high resolution (Hamilton *et al.*, 1998). This so-called coffee-can method allows researchers to measure rates of vertical velocity very accurately. Comparison of that velocity with rates of surface accumulation provides researchers with accurate figures for local mass balance (Box 7.3).

Where longer-term accumulation data are sought, it is possible to count annual layers along sections of snow, firn and (cold) ice core. This may be done from measurements of any seasonally variable physical property, the most commonly used of which are density, electrical conductivity, and stable-isotopic, ionic and elemental chemistry (section 3.4.1). However, the deeper into an ice mass the layers are identified, the greater the density of that material and the thinner the layers. This densification, along with ice advection, needs to be corrected in reconstructing the original accumulation rates.

Apart from counting annual layers it may be possible to identify specific reference horizons within ice cores and core holes. At the decadal scale, distinctive reference horizons have been deposited worldwide by radioactive fallout associated with the Chernobyl incident in 1986 and nuclear bomb testing through the period from 1954 to 1970. These two reference horizons

---

Box 7.3   Use of the 'coffee-can' method to measure accumulation rates continually at high resolution: Siple Coast, Antarctica

Hamilton *et al.* (1998) developed and reported the first application of the 'coffee-can' or 'submergence velocity' method to measure rates of local ice-sheet thickening or thinning. This method relies on the requirement for the maintenance of mass balance that the speed at which local firn or ice moves downwards needs matched by the rate of accumulation of new snow at the surface. If the former exceeds the latter then the ice-mass thins and mass balance is negative and

Box 7.3 (Continued)

conversely. The coffee-can method provides a simple means of measuring local vertical velocity within the firn layer. To achieve this, an anchor (originally an actual coffee can, hence the name) is installed at the base of a hole drilled in the firn and the precise movement of a wire, attached to the can and run up the hole to the surface, is monitored by intermittent global positioning system (GPS) survey or data logger (Figure B7.3). The resulting vertical velocities are compared with the rate of snow accumulation at the surface (from the study of layers in firn cores; section 3.4.1) to determine local mass balance.

In their study, Hamilton *et al.* (1998) applied this technique to sites in the Siple Coast region of Antarctica. Their results indicated that Byrd Station was approximately in balance, yielding -0.004 (± 0.022) $m\,a^{-1}$, while the 'Dragon' (an area between two ice streams) had slightly negative mass balance at -0.096 (± 0.044) $m\,a^{-1}$. Since this initial study, several improvements to the submergence velocity method have been made, reducing these error terms markedly.

(a)

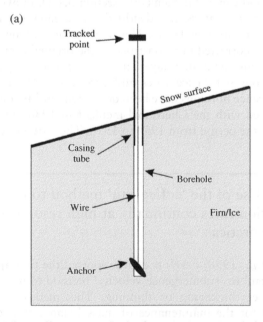

Figure B7.3 The coffee-can method of measuring mass balance by local thickness changes: (a) schematic illustration of the experimental set-up; and (b) field researchers from the Byrd Polar Research Institute installing a field experiment (Image: Gordon Hamilton)

(b)

Figure B7.3   (Continued)

Hamilton, G.S., Whillans, I.M. and Morgan, P.J. 1998. First point measurements
of ice-sheet thickness change in Antarctica. *Annals of Glaciology*, **27**, 125–129.

can be readily identified in snow sampled from cores by their elevated beta
radioactivity (e.g. Hamilton, 2002), and specifically in terms of ^{137}Cs concentrations, or in the field through the use of a neutron probe, which
measures gross beta radioactivity as it is lowered along a hole bored in the
snow (e.g. Dunphy *et al.*, 1994).

Rough estimates of long-term accumulation rates and their spatial variability can be obtained from ice-radar-based records of internal reflecting
horizons in large ice masses (section 6.7.1). In this case, since IRHs act as
passive isochrones, spatial variations in their depth may be equated to
spatial and/or temporal variations in accumulation rate. The derivation of
these rates, however, is complicated not only by variations in compression,
but also by variations in ice flow and in mass gains or losses at the ice-mass
base.

Finally, glaciers can gain mass through the accumulation of superimposed ice, which forms from the refreezing of meltwater on or near its
upper surface. Importantly, superimposed ice can form either above or
below the end-of-melt-season snow line. Thus, in cases where superimposed ice is present below the snow line, the common assumption that this
line separates the accumulation area from the ablation area is incorrect.
Superimposed ice can also form on the snow or ice surface or within the
snowpack. In the latter case, although the ice may not be strictly superimposed, refrozen meltwater can form spatially extensive ice layers or
lenses that are generally some centimetres thick and which can be several
metres in lateral extent. Field-based measurements of superimposed ice
thickness and extent are rare (cf. Fujita *et al.*, 1996). Where superimposed
ice is present at the ice surface its extent may be mapped by GPS or optical
survey, and its thickness may be ascertained by retrieving a core or, where
the layer is in the order of tens of centimetres in thickness (as is often the
case), by breaking it up with an ice axe. Sections can also be viewed in such
samples. It may also be possible to record surface superimposed ice thickness by high-frequency GPR (Box 6.6), as long as there is a sufficient
thickness of superimposed ice and as long as there is a detectable interface
between it and the underlying ice surface.

In addition to supraglacial superimposed ice, proglacial icings have been
described, particularly at polythermal, high-Arctic glaciers (Wadham *et al.*,
2001). These ice deposits, which are several metres thick and may be tens to
hundreds of metres in lateral extent, appear to form during the winter by the
refreezing of percolating meltwater as it emerges through proglacial sediments near to the glacier terminus. This water has most probably followed
a deep subglacial pathway from the warm interior of these polythermal
glaciers.

Very little is known about the internal structure of superimposed ice,
or about temporal and spatial variations in its mass balance. While the
physical properties of superimposed ice could be studied in the field by
geophysical methods or subsequently in the laboratory on cores, its mass-
balance properties would be best studied by a combination of field survey
and remote sensing.

7.4   GLACIER MOTION AND ICE VELOCITY

7.4.1   Principles and models

Glacier ice forms at high altitudes and is lost by ablation at lower altitudes. It moves from the former position to the latter position through motion within the ice mass or focused near the base of the ice mass.

7.4.1.1   Internal motion

Most internal motion is considered in terms of deformation, or ice creep, which considers the strain response of ice to an applied stress. Here, *stress* is defined as the force acting on the ice, and has units of newtons per square metre ($N m^{-2}$), commonly termed pascals (Pa) or, for more manageable magnitudes at glaciers, bars (bar) where 1 bar is 100 kPa (10^5 Pa). *Shear stress* (denoted τ) is the type of stress that most commonly acts to move an ice mass down-slope. *Strain* (denoted ε) is a measure of ice deformation under an applied stress. It is expressed as a dimensionless ratio of change in length to initial length. Strain rate is therefore the rate at which a body deforms in response to stress, and has units of *time*$^{-1}$ (and usually represented by $\dot{\varepsilon}$). The rheology of ice, which describes its strain rate as a function of applied stress, is expressed by its *constitutive relation* or *flow law*. The most widely accepted flow law for ice was derived by Glen (1955) on the basis of laboratory deformation studies. In tensor notation with $i, j = x, y, z$, the three axes of the Cartesian coordinate system, it takes the form:

$$\dot{\varepsilon}_{ij} = A\tau_e^{n-1}\tau_{ij} \qquad (7.5)$$

Here, A is a rate factor that reflects ice hardness (principally considered to be temperature-dependent), τ_e is the effective stress (a measure of the total stress state of the ice), and the exponent n is generally taken to have a value of ~3 (Hooke, 1981). Stress is a function of ice thickness and ice surface slope, and can be transferred longitudinally or laterally across an ice mass.

In applying Glen's flow law to real ice masses through computer-based models it is important to understand that a number of simplifications of reality may have been assumed. These include unconstrained factors that influence either the distribution of shear stress in the ice or the rheology of that ice. The model may, for example, not solve the first-order terms of the momentum equations on which they are based and may not therefore account for the transfer of longitudinal stresses. Although these may not be important at large ice masses (where transmitted stresses caused by variations in boundary traction may not represent a significant proportion of the overall stress field), they can be very important at the margins of ice streams and at smaller valley glaciers (Box 7.4).

Ice softness can also vary in response to numerous physical properties other than temperature. These include ice crystal size and orientation, liquid water content, solute content, bubble content and incorporated debris. The evolution of a preferred crystal alignment over long periods of strain may be very important in this context (see Paterson, 1994).

Finally, in addition to ductile deformation, ice motion can involve more discrete, brittle fracture. For example, surface crevasses, caused by tensile

---

Box 7.4 Use of field-measured 3D ice velocities to calibrate and validate a numerical model of valley glacier motion: Haut Glacier d'Arolla, Switzerland

Hubbard *et al.* (1998) applied a 3D first-order numerical model, based on the algorithms of Blatter (1995), of ice-mass motion to Haut Glacier d'Arolla, Switzerland. Using the glacier's 3D geometry as a basic boundary condition, the model converged successfully at a finite difference dimension of down to 70 m and with 40 vertical layers throughout. The model was then run iteratively, minimizing the difference between the modelled surface velocity field and that measured at the glacier during the winter (when basal motion is considered to be negligible), in order to derive the optimum rate factor, A, in Glen's flow law. This tuning produced a best-fit value of A of $0.063\,a^{-1}\,bar^{-3}$, which resulted in a good match between measured and modelled surface velocities. Once the ice deformation motion field was tuned in this way the model was applied to two further experiments. First, zones of particularly high surface tensile strain (actually, the second invariant of the surface stress tensor) were identified in the model output and compared with observed zones of crevassing at the glacier. These matched very well, both in terms of the locations of crevasse fields and in terms of their orientation (i.e. with crevasses being aligned at right angles to the direction of tensile stress). Second, this deformation flow field was supplemented by two different patterns of basal motion applied over different periods of the year. These were a short period of sliding focused along a preferential drainage axis to mimic the so-called spring event measured at the glacier, and an extended period of more general sliding to mimic that induced by the flux of large quantities of meltwater across the glacier bed during the summer melt season. Integrating the three resulting flow regimes (winter, no

(a)

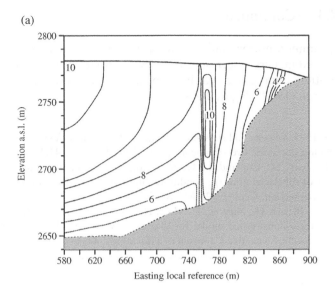

(b)

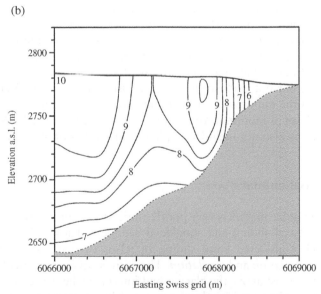

Figure B7.4 Distributions of annually averaged down-glacier velocity through a half cross-section of Haut Glacier d'Arolla, Switzerland: (a) measured by repeat borehole inclinometry; and (b) modelled as a composite of three flow regimes. Reproduced from: (a) Harbor *et al.* (1997) with the permission of the Geological Society of America; and (b) Hubbard *et al.* (1998) with the permission of the International Glaciological Society

Box 7.4   (Continued)

sliding; summer, normal sliding and spring, enhanced sliding) over a
year resulted in a cross-sectional velocity distribution that agreed well
with motion data generated by repeat inclinometry, reported by
Harbor *et al.* (1997) (Figure B7.4).

Blatter, H. 1995. Velocity and stress-fields in grounded glaciers: A simple algorithm
for including deviatoric stress gradients. *Journal of Glaciology*, 41(138), 333–344.
Harbor, J., Sharp, M., Copland, L., Hubbard, B., Nienow, P. and Mair, D. 1997.
Influence of subglacial drainage conditions on the velocity distribution within a
glacier cross section. *Geology*, 25(8), 739–742.
Hubbard, A., Blatter, H., Nienow, P., Mair, D. and Hubbard, B. 1998. Compari-
son of a three-dimensional model for glacier flow with field data from Haut
Glacier d'Arolla, Switzerland. *Journal of Glaciology*, 44(147), 368–378.

strain exceeding local ice strength, are commonplace on most ice masses.
Thrusting along high-angle slip planes is, however, perhaps of even greater
importance to overall ice motion at many ice masses. Such motion has been
inferred from geomorphic evidence near the frontal margin of numerous
glaciers, particularly in polythermal, Svalbard glaciers (Hambrey *et al.*,
1997, 1999), but not exclusively so (e.g. Tison *et al.*, 1989). Very little is
known about the importance of brittle failure to net ice motion at different
scales, and rates of motion have not yet been measured directly. Motion by
brittle failure is also not included explicitly in models of ice-mass motion.

7.4.1.2   Basal motion

Basal motion at ice masses is very poorly understood, although it is generally
divided into basal sliding and subglacial sediment deformation. Weertman
(1957, 1964) considered basal sliding in terms of two discrete processes:
enhanced deformation and *regelation*. The former involves particularly rapid
ice deformation just above the glacier–bed interface where already high basal
stresses are supplemented by those generated by ice impinging on the stoss
face of bedrock hummocks. The later process is also driven by local stresses
caused by bedrock hummocks, but in this case it only applies to temperate
glacier beds as enhanced, stoss-face stresses drive the melting temperature
down and cause local ice melting. The meltwater thereby produced flows
around the hummock concerned and refreezes at its lee face, where the stress
is reduced and the melting point returned to its (higher) ambient level.

Theoretically, these two processes are considered to combine at temperate- and hard-based ice masses to result in a basal sliding rate (U) that is principally dependent on the driving stress (τ_b) and on bedrock roughness (r), in particular the predominance of an intermediate, or controlling, obstacle size (which has a length scale of ~ 0.1–1.0 m) that most effectively inhibits both sliding processes. Thus, following Weertman's theory,

$$U \propto \frac{\tau_b^2}{r^4} \qquad (7.6)$$

Basal sliding rate, however, also varies through time due to the additional control exerted by the variable presence of water-filled cavities at the ice–bedrock interface. The presence of such cavities, and the pressure of water within them, enhances sliding speed through separating ice from its bed, increasing the shear stress on the locations remaining in contact with the bed, and by exerting a net down-glacier force on the overlying ice. Although these processes have not been incorporated formally into sliding theory, the most applicable relation expresses sliding speed U as an inverse function of effective pressure N (defined as ice pressure minus basal water pressure):

$$U \propto \frac{\tau_b^p}{N^q} \qquad (7.7)$$

The values of p and q may be determined empirically.

Basal motion by the deformation of subglacial sediments is now widely considered to be responsible for the fast flow of many of the Earth's ice streams and glaciers. The most common model used to describe this motion process is a viscous approximation (above a yield strength) with an inverse dependence on effective pressure, proposed by Boulton and Hindmarsh (1987):

$$\dot{\varepsilon} \propto \frac{(\tau - \tau_0)^a}{N^b} \qquad (7.8)$$

Here, the yield strength (τ_0) is defined by the Coulomb failure criterion:

$$\tau_0 = C_0 + N \tan \theta \qquad (7.9)$$

where C_0 is sediment cohesion, θ is the angle of internal friction, and the exponents a and b are determined empirically. More recently, a highly non-linear rheology has been proposed, more closely approximating plastic failure than viscous flow. Hooke *et al.* (1997), for example, characterized this rheology as:

$$\dot{\varepsilon} \propto e^{k\tau} \qquad\qquad (7.10)$$

for values of $\tau \geq \tau_0$. The constant k has a value of between 10 and 60.

The reality of subglacial sediment deformation, however, is far more complex than either of these models implies. For example, a subglacial sediment layer may be of variable depth and may also be water-saturated to a highly variable degree, creating corresponding spatial variability in strength. Subglacial sediments are also often of polymodal grain-size texture (Figure 7.5) and characterized by large spatial variations in that texture, again inducing corresponding variations in strength. Indeed, the presence of large clasts within subglacial sediments and entrained within the basal ice has led to some researchers considering subglacial sediment deformation in terms of the interactions between such clasts, termed 'ploughing' (e.g. Hooyer and Iverson, 2002).

Figure 7.5 A typical interface between basal ice subglacial debris: Haut Glacier d'Arolla, Switzerland. Note the polymodal nature of the debris, dominated by coarse gravel-sized material, both within and beneath the ice

7.4.2 Measurement

7.4.2.1 Surface motion

Ice-mass surface motion can be measured by a variety of field-based survey techniques. Simple 2D velocity measurements can be made from aerial photographs by monitoring the movement of surface structures through time (e.g. Krimmel and Rasmussen, 1986; Harrison et al., 1992). Such optical remote sensing can only be used to reconstruct glacier velocities if features on the glacier surface can be identified in two images. Multispectral satellite data have been used successfully to do this, limited in its accuracy only by the spatial resolution of the sensor. For example Rolstad et al. (1997) applied a feature-tracking algorithm to crevasse structures on three consecutive SPOT and Landsat TM images to derive velocity fields over a glacier surge in Svalbard. These authors reported an overall ice displacement of 1 km over a 12-month period. Daily velocity vectors were also constructed, varying from 2.6 to 6 m d^{-1}. More recently, sophisticated software packages have been developed to allow several vertical or oblique images to be combined and overlain to provide multi-parameter spatial data, including elevation and velocity, termed multi-model photogrammetry (e.g. Jiskoot et al., 2001). Increasingly, ice surface velocity fields are calculated from repeat synthetic-aperture radar (SAR) images recovered from satellite platforms, including ERS1, ERS2, JERS, IRS and Radarsat. Comparison of the intensity and phase of radar signals from pairs of images through synthetic-aperture radar interferometry (InSAR) allows both the surface topography and the vector velocity field of the area common to both images to be quantified precisely at a high spatial resolution. InSAR has been applied successfully in numerous glaciological settings, including the Earth's major ice sheets, ice streams, ice shelves, sea ice and rock glaciers. Although the technique falls outside the remit of this text, many other texts dedicated to the technique are available (Massonnet and Feigl, 1998).

Despite the evident utility of remote-sensing techniques to reconstructing large-scale ice velocities, high-resolution surveys conducted over time scales of hours to months are still best conducted on the ground by either optical survey or, increasingly, differential global positioning survey (DGPS; Figure 7.6a). Both techniques are based on measuring repeatedly the position of sets of stakes mounted on the glacier surface. Often these double as ablation stakes (section 7.3.2.3). For optical survey it is necessary to mount reflective prisms onto these poles, where possible backing them with a board to ensure visibility from the survey station, which should be located on a stable surface (preferably bedrock) off the glacier (Figure 7.6b). If the purpose of the survey is to record glacier motion or surface topography then it may not be essential to record positions according to a national or international reference system, and a local reference system may be used.

For example, the principal base station may be considered $(0, 0)$ in space and an artificial north bearing may be defined. However, this approach should only be adopted if absolute referencing is not possible as positional data within a recognized grid system can be compared more easily with future studies. It is also important to remember that the exact position and

(a)

Figure 7.6   Survey base stations located on solid bedrock adjacent to Tsanfleuron Glacier, Switzerland: (a) DGPS, and (b) total station

(b)

Figure 7.6   (Continued)

orientation of a survey station must be located at the time of each relative survey for data to be even internally comparable – and therefore used to calculate velocities. Two further problems with optical surveying may usefully be noted: that of maintaining a prism whose movement is the same as the glacier surface and that of surveying overnight. The former requires solid prism placement onto a rigid pole that is inserted vertically and stably, to some depth, within the ice. Ice-mass surfaces may also experience high winds and it is important that survey poles are prevented from rotating the prism orientation away from the view of the survey station. Where high winds are anticipated, and survey poles cannot be guaranteed to freeze in, it may be preferable to use a survey pole that is of irregular cross-section, square for example, to inhibit such rotation. Alternatively, prisms may be attached to surface-mounted survey poles, based around a tripod – although some experimentation is required in the design and implementation of such apparatus in specific field conditions. Night-time survey is hindered by the surveyor's inability to locate accurately the centre of the target prisms. If surveying in darkness is desired, it is possible to sight onto a light mounted adjacent to the centre of the prism. Small LEDs are ideal for this purpose as they are luminous, lightweight and have low power consumption (Figure 7.7). A sufficiently luminous LED may have a rating of 50–75 mA, meaning

(a)

Figure 7.7 The upper section of *in situ* velocity stakes housing reflective prisms screwed into the centre of high-visibility targets: (a) standard stake set-up; and (b) set-up including a LED adjacent to the reflective prism for night-time surveying (the additional prism has been added to increase the surveying range)

(b)

Figure 7.7   (Continued)

that it can be powered for a full night, even at sub-freezing temperature, by a single AAA-sized battery (~750 mAh).

For DGPS survey the base station is also mounted on a stable surface located off the glacier, while roving stations are mounted either permanently or intermittently at the glacier surface. The permanent location of GPS units at the glacier surface requires a continuously powered GPS unit to be dedicated to each survey location, correspondingly resulting in a continual, high-resolution record of surface movement at each location. In contrast, using a roving unit to record position intermittently at fixed stakes does not require continuous power (although over an hour may be needed at each site for a reading to have sub-centimetre accuracy, depending on satellite and atmospheric conditions) and requires only one roving unit to cover several survey locations. However, the method produces a velocity record that has a temporal resolution dictated by the frequency of survey, commonly one or more days. DGPS surveying also generates positions in absolute space according to any predetermined datum.

Transforming sets of spatial coordinates into velocities can be achieved through bespoke programming or in a spreadsheet such as Microsoft Excel. An example of the latter might be as follows. A worksheet is set up with input columns for: (i) station identifier, (ii) survey date (decimal day of year is probably most suitable), (iii) position (east), (iv) position (north), and (v) position (elevation). These positional data can be in any reference system as long as all records are consistent, and the user must be sure of the units being used. A new row is ascribed for each survey record, and survey stake redrilling should have two records, one for immediately before and one for immediately after repositioning. Subsequent, derived columns should include: (vi) distance travelled (east), (vii) distance travelled (north), (viii) distance travelled (vertical), (ix) distance travelled (net), (x) net horizontal velocity, (xi) net vertical velocity, (xii) net horizontal bearing, and (xiii) net vertical bearing. Separate easting, northing and vertical components are calculated here to simplify the calculations of the net distances. Net bearings can be calculated from the geometry of the separate components (Figure 7.8), and net velocities are then calculated by dividing the total distances travelled by the time interval separating the readings.

Since each survey has an associated distance (and therefore positional) error, the larger the distance travelled between surveys the smaller the velocity error. Although total stations and DGPS have published errors, it is important to remember that these are essentially instrumental errors and that they may be supplemented by field-based and operator-based errors associated with carrying out the surveys under field conditions. These errors can only be assessed empirically by, for example, repeatedly surveying, under representative conditions, a fixed point or a range of points, carrying out the full suite of operations for each reading that would normally be carried out for separate surveys. If, under these conditions, a point is surveyed 30 times then a sample set of 30 distances, including errors, will result. These data can be normalized, by expressing each reading as a deviation from the mean distance value, and plotted as a histogram whose distribution can be described statistically. This allows, for example, the standard error of the distances to be quantified exactly. Importantly, the resulting error includes instrument-related, method-related and operator-related errors and is therefore more representative of the actual errors in the final data than the published instrument error. However, further errors could be associated with surveying, resulting for example from misreading a number. In such cases, an anomalous reading may result which should be removed from the analysis. Any displacement of the tripod during survey should also show by analysis of readings to fixed reference points, which should be made at the beginning, during and at the end of each individual survey session.

(a) Planform view

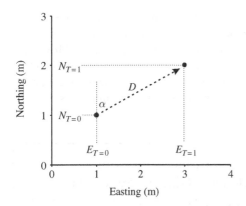

$$D = \sqrt{(N_{T=1} - N_{T=0})^2 + (E_{T=1} - E_{T=0})^2}$$
$$= \sqrt{(2-1)^2 + (3-1)^2}$$
$$= \sqrt{1+4} = 2.24\,\text{m}$$

$$\alpha = \tan^{-1}\left(\frac{(E_{T=1} - E_{T=0})}{(N_{T=1} - N_{T=0})}\right)$$

$$= \tan^{-1}\left(\frac{(3-1)}{(2-1)}\right)$$

$$= \tan^{-1}\left(\frac{(2)}{(1)}\right) = 63.3°$$

(b) Vertical section along travel path view

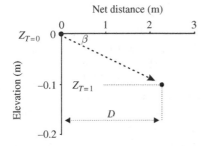

$$\beta = \tan^{-1}\left(\frac{(Z_{T=1} - Z_{T=0})}{D}\right)$$

$$= \tan^{-1}\left(\frac{0.1}{2.24}\right) = 2.56°$$

Figure 7.8 Illustration of geometric calculations made to determine positional changes resulting from surface velocity survey: (a) planform calculations (to determine easting and northing); and (b) vertical calculations (to determine elevation)

7.4.2.2   Internal deformation

The internal motion of ice masses is principally measured by tilt sensors installed along boreholes and by the repeat inclinometry of boreholes, the former providing excellent spatial resolution integrated over specific time windows and the latter providing excellent temporal resolution from specific spatial locations. Both techniques are described fully in section 5.4.1. Often the internal motion field of an ice mass is not required so much for its own sake, but to subtract from the measured surface velocity field in order to determine the basal motion field. If, in such cases, the internal motion field is not available, it may be approximated (subject to numerous assumptions) from the output of a spatially distributed flow model.

7.4.2.3   Basal motion

The basal motion of ice masses is most commonly measured through the use of sensors installed at the base of boreholes. These sensors include drag spools, ploughmeters and tilt cells, described fully in section 5.4.3. However, basal motion may also be measured *in situ* in situations where the ice–bed interface can be accessed by researchers (Box 7.5). Such access is generally possible: (i) in natural cavities formed in the lee of bedrock hummocks beneath

---

Box 7.5   *In situ* investigations of basal motion: Engabreen, Norway, Tsanfleuron Glacier, Switzerland, Meserve Glacier, Antarctica and Suess Glacier, Antarctica

Several researchers have studied basal motion processes and rates by accessing the glacier bed directly, either in natural cavities that open up in the lee side of bedrock hummocks or by melting temporary cavities at the basal interface. For example, Cuffey *et al.* (1999) excavated a tunnel beneath Meserve Glacier, Antarctica, by chainsaw to access a natural cavity in the lee of a bedrock boulder. Here the authors installed very high-resolution linear variable differential transformers (LVDTs) $\sim$1 mm above the ice–bedrock interface and measured smooth sliding at rates of between 2 and 8 mm a^{-1}. This result was particularly interesting because the temperature of the study site was $-17\,^{\circ}$C supporting earlier observations of sliding at sub-freezing temperatures in Urumqui Glacier #1, China, by Echelmeyer and Zhongxiang (1987). Similarly, Fitzsimons *et al.* (1999) excavated a tunnel by chainsaw beneath Suess Glacier, Antarctica (see Box 3.1), and reported complex structures and material contrasts in the frozen subglacial sediments underlying the glacier. On the basis of these contrasts, the authors argued for the existence within these sediments of planes of weakness that could be exploited by preferential shear or slip.

Basal motion has also been measured directly at the base of temperate glaciers. For example, Hubbard (2002) accessed the frontal margin of Tsanfleuron Glacier, Switzerland, and installed a vertical array of drag spools to measure sliding and ice deformation in the lowermost $\sim$300 mm of the basal ice of the glacier (Figure B7.5a). In this frontal zone of the glacier, beneath a few metres of ice, motion was found to be jerky, consistent with stick-slip motion, at the timescale of minutes. Ninety percent of the overall $\sim$12 mm d^{-1} of motion in the basal zone was either pure slip at the ice–rock interface or confined to the lowermost 25 mm of the ice. In one of the most ambitious studies to date, Cohen *et al.* (2000) installed an artificial bedrock hummock instrumented with a variety of sensors into a temporary cavity melted by hot water beneath 210 m of ice at the bed of Engabreen, Norway. The authors then left the bump in place and recorded data from it following cavity re-closure, providing a unique insight into conditions at a closed ice–bedrock basal interface. Amongst other instruments,

(a)

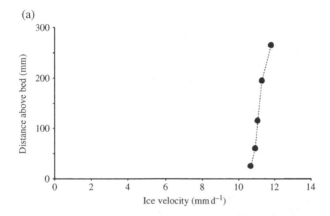

(b)

Figure B7.5 Measuring basal sliding at the glacier bed: (a) net ice velocity plotted against distance above the ice–bed interface measured by five adapted drag spools near the frontal margin of Tsanfleuron Glacier, Switzerland. (a) Reproduced from Hubbard (2000) with the permission of the International Glaciological Society. (b) the instrumented artificial (concrete) bump installed by Cohen *et al.* (2000) beneath the interior of Engabreen, Norway, with glass viewing portals for the video camera clearly visible. (b) Photograph Neal Iverson

---

Box 7.5   (Continued)

the authors installed video cameras in the bump to measure ice velocity (Figure B7.5b). The resulting records of the passage of small debris particles past the object indicated a sliding speed at the ice–bed interface of between $\sim$25 and $\sim$150 mm d^{-1}. This contrasted sharply with the $\sim$800 mm d^{-1} measured at the glacier surface.

Cohen, D., Hooke, R.L., Iverson, N.R. and Kohler, J. 2000. Sliding of ice past an obstacle at Engabreen, Norway. *Journal of Glaciology*, **46**(155), 599–610.

Cuffey, K.M., Conway, H., Hallet, B., Gades, A.M. and Raymond, C.F. 1999. Interfacial water in polar glaciers and glacier sliding at −17 °C. *Geophysical Research Letters*, **26**(6), 751–754.

Echelmeyer, K. and Zhongxiang, W. 1987. Direct observation of basal sliding and deformation of basal drift at sub-freezing temperatures. *Journal of Glaciology*, **33**(113), 83–98.

Fitzsimons, S.J., McManus, K.J. and Lorrain, R.D. 1999. Structure and strength of basal ice and substrate of a dry-based glacier: Evidence for substrate deformation at sub-freezing temperatures. *Annals of Glaciology*, **28**, 236–240.

Hubbard, B. 2002. Direct measurement of basal motion at a hard-bedded, temperate glacier: Glacier de Tsanfleuron, Switzerland. *Journal of Glaciology*, **48**(160), 1–8.

---

thin, marginal ice, (ii) in tunnels cut or melted in ice accessed from the glacier margins, (iii) in natural cavities formed in the lee of bedrock hummocks and above subglacial channels beneath thick interior ice (accessible only by tunnels cut beneath the ice), and (iv) in temporary cavities melted or cut by researchers beneath thick ice in the interior of glaciers (accessible only by tunnels cut beneath the ice). Each of these locations has advantages and disadvantages for studying basal motion. While research conducted within temporary cavities (and by instrumentation left to record following cavity closure) is probably the most representative of basal conditions over the majority of the interior of ice masses, access is extremely logistically demanding and requires bespoke tunnels and subglacial risers. In contrast, conditions within natural basal cavities located near the glacier margin may only be locally representative. However, access to such cavities is straightforward (Figure 7.9), allowing long-term or repeated experiments. Between these two extremes, it may be possible (particularly in cold ice) to use a chainsaw (section 3.3.3) to cut an access tunnel to the immediate interior of an ice mass from its margin. This technique has been used successfully in the Antarctic, but has met with limited success in warm glacier ice.

(a)

(b)

Figure 7.9 Frontal margin of Tsanfleuron Glacier, Switzerland, showing: (a) opening to subglacial cavity; and (b) a person working within the cavity. While such cavities are commonly stable over periods of weeks to months, thin ice at the glacier margin are prone to collapse. Work within such cavities should therefore only be conducted by well-equipped, experienced researchers having considered all health and safety issues

7.5   STUDENT PROJECTS

7.5.1   Surface energy balance and glacier mass balance

- Quantifying relationships between rates of ice or snow surface lowering and surface energy balance components. Components measured would depend on equipment available, but the simplest case would involve calibrating and investigating the degree-day factor for particular conditions. Factors could be derived and compared for different surfaces (e.g. ice as opposed to snow) and surface conditions (e.g. ice roughness or albedo).
- Relating variations in rates of glacier ablation to physical conditions at the glacier surface, for example by investigating the relationships between ice roughness and turbulent heat transfers.
- Investigating the role of energy uptake by cold ice (i.e. energy used to raise the temperature of ice up to the melting point), for example by measuring rates of ice warming in the morning following overnight cooling.
- Investigating spatial variations in snowpack thickness and density and relating these to surface topography.
- Investigating the presence and nature of ice lenses and layers within the surface snowpack, and perhaps the influence of this ice on snowpack hydrology.

7.5.2   Glacier motion

- Measuring spatial and/or temporal variations in glacier surface velocity. Where other data are available, spatial variations could be related to controlling factors such as local ice thickness and surface slope (and possibly meltwater routing) and temporal variations could be related to factors such as temporal variations in meltwater generation.
- Measuring spatial and/or temporal variations in glacier basal motion and relating them to controlling factors such as local ice thickness and slope, up-glacier ice velocities and local basal conditions. In the first instance these could be the character of the bed (i.e. bedrock as opposed to sediments) and for a more advanced analysis the nature of the substrate could be characterized in more detail (e.g. sediment grain-size texture and fabric (see Chapter 8).

8

Glacigenic sediments

8.1 AIM

The aim of this chapter is to explain the techniques used in the field description of glacigenic sediments and to provide guidance on the types of follow-up activities, including laboratory analysis, that can be used to make inferences about depositional processes (Table 8.1). Sediment descriptions are used in contemporary glacial environments to make inferences about the processes operating (e.g. erosion, transportation and deposition). In the glaciated environment sedimentary descriptions are used mainly to make interpretations about the depositional environment based on sediment properties such as lithology, texture, bedding, sedimentary structures and palaeocurrent data (Table 8.2). This chapter is intended to be a guide to the types of data that can be collected and provides an outline of the procedures normally adopted for sediment description in the field.

8.2 INTRODUCTION TO FIELD SEDIMENTOLOGY

Sedimentology is a wide-ranging and complex discipline, which includes all aspects of sediment transport and deposition across a diverse range of environments. Tucker (1988, 1996) and Leeder (1983) give the general background to the principles of sedimentology, whilst Ashley *et al.* (1985), Hambrey (1994) and Benn and Evans (1998) provide extensive reviews of the processes and sediments that dominate glacial environments. Gale and Hoare (1991) and Jones *et al.* (1999) provide guidance about the field description and laboratory analysis of Quaternary sedimentary successions.

Field Techniques in Glaciology and Glacial Geomorphology   Bryn Hubbard and Neil Glasser
© 2005 John Wiley & Sons, Ltd

Table 8.1 A summary of the principal field and laboratory techniques commonly used in the description and analysis of glacigenic sediments. The column headed 'Where can I find other studies to compare with my data?' is intended as an indicative list of published studies. It is by no means an exhaustive list – there are numerous other published studies that make use of these techniques to a greater or lesser extent

Technique	What do I measure/look for?	What do I do with the data collected?	What will it tell me?	Where can I find other studies to compare with my data?
Clast macrofabric measurements	Measure the dip angle and dip direction of 50 prolate clasts (each clast must have an elongation ratio of at least 1.5:1).	Plot data on rose diagrams (2D), polar diagrams or equal-area stereonets (3D). Calculate fabric strength using eigenvalue analysis and plot data on triangular eigenvalue plots (Benn, 1994) or on bivariate eigenvalue plots (Dowdeswell and Sharp, 1986).	Nature of deposition, for example lodgement, meltout, subglacial deformation, glacigenic debris flow. Clast macrofabric can also be used to indicate relative strain within deforming sediment or basal ice layers.	Mark (1974) Dowdeswell and Sharp (1986) Dowdeswell et al. (1985) Hart (1994) Benn (1994, 1995) Hicock et al. (1996) Bennett et al. (1999) Kjaer and Krüger (1998) Hooyer and Iverson (2000) Benn and Ringrose (2001) Benn (2002)
Microstructural analyses	Make impregnated thin sections of sediment for viewing under microscope. Describe the deformation structures (e.g. folds, shears and faults) present in thin sections. Measure 2D orientation of clasts (if present) in orientated thin sections.	Make detailed descriptions and interpretations of structures present. 2D clast microfabric data can be plotted in the same manner as clast macrofabrics (see above).	Sediment genesis Evidence of subglacial deformation.	Menzies and Maltman (1992) Van der Meer (1993) Menzies (2000) Van der Wateren et al. (2000) Phillips and Auton (2000) Khatwa and Tulaczyk (2001) Lachniet et al. (2001) Carr et al. (2000) Carr (2001)

Parameter	Method	Interpretation	References
Clast size and shape measurements	Measure the three orthogonal axes (a-, b- and c-axes) of samples of 50 gravel-sized clasts (see Krumbein, 1941; Blair and MacPherson, 1999). Plot data on general shape triangle using TRI-PLOT spreadsheet of Graham and Midgley (2000). Calculate RA and C_{40} indices. Plot data on RA/C_{40} diagram (Benn and Ballantyne, 1994).	Clast shape can be used to reconstruct the transport paths of debris through the glacier system.	Wentworth (1935) Holmes (1960) Barrett (1980) Benn and Ballantyne (1993, 1994) Huddart (1994) Illenberger (1991) Bennett et al. (1997)
Clast roundness measurements	Estimate roundness of samples of 50 clasts on Powers (1953) scale or measure radius of curvature of sharpest corner on Cailleux diagram. Plot data as histograms of roundness class against percentage frequency.	Clast roundness can be used to reconstruct transport history, in particular modification by subglacial processes and degree of reworking by (glacio)fluvial processes.	Boulton (1978a) Mills (1979) Bennett et al. (1997) Huddart (1998)
Clast or matrix lithology	Identify clast lithology in samples of 100 clasts (Bridgland, 1986). Identify mineral composition and magnetic susceptibility using laboratory techniques (see Walden et al. 1999). Plot as histograms of lithology against percentage frequency or display data as pie charts.	Provenance of material. Erratic lithologies indicate transport directions, changes in transport paths over time and degree of mixing during transport.	Walden et al. (1987) Walden et al. (1996) Kjaer (1999)
Clast surface features	Record the occurrence of striated, grooved, faceted or chipped clasts in samples of 50 clasts. Examine and describe surface features on sand-sized grains scale using Scanning Electron Microscope (SEM). Plot as histograms of surface feature type against percentage frequency or display data as pie charts. Produce surface texture variability plots for samples examined (Fuller and Murray, 2002).	Proportion of striated, grooved, faceted or chipped clasts in samples of 50 clasts provides evidence of basal glacial transport. Surface features on sand-sized grains such as fracture surfaces, edge rounding, striations and chips provide evidence of subglacial transport.	Drake (1972) Whalley and Krinsley (1974) Boulton (1978a) Krüger (1984) Sharp and Gomez (1986) Gomez et al. (1988) Mahaney and Kalm (2000) Fuller and Murray (2002)

Table 8.1 (Continued)

Technique	What do I measure/look for?	What do I do with the data collected?	What will it tell me?	Where can I find other studies to compare with my data?
Particle-size analysis	Laboratory sieving and sedigraph analysis of bulk samples of c.1.5 kg of matrix and clasts or samples of c.100 g if matrix only (McManus, 1988).	Plot as cumulative frequency diagrams, percentage frequency histograms or ternary plots. Calculate descriptive statistics, including mean, median, mode, sorting, skewness and kurtosis using GRADISTAT spreadsheet of Blott and Pye (2001). Produce bivariate plots of descriptive statistics for environmental discrimination. Consider plotting data as double logarithmic plots (Hooke and Iverson, 1995).	Descriptive statistics can be used to determine the degree of sorting (or otherwise) in matrix for use in environmental discrimination. Double logarithmic plots can indicate degree of grain crushing and therefore subglacial transport (Hooke and Iverson, 1995).	Folk and Ward (1957) Landim and Frakes (1968) Greenwood (1969) Peach and Perrie (1975) Boulton (1978a) German et al. (1979) Haldorsen (1981) Dreimanis and Vagners (1971) Humlum (1985) Barrett (1989) Duck (1994) Iverson et al. (1996) Kjaer (1999) Khatwa et al. (1999) Taylor and Brewer (2001) Benn and Gemmell (2002)
Geotechnical properties	A variety of laboratory tests can be used, including ring-shear devices, shear boxes and triaxial tests. Preconsolidation stress can be measured using an oedometer.	Plot data according to tests applied (Bowles, 1992). Plot oedometer data following methods outlined by Tulaczyk et al. (2001).	Tests provide information on water content, liquid and plastic limits, undrained compressive strength, shear strength, density, porosity and modulus of elasticity. Preconsolidation stress in sediment samples provides information about former subglacial effective stress.	Milligan (1976) Boulton and Paul (1976) Boulton (1978b) Boulton and Dobbie (1993) Clarke et al. (1998) Boone and Eyles (2001) Tulaczyk et al. (2001)

Table 8.2 Commonly recorded field characteristics of sediments

Characteristic	Specific features
Lithology	Mineralogy/composition and colour of the sediment
Gravel-sized clast lithological analysis	Mineralogy/lithology of gravel-sized clasts only
Texture	Particle size, particle shape, roundness, sorting and fabric
Bedding	Designation of beds and bedding planes, bed thickness, bed geometry, contacts between beds
Sedimentary structures	Internal structures of beds, structures on bedding surfaces and structures involving multiple beds
Deformation structures	Evidence of folding, thrusting and faulting
Palaeocurrent data	Orientation of palaeocurrent indicators

Reineck and Singh (1980), Compton (1985), and Graham (1988) also provide overviews of the collection and analysis of field data from a general sedimentological perspective. The term 'sedimentary facies' or 'lithofacies' is often used to describe a body of sediment or rock that forms under certain conditions of sedimentation, reflecting a particular process or environment (Hambrey, 1994). Sedimentary facies have specified characteristics such as colour, bedding, geometry, texture, fossils, sedimentary structures and types of external contacts. Facies may be subdivided into **subfacies**, grouped into **facies associations** or considered on a regional scale in terms of **facies architecture**.

The main skills required are those of basic observation, the ability to describe concisely, think laterally, to visualize in three dimensions, and to make logical deductions and inferences. The key to producing good sedimentological descriptions is the ability to separate description from interpretation. The task of describing sedimentary exposures in the field in vertical sections is called **sedimentary logging** and the recording of this information visually onto paper is known as **graphic logging** or **field logging**. Graphic logs are simply diagrams of measured vertical sections through a sedimentary succession. Hanging a 30-m tape measure over the section will help with continuous logging. If it is impossible to log one single continuous section, for example because of overhangs or slumped material, horizontal offsets can be used as long as you are careful to follow a distinctive marker horizon sideways at the point where an offset is required. Graphic logs allow information collected in the field to be recorded in a standard manner, enabling comparison between exposures at different locations and therefore enabling comparison with other published descriptions.

The techniques of sedimentary logging can be applied to contemporary (glacierized) and ancient (glaciated) environments, whether these are exposed in sections or in borehole material. The interpretation of contemporary glacigenic sediments is usually much easier because the sedimentary processes can often be observed in action, whereas in formerly glaciated areas a greater level of interpretation is required to establish the precise depositional environment. In this section we describe the procedures used in sedimentary logging from a personal perspective – do not be surprised if other practitioners suggest other procedures. A suggested sequence of events is provided in Figure 8.1. Each will be described in turn.

8.2.1   Preparation for fieldwork

Most fieldworkers record their data on **logging sheets,** which consist of a number of columns for recording data or information (Figure 8.2). The entire log should be placed on one sheet wherever possible. The width of the grain-size column is determined by the average grain size of the sediment. Poorly sorted sediments (gravel/diamictons/diamictites) are represented by their coarsest components. The depth of individual units is given by the vertical scale (m). Contacts between units are marked as horizontal lines, using standard symbols for the nature of this contact (e.g. solid lines for sharp contacts and dashed lines for transitional contacts). Further columns can be added to record written descriptions of the sediment and an interpretation of its depositional environment. You should also ensure that you have the correct equipment for the job (Box 8.1).

8.2.2   Field procedures

Glacigenic rocks and sediments are found in naturally occurring exposures such as coastal or river cliffs, upland and mountainous terrain and in artificial exposures (active or disused pits, quarries or road cuttings). Access permission (if applicable) should always be obtained from the owner/occupier of the land in question. Be aware of health and safety risks, especially if the section is adjacent to a tidal area, has steep or high cliffs, contains loose boulders or is located in a working quarry. A suggested field procedure for field site descriptions is as follows:

- Record the grid reference of the site or take a GPS reading so that others can subsequently relocate the section. If the site is not named on existing maps, give it a name or unique location code.
- Make detailed notes of the site including its overall orientation and dimensions, as well as the nature of the surrounding topography and vegetation. Record the relationship of the section to other glacial features in the immediate vicinity. If the section is part of a larger landform, describe the landform in which it occurs.

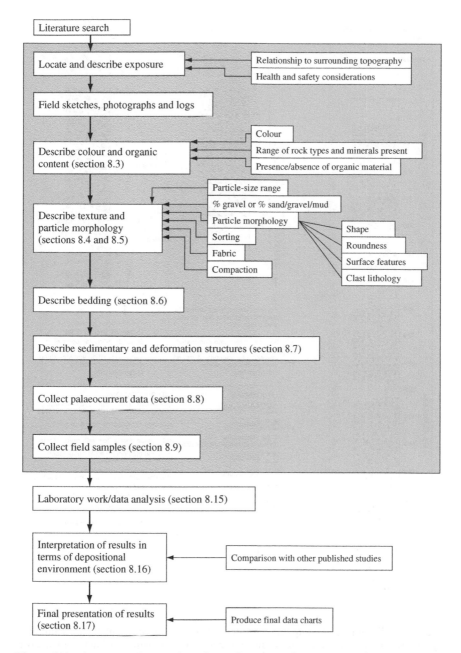

Figure 8.1 A suggested procedure for sedimentary description and interpretation. Activities outside the shaded area are normally carried out as a desk-based exercise or in the laboratory; those within the shaded area are carried out in the field. Note that this procedure describes the 'perfect situation', which of course rarely exists in reality. Therefore, not all steps will be applicable in all situations

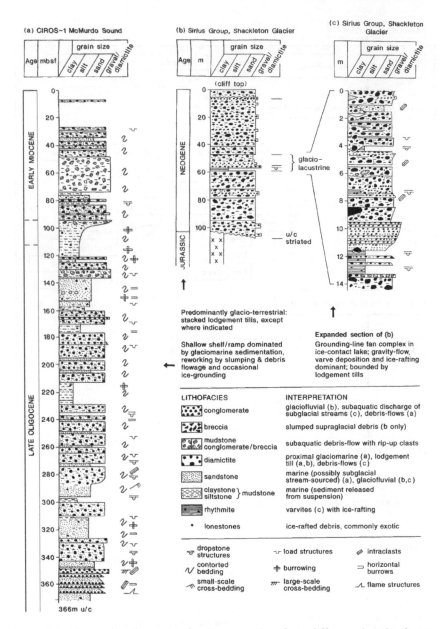

Figure 8.2 Examples of completed sedimentary logs from different glacial sediment-ary environments: (a) late Oligocene–Miocene glaciomarine sediments in a core drilled on the Antarctic continental shelf; (b) predominantly terrestrial glacigenic sediments from sections in the Sirius Group (Neogene), Shackleton Glacier, Trans-antarctic Mountains; and (c) detail of glacial lake sediments in the middle of the section shown in (b). Diagram from Figure G16 of Hambrey and Glasser (2003), reproduced with kind permission of Springer Science and Business Media

---

Box 8.1   Equipment required for sediment logging

The following equipment is desirable for field investigations of glacigenic sediments:

Hard hat
30-m tape for measuring sections
Grain-size card with roundness and sorting scales (laminated)
Munsell colour chart for descriptions of sediment colour
Ruler or pocket tape measure
Compass clinometer
Large callipers
Spade or trowel (to clean the section face)
Geological hammer
Plastic sample bags and ties (for sample collection)
Permanent marker pen (to label sample bags)
Notebook or clipboard (and plastic sheet or bag in case of rain)
Logging sheets
Camera and film/digital camera.

- Stand back from the section and make a detailed field sketch of the exposure. The simplest way to do this is to imagine that the section face is divided into an arbitrary number of individual grid squares. String or tape can be stretched across the whole section face to mark the grid squares. By sketching the content of each of the grid squares, it is possible to build up a scaled sketch of the whole section face. If the section is internally layered, record if these layers are parallel to any of the bounding surfaces. Study the boundaries between internal layers from a moderate distance and again close up. Decide if the boundaries are sharp or gradational, and note if any boundaries cut across one another.
- Describe the lateral continuity (or otherwise) of beds and units. Decide how many sedimentary units are present (sedimentary units may be obvious even from a distance on the basis of large-scale differences in colours, texture, particle size or sorting), looking especially for structures that establish the tops and bottoms of layers. Add to the sketch any obvious boundary relationships and large-scale deformation (glaciotectonic) structures such as folds or faults that might influence the distribution of the sedimentary facies present.
- Photographing the exposure is always advisable, especially if the section is large. Remember to include an object of known size (e.g. a person, ranging pole, spade, trowel or lens cap depending on the area

of the photograph) as a visible scale marker. Multiple photographs can later be made into a montage or photomosaic to record facies changes and to obtain sectional data that might be missing from simple sketches. Photography can also be used in the field to capture information about sedimentary exposures. For example, Boulton *et al.* (1999) took a continuous series of Polaroid photographs to survey sections through a push moraine in front of the glacier Holmstrømbreen in Svalbard. These authors then overlaid the Polaroid photographs, in the field, with transparent sheets on which details of the section were drawn. By mapping the location from which each photograph was taken and the distance to the base of the cliff, it was possible to correct the photographic distortion and to project each sketch section onto any given plane.

• Before starting detailed investigations it is vital to check that what you can see is a true section and not merely slumped or disturbed material that has fallen down the face. Often it will be necessary to clean parts of the section with a spade or trowel, in order to remove loose debris and create a fresh exposure.

Sedimentary facies are described in the field according to lithology, texture, bedding and sedimentary structures. The general rule is to describe first, and interpret later. Sediments can be readily classified in the field using the facies schemes developed by Miall (1978) and Eyles *et al.* (1983). For the poorly sorted sediments such as diamictons that are characteristic of many glacial environments, some workers have adopted the Hambrey (1994) modification of the Moncrieff (1989) classification (Figure 8.3). Fine-grained facies are classified according to the scheme illustrated in Figure 8.4.

Logs that are drawn up roughly in the field may be supplemented by further laboratory and data analysis, together with a final interpretation of the depositional environment (sections 8.16 and 8.17).

8.3   COLOUR AND ORGANIC CONTENT

8.3.1   Colour

Describe the colour of the sediment, either by using a verbal description or by comparing samples to a Munsell soil colour chart. The Munsell soil colour chart is the most widely used method of colour quantification. It enables similar colours to be discriminated, provides a useful shorthand form of colour description and allows the universal communication and correlation of colours between different studies. The Munsell soil colour chart, originally designed for the field description of soils, is an abridged

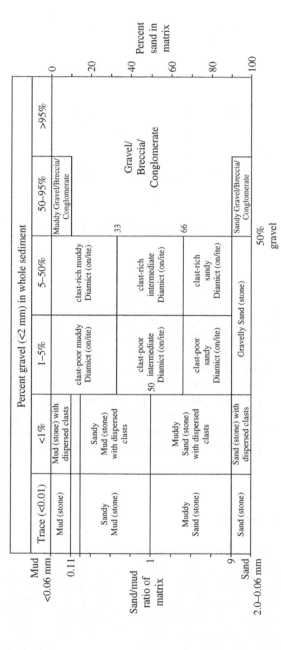

Figure 8.3 Particle-size classification for poorly sorted sediments. This scheme is the Hambrey (1994) modification of the Moncrieff (1989) scale

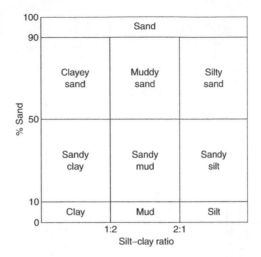

Figure 8.4   Particle-size classification for materials of sand grade and finer (all particles <2 mm)

version of the much larger and expensive Munsell book of colour (which has some 100 000 colours). The soils version has 199 colours on 7 charts, covers most of the colours likely to be encountered in the field, and is available in a pocket-sized wallet. The scheme works by classifying the colour of a sample in the field within the colour scheme. Samples should not be dried or crushed. Compare the colour of your unweathered, unsmeared sample with those in the book in two stages: (1) match the dominant colour of your sample to the nearest colour square in the chart and note the one which most closely matches your sample (e.g. 5YR 6/3 for a light reddish brown); and (2) record the extreme and/or dominant colours if your sample is variegated or highly variable in colour over short distances.

8.3.2   Organic content

Note the presence or absence of organic matter (e.g. rootlets or broken twigs) in the sample. If the sample is particularly black, brown or 'dirty'-looking, then it probably contains some organic matter. If the sediment looks 'clean', then it probably does not contain any organic matter. If establishing the presence or absence of organic material is of vital importance to your research design, this simple field description will not suffice and you will need to collect samples for more precise laboratory analysis (e.g. the determination of its chemical composition, see section 8.15.2).

8.4   SEDIMENT TEXTURE

8.4.1   Particle size

Measure the **particle-size range** (i.e. the extremes in particle size) of the sediments present. This can be done quickly with a ruler or callipers by measuring the grain size (the average of ten measurements of the intermediate axis (*b*-axis) of the largest particles present in the sediment). A more accurate determination of maximum grain size is achieved by recording the size of each of the three orthogonal axes (*a*-, *b*- and *c*-axes) of the largest particles present using the approach outlined in section 8.5.1. Other authors (e.g. Billi, 1984) noted that there are two methods for determining the mean diameter of particles in the field. The first method is to measure the *a*-, *b*- and *c*-axes and to calculate the mean diameter (*d*) using the formula $d = (a + b + c)/3$. The second method is to measure the sieve diameter of the *b*-axis in the field using a template with a graded series of openings (a 'gravelometer'). The procedure is to pass pebbles through the openings to find the smallest openings through which they cannot pass. Most gravelometers have square openings, but Billi (1984) advocated the use of those with round holes, concluding that templates with round openings provide greater reliability when a rapid method of sizing unconsolidated gravels is required. Use simple tests (i.e. rubbing a sample between your fingertips) to determine the approximate size of the finest sediment, i.e. whether this is fine sand, silt or clay. Sand feels gritty when rubbed between thumb and forefinger, silt feels silky to the touch, and clay feels sticky and will normally roll into a ball. Use a **grain size classification scheme** such as the one suggested in Figure 8.5 to classify your measurements. The most commonly used classification scheme is probably that of Wentworth (1922) but the scheme proposed by Friedman and Sanders (1978) has a lot to commend it because it offers subdivisions for categories such as silt, sand, pebbles, cobbles and boulders. Note that the term 'mud' is also in common usage – this term simply describes a mixture of silt and clay. The term 'gravel' is also commonly used to describe all particle sizes above 2 mm or −1 phi (ϕ) (including boulders). This classification scheme has recently been extended by Blair and MacPherson (1999) to include particles larger than 2048 mm. The average grain size is depicted on the graphic log, and thus determines the width of the log column.

8.4.2   Overall sediment texture

Assess the overall **texture** of the sample, that is the overall proportion in percentage terms of each sediment size fraction. Some workers do this by estimating the percentage of silt and clay (i.e. mud); percentage of sand;

Grain size		Descriptive terminology		
phi (ϕ)	mm/μm	Udden (1914) and Wentworth (1922)	Friedman and Sanders (1978)	Blott and Pye (2001)
−11	2048 mm		Very large boulders	
−10	1024		Large boulders	Very large
−9	512	Cobbles	Medium boulders	Large
−8	256		Small boulders	Medium
−7	128		Large cobbles	Small
−6	64		Small cobbles	Very small
−5	32		Very coarse cobbles	Very coarse
−4	16	Pebbles	Coarse pebbles	Coarse
−3	8		Medium pebbles	Medium
−2	4		Fine pebbles	Fine
−1	2	Granules	Very fine pebbles	Very fine
0	1	Very coarse sand	Very coarse sand	Very coarse
1	500 μm	Coarse sand	Coarse sand	Coarse
2	250	Medium sand	Medium sand	Medium
3	125	Fine sand	Fine sand	Fine
4	63	Very fine sand	Very fine sand	Very fine
5	31		Very coarse silt	Very coarse
6	16	Silt	Coarse silt	Coarse
7	8		Medium silt	Medium
8	4		Fine silt	Fine
9	2	Clay	Very fine silt	Very fine
			Clay	Clay

(Right-hand brackets: Boulders, Gravel, Sand, Silt)

Figure 8.5 Particle-size classification for description of sediments in the field. Diagram assembled from various sources including Udden (1914), Wentworth (1922), Friedman and Sanders (1978) and Blott and Pye (2001)

and percentage of granules, pebbles, cobbles and boulders (i.e. gravel). They then quote this as a mud/sand/gravel percentage (e.g. 40% mud, 10% sand, 50% gravel). At a bare minimum you should record the percentage of gravel by visually estimating the gravel content over exposed areas of between 0.5 and 2 m^2. Although sediment texture is normally described visually in the field, automated methods also exist. Shoshany (1989) describes the construction of a portable device for measuring and recording the surface microtopography and texture of sediments in the field. The device can be used at a scale of millimetres to centimetres to measure and record the size, shape and roundness of individual clasts, and for measuring the spatial arrangement and orientation of fabric elements. Although this is an objective method, it is not yet in widespread usage.

8.4.3   Sediment sorting

Make an estimate of the **sorting** in each unit using standard sorting indices (Figure 8.6) to define the level of sorting. The standard terms are: 'very well sorted', 'well sorted', 'poorly sorted', 'very poorly sorted'. You can also make descriptions of whether the sediment is **unimodal** (one particle-size fraction dominates), **bimodal** (extremes of particle size are present, e.g. sand and cobbles), or **polymodal** (a range of sizes are present and no single fraction dominates).

Sediments can also be described as **clast-supported** (predominantly coarse particles, where individual coarse particles touch one another but with an infill of fines) or **matrix-supported** (predominantly fine material, with coarse particles 'floating' in the fines). You should record a brief description of the **fabric** of the sediment. Fabric refers to the orientation of particles in sediment. Sometimes this is obvious, for example where gravel particles appear to have a strong preferred orientation or where there is well-developed imbrication. On other occasions, no preferred orientation is apparent. If the sediment is a diamicton it may be worth recording a clast fabric as part of your sedimentological description. This can be a clast **macrofabric, mesofabric or microfabric.** Clast macrofabrics are recorded in the field, whilst mesofabrics and microfabrics require laboratory facilities such as microscopes (section 8.12).

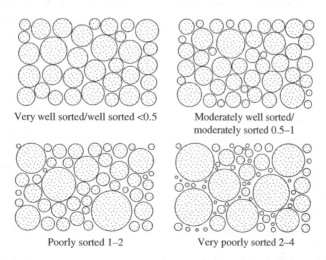

Figure 8.6   Comparison chart for describing sediment sorting

8.4.4 Physical state

You should also determine the physical state of each bed by testing its compaction and cementation by the reaction to breaking in the hand (Table 8.3).

8.5 PARTICLE MORPHOLOGY: THE SHAPE AND ROUNDNESS OF SEDIMENTARY PARTICLES

Morphology encompasses both the **shape** and the **roundness** of sedimentary particles. Note that the terms 'shape' and 'roundness' are commonly confused, but are in fact entirely different parameters.

8.5.1 Clast shape

Clast shape is a measure of the overall shape of a particle. This requires the measurement of the three orthogonal axes (called the a-, b- and c-axes) of particles to quantitatively describe the shape of a particle (Figure 8.7). The a-axis is the longest axis of a particle, the b-axis is the intermediate axis, and the c-axis is the shortest axis. Measurements of these three axes must be orthogonal (i.e. at right angles to one another). In a fieldwork setting, it is usually the gravel fraction (>2 mm) that is analysed in this way, since this size fraction can be readily handled and measured with a ruler. Clasts are chosen with a-axis between 20 and 100 mm. To be representative, samples must comprise at least 50 clasts. Some fieldworkers prefer to collect samples of 50 clasts and analyse these later, rather than perform measurements in the field. The advantage of this approach is that it reduces the time spent in the field and allows time to be dedicated to clast measurement in the

Table 8.3 A scheme for assessing the compaction of sediments in the field

Description	Terminology
Loose	Unconsolidated
Crumbles easily between fingers	Very friable
Rubbing with fingers frees numerous grains; gentle blow with geological hammer disintegrates sample	Friable
Grains can be separated from sample with a steel probe; breaks easily when hit with a geological hammer	Hard
Grains are difficult to separate with a steel probe; difficult to break with a geological hammer	Very hard
Sharp hammer blow required to break sample	Extremely hard

1. A well-rounded particle

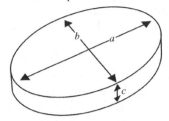

2. Orthogonal arrangement of the *a*-, *b*-, *c*-axes

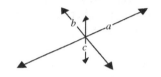

3. A very angular particle

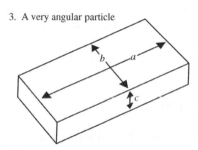

Figure 8.7   How to measure the three orthogonal axes (*a*-, *b*-, *c*-axes) of a particle or a clast. The *a*-axis is the longest axis of a particle, the *b*-axis is the intermediate axis, and the *c*-axis is the shortest axis. This diagram also illustrates the difference between particle shape and roundness: measuring the three orthogonal axes of both particles will produce the same values for *a*, *b* and *c* in both cases (**shape**), but in the top example (1) the particle is well rounded and in the bottom example (3) it is very angular (**roundness**)

laboratory. The disadvantage is that it can mean sampling and transporting large numbers of clasts out of the field – often this is simply not possible. Once the three axial measurements have been made, data can be analysed using a variety of different methods:

- Ratios of *b/a* and *c/b* axes can be calculated and transferred to bivariate plots to indicate clast shapes as discs, blades, rods and spheres/cubes according to the method outlined by Zingg (1935).
- Data can be plotted on shape triangles, following the method of Sneed and Folk (1958) and Benn and Ballantyne (1993). These triangles or

ternary plots can easily be produced using the TRI-PLOT spreadsheet of Graham and Midgley (2000).

- The flatness ratio for particles can be calculated using the simple formula $(a + b)/2c$.
- These data can be plotted on roundness/sphericity diagrams, where roundness (see section 8.5.2) is plotted against sphericity (the ratio of the surface area of a sphere to the actual surface area of the particle, equal to $3\sqrt{(b \times c)}/a^2$ (Wadell, 1932, 1933; Boulton, 1978a).
- Clast shape data can be plotted on RA/C_{40} bivariate plots (Benn and Ballantyne, 1994), in which the RA index (% of angular and very angular clasts; see section 8.5.2) is plotted against the C_{40} index (% of clasts with c/a axial ratio ≤ 0.4). This method has been shown to provide good discrimination between glacial facies in the high-arctic environment (Bennett *et al.*, 1997).

The shape of sedimentary particles can also be described by other approaches, but most of these are too time-consuming to carry out in the field and are suitable only for laboratory investigation. For example, the method of Ehlrich and Weinberg (1970) requires the operator to measure a series of points (position co-ordinates) around the outside of each particle. Shape is then calculated by an expansion of the periphery radius as a function of the angle about the particle's centre of gravity by a Fourier Series.

8.5.2   Clast roundness

The **roundness** of a particle is defined by small-scale directional changes of the particle surface – its roundness at one end of a scale and its angularity at the other (Figure 8.8). Roundness is normally analysed for samples of at least 50 particles by comparing each particle to the Powers (1953) scale (Figure 8.9). This scale requires the user to make a visual assessment of the roundness of the corners of individual particles by assigning it to one of six class intervals ranging from very angular to well-rounded (Table 8.4). The scale begins at 0.12 rather than at zero because particle roundness less than 0.12 is not readily differentiated by the observer (Powers, 1953). The class ranges are designed such that the ratio of the upper limit to the lower limit of any interval is 0.7, and the scale is deliberately designed to provide increasingly fine divisions at its lower end. In glacial environments, field-work normally involves measuring clasts with a-axis between 20 and 100 mm. Some subjectivity is inevitable in using this scale in glacial environments, for example in dealing with a rounded clast that has one broken corner (this clast should still be classified as rounded). Data are usually drawn as histograms showing the percentage of a sample that falls in each of the six roundness classes.

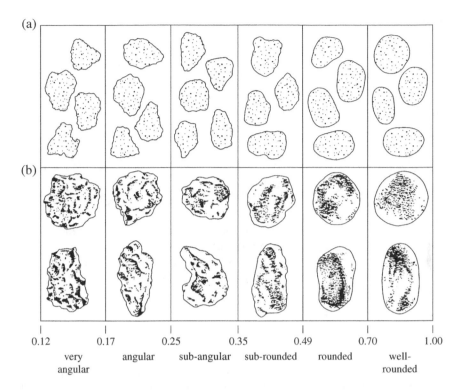

Figure 8.8 Visual images for the determination of roundness of grains or clasts: (a) 2D outlines modified from Krumbein (1941), and (b) 3D images modified from Powers (1953). Roundness classes and boundaries are those of Powers (1953)

Other more time-consuming measures of roundness also exist. For example, Wadell's (1935) method for determining the roundness of a particle involves dividing the average of the radii of the corners of a particle by the radius of the maximum inscribed circle. Lees (1964) criticised the measures of roundness adopted by Krumbein (1941) and Powers (1953) because these classification schemes emphasize only *roundness* and fail to deal adequately with angular and very angular particles. Lees (1964) argued that angularity is not just the absence of roundness but a distinct concept in its own right. He introduced a new classification scheme for angularity, based on the degree of angularity as defined by the angle between faces on a particle, the projection of its corners relative to the largest internal spherical mass of the particle, and the number of angular corners. He also provided visual comparators ranging from 0 to 1599 for the rapid visual estimation of degree of angularity. Unfortunately this scheme has failed to meet with universal approval and most fieldworkers still adopt the Powers scale. Olsen (1983) further criticized the Powers roundness scale because it deals only

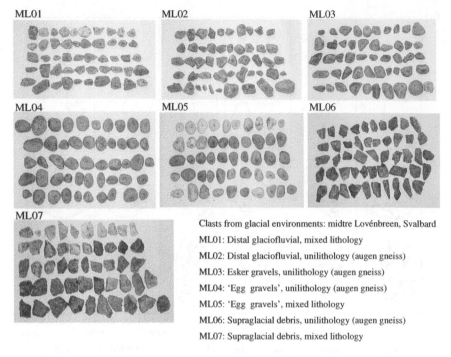

ML01

ML02

ML03

ML04

ML05

ML06

ML07

Clasts from glacial environments: midtre Lovénbreen, Svalbard

ML01: Distal glaciofluvial, mixed lithology

ML02: Distal glaciofluvial, unilithology (augen gneiss)

ML03: Esker gravels, unilithology (augen gneiss)

ML04: 'Egg gravels', unilithology (augen gneiss)

ML05: 'Egg gravels', mixed lithology

ML06: Supraglacial debris, unilithology (augen gneiss)

ML07: Supraglacial debris, mixed lithology

Figure 8.9 Clast roundness and its relationship to the environment of deposition. The photographs show samples of 50 clasts collected on the forefield of the glacier midre Lovénbreen in Svalbard. (Photograph: N.F. Glasser)

Table 8.4 Roundness grades (modified from Powers, 1953)

Grade term	Abbreviation	Class interval	Geometric mean
Very angular	VA	0.12–0.17	0.14
Angular	A	0.17–0.25	0.21
Sub-angular	SA	0.25–0.35	0.30
Sub-rounded	SR	0.35–0.49	0.41
Rounded	R	0.49–0.70	0.59
Well-rounded	WR	0.70–1.00	0.84

with abrasion of particles and ignores crushing, which is common in glacial environments. He advocated a new approach using only four categories (A, SR, R, WR) to describe roundness due to abrasion with a crushing index (Cr) defined as the percentage of the rounded and well-rounded material which has breakage zones with sharp-edged to very weakly subrounded limits. Again, although this approach is good for glacial environments, it has not been universally adopted.

8.5.3   Surface features on clasts

It is important to record the presence or absence of **surface features on clasts** because these features may give an indication of the transport history (basal glacial or otherwise) of a sedimentary facies. In the field this is normally restricted to observations of the surface features on clasts in the gravel fraction (>2 mm). Analysis of surface features on clasts below this size requires magnification and is normally possible only by using techniques such as scanning electron microscopy (SEM; see section 8.15.1). Features commonly recorded in the field are the occurrence of faceted clasts (flat surfaces with rounded edges), bullet-shaped clasts and those with a stoss-and-lee form, as well as clasts with striations, grooves, gouges and cracks. All of these features indicate basal glacial transport. Normally the percentage of clasts displaying these features in samples of 50 clasts is recorded. Note that the development of striations on clasts is strongly dependent on lithology. Hard crystalline rocks such as quartzite, granite, gneiss and schist rarely display striations, whereas fine-grained igneous rocks, carbonates and mudstones commonly do. Striations are most obvious when a clast is wet, and a common technique is to moisten the clast and hold it up to the light to check for striations. Facets may develop on all rock types subjected to basal glacial transport. Finally, it is worth considering the overall morphology of the clasts: for example Krüger (1984) has argued convincingly that clasts with a stoss-and-lee form are indicative of basal glacial transport.

8.5.4   Clast lithological analysis

The purpose of **clast lithological analysis** is to identify the source areas (provenance) of a sedimentary facies from the range of clast lithologies present (Box 8.2). Two important considerations here are the size range to be sampled and the number of clasts that need to be identified to ensure a representative sample. To avoid bias in the sizes analysed, clasts should be selected so that they are all within a half phi-unit, and preferably within a quarter phi-unit, range. For example, in their study of the provenance of tills in Eastern England, Lee *et al.* (2002) studied only pebbles from the 4–8-mm and 8–16-mm fractions by crumbling the till into a 4-mm sieve and washing out the matrix. Although many studies have used samples of less than one hundred clasts, a minimum sample of 250–300 clasts is required to produce meaningful results (Bridgland, 1986). This technique is based on the field identification of hand specimens and therefore requires some basic geological knowledge in order to correctly identify and classify different rocks and minerals. Figure 8.10 provides some guidance on how to identify different lithologies in the field from hand specimens, which can be used in

Box 8.2   Provenance studies of glacigenic deposits

Provenance studies involve using characteristic suites of clast lithologies or heavy mineral assemblages in the matrix of glacial deposits
to identify the source areas of these deposits. These provenance techniques can be used in both glaciological studies of existing glaciers to

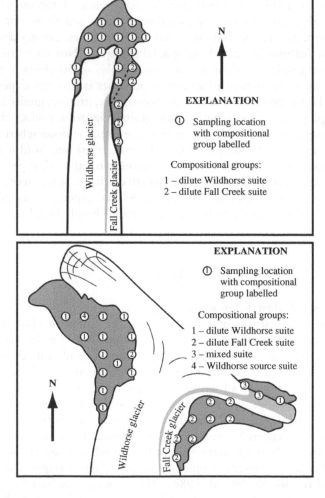

Figure B8.2   The terminus geometries of the former Wildhorse and Fall Creek
Glaciers during two different stages of their existence according to Brugger
(1996). Reprinted from the *Annals of Glaciology* with permission of the
International Glaciological Society

identify transport paths through a glacier and in geomorphological studies to identify the source areas of former glaciers. Brugger (1996) used these techniques to reconstruct transport paths through a Late Pleistocene glacier in the Wildhorse Canyon area of Idaho. During this period of glacier expansion, a compound glacier, fed by two tributary glaciers (the Wildhorse and Fall Creek Glaciers), existed in the mountains. The catchments of the two tributary glaciers were underlain by different bedrock lithologies, allowing Brugger to identify with confidence the relative contribution of the components of the compound glacier (Figure B8.2). Gneisses, schists and quartzites underlay the catchment area of Wildhorse Glacier whilst granitic rocks underlay the catchment area of the Fall Creek Glacier. Having mapped the distribution of these lithologies in both the pebble-sized fraction and the matrix of tills deposited by the former compound glacier, Brugger found he could reconstruct the dimensions and terminus geometries of the glacier. The position of the medial moraine that marked the flow unit boundary between the glaciers was particularly useful in determining the relative contribution to flow made by the tributary glaciers. Once this was established, Brugger was able to reconstruct the surface contours of the glacier and to make glaciological reconstructions of the palaeoglaciers, including the equilibrium line altitude (ELA) and values for former basal shear stresses. This study is a good illustration of how till-provenance studies can be used as an independent verification of glaciological reconstructions of palaeoglaciers.

Brugger, K.A. 1996. Implications of till-provenance studies for glaciological reconstructions of the paleoglaciers of Wildhorse Canyon, Idaho, USA. *Annals of Glaciology*, **22**, 93–101.

conjunction with knowledge of the local and regional geological conditions to relate the lithologies to their provenance.

8.6   BEDDING

The term 'bedding' refers to the designation of beds and bedding planes, bed thickness, bed geometry, and the nature of the contacts between beds. In descriptive terms, a bed is simply a layer that is sufficiently distinct from adjoining layers, while in genetic terms, it is a layer that represents a depositional environment during which conditions were relatively uniform (Graham, 1988). Those sediments where no bedding is apparent are termed

(a)

Questions

1. Is the rock made up entirely of interlocking crystals? Note that in some cases these may not be visible to the naked eye and the use of a hand lens or microscope will be required for them to be seen. Note also that glassy and cryptocrystalline rocks are made up entirely of interlocking crystals, but that these will not be visible unless the rock is viewed in thin section.
2. Is the rock both massively crystalline and monomineralic; that is, is it made up of large interlocking crystals of a single mineral?
3. Is the rock phaneric; that is, are more than 50% of the particles of which it is composed visible to the naked eye or with a hand lens? Note that in the case of igneous rocks a phaneric texture is generally indicative of an intrusive origin.
4. Is the rock aphanitic; that is, are less than 50% of the particles of which it is composed visible to the naked eye or with the hand lens? Note that in the case of igneous rocks an aphanitic texture is generally indicative of an extrusive origin.
5. Is the rock glassy in appearance, with a fresh fracture surface smooth to the touch, and no trace of grittiness? Note that in the case of igneous rocks a glassy texture is indicative of an extrusive origin.
6. Is the rock porphyritic (in the case of igneous rocks) or porphyroblastic (in the case of metamorphic rocks); that is, does it possess minerals that are conspicuously larger than those surrounding them?
7. Is the igneous rock leucocratic; that is, is it light-coloured, having less than 30% dark-coloured minerals? Note that a leucocratic colour index is generally indicative of an acidic composition.
8. Is the igneous rock mesocratic; that is, is it medium-coloured, having 30–60% dark-coloured minerals? Note that a mesocratic colour index is generally indicative of an intermediate composition.
9. Is the igneous rock melanocratic; that is, is it dark-coloured, having 60–90% dark-coloured minerals? Note that a melanocratic colour index is generally indicative of a basic composition.
10. Does the rock possess a parallel structure (not necessarily of crystals)?
11. Does the rock possess vesicles: small, rounded gas cavities often lined or infilled by minerals?
12. Does the rock split into thin parallel sheets?
13. Is the rock schistose; that is, does it possess an undulatory foliation (parallel banding formed by flat-lying minerals) with a tendency for the folia to part from one another, with the surfaces of the foliation planes shimmering as a result of the dominant presence of mica crystals?
14. Is the rock gneissose; that is, is it banded as a result of the concentration of different minerals in successive layers? Note that the bands are generally alternately light and dark, the dark bands being narrower than the light.
15. Is the rock calcareous; that is, does the bulk of the mass of the rock effervesce vigorously upon contact with cold hydrochloric acid? If only the matrix effervesces, then the rock has a calcareous cement. A less likely occurrence is that only particular particles or minerals effervesce: this should be taken only to demonstrate the presence of calcite particles or minerals. Note that care is required to distinguish effervescence caused simply by reaction with calcite precipitated on the surface of the clast or with traces of calcareous material adhering to the clast. Note also that some clasts, particularly those that are weathered, may produce air bubbles upon contact with dilute hydrochloric acid. This should not be confused with the vigorous reaction characteristic of calcareous rocks.
16. Is the rock made up of non-interlocking grains of rock particles of mean diameter greater than 2 mm bound together by a matrix of finer material?
17. If the answer to question 16 is yes, are the grains angular (A) or rounded (R)?
18. Is the rock made up of non-interlocking sand-sized grains bound together by a cement or by a matrix of finer material?
19. Is the rock made up of non-interlocking fine grains, only visible with the hand lens?
20. Are fossils visible in the rock?
21. Is the rock made up almost completely of a single mineral: quartz (Q), calcite (Ca), coal (C), gypsum (G) or halite (H)?
22. Does the rock have a brown to black streak?
23. Is the rock denser than 'normal' rocks?
24. Does the rock break along a conchoidal fracture plane?

Figure 8.10 (a) A scheme to aid the recognition of different lithologies in hand specimens. Twenty four lithologies can be identified on the basis of answers to 24 questions. With the exception of questions 17 and 21 the questions only require yes/no answers. (b) Diagram modified from Gale and Hoare (1991)

(b)

Question Number

Igneous	1	2	3	4	5	6	7	8	9	10	11	12	13	14	15	16	17	18	19	20	21	22	23	24
Granite[a]	Y		Y			S	Y																	
Diorite	Y		Y			S		Y		S														
Gabbro	Y		Y			S			Y	S													S	
Rhyolite[b]	Y			Y	S	S	Y			S	S													S
Andesite[b]	Y			Y	S	S		Y		S	S													S
Basalt[c]	Y			Y	S	S			Y		S												S	S
Metamorphic	**1**	**2**	**3**	**4**	**5**	**6**	**7**	**8**	**9**	**10**	**11**	**12**	**13**	**14**	**15**	**16**	**17**	**18**	**19**	**20**	**21**	**22**	**23**	**24**
Slate	Y			Y		S	–	–	–	Y		Y								S				
Schist	Y		Y			S	–	–	–	Y			Y										S	
Gneiss	Y		Y				–	–	–	Y				Y									S	
Granulite[a]			Y				–	–	–														S	
Quartzite	Y		Y				–	–	–												Q		S	S
Hornfels[c]				Y		S	–	–	–	S													S	S
Marble	Y		Y				–	–	–						Y					S	Ca			
Sedimentary	**1**	**2**	**3**	**4**	**5**	**6**	**7**	**8**	**9**	**10**	**11**	**12**	**13**	**14**	**15**	**16**	**17**	**18**	**19**	**20**	**21**	**22**	**23**	**24**
Breccia[d]			Y			–	–	–	–	S					S	Y	A							
Conglomerate			Y			–	–	–	–	S					S	Y	R			S				
Sandstone			Y			–	–	–	–	S					S			Y		S				
Mudrock				Y		–	–	–	–	S		S			S				Y	S		S		
Coal	S	S	Y			–	–	–	–	S										S	C	Y		S
Limestone	S	S	S	S		–	–	–	–	S					Y	S		S	S	S	Ca			
Gypsum	S	S	S	S		–	–	–	–												G			
Halite	Y	S	Y			–	–	–	–												H			
Chert	Y			Y	Y	–	–	–	–	S										S	Q			Y
Mineral veins	**1**	**2**	**3**	**4**	**5**	**6**	**7**	**8**	**9**	**10**	**11**	**12**	**13**	**14**	**15**	**16**	**17**	**18**	**19**	**20**	**21**	**22**	**23**	**24**
Quartz	Y	Y	Y				–	–	–	S											Q			
Calcite	Y	Y	Y				–	–	–	S					Y						Ca			

Y = yes, S = sometimes, no symbol = no,– = question not applicable.

[a] Many granulites are indistinguishable from granite in hand specimen.

[b] Rhyolites and andesites may be difficult to differentiate in hand specimen.

[c] Some hornfelses may be difficult to distinguish from basalt.

[d] Some breccias may be igneous or metamorphic rather than sedimentary.

Figure 8.10 (Continued)

'**massive**'. Individual beds are often difficult to recognize in the field, especially since thicknesses can range from a few millimetres to several metres
(Table 8.5).

A single bed may be internally homogeneous, show continuous gradational variation, or be internally layered in thin or very thin beds (those
under 100 mm in thickness). Thin beds are called laminae and are also
classified according to thickness (Table 8.6).

When recording the nature of the contact between beds, features to look
for include burrowed surfaces, hardgrounds, evidence of periods of soil
formation and erosional contacts. The first two of these are relatively
uncommon in terrestrial glacigenic sediments, while the latter two are more
common in such environments.

At a bare minimum you should record for each bed its thickness, structural attributes (dip and strike) and make observations of bed geometry and
the nature of their lateral termination. Beds can terminate laterally by:

1. Convergence and intersection of bedding surfaces.
2. Lateral gradation of the material into another bed in which bedding
   surfaces become indistinguishable.
3. Abutting a fault, unconformity or cross-cutting feature such as a
   channel.

Table 8.5   Terminology for thickness of beds

Bed thickness (mm)	Terminology
>1000	Very thick bed
300–1000	Thick bed
100–300	Medium bed
10–100	Thin bed
<10	Very thin bed

Table 8.6   Terminology for thickness of lamina

Lamina thickness (mm)	Terminology
>30	Very thick lamina
10–30	Thick lamina
3–10	Medium lamina
1–3	Thin lamina
<1	Very thin lamina

8.7   SEDIMENTARY STRUCTURES

Sedimentary structures can be important clues to environmental interpretation and these should therefore be recorded during sedimentary logging (Table 8.7).

8.7.1   Deformation structures and folds

Deformation structures can be found by examining each sedimentary unit for grains or other small bodies (inclusions) that have been deformed into planar or linear shapes. Where such structures exist, it is important to make a detailed description of their appearance, and to record their orientation relative to other structures in the sediment and with regard to the overall geometry of the sedimentary exposure. For a full discussion of deformation processes and a classification of the deformation structures present in subglacially deformed materials see Benn and Evans (1996). Describe any large-scale structures such as folds (structures produced when an originally planar surface becomes bent or curved as a result of deformation) that are apparent within either individual beds or groups of beds. Folds are relatively common in glacial environments, especially in sediments deposited at or near the ice margin, and in sediments that have been deformed by glacier movement. Suggested terminology for the field description of folds is shown in Figure 8.11.

Table 8.7   Sedimentary structures commonly encountered in glacial environments (modified from Graham, 1988)

---

1. **Structures occurring primarily within beds**
   Cross-stratification
   Lamination
   Grading
   Convolute bedding or laminations (soft sediment deformation)
   Pedogenic (soil) horizons

2. **Structures occurring primarily on bedding surfaces**
   (i) Bottom surfaces
   Flute marks
   Tool marks
   Load casts
   Geometry due to scour or topographic fill

   (ii) Top surfaces
   Surface topography
   Bedforms such as ripples, dunes and hummocks
   Soft sediment striations (e.g. from iceberg keels)
   Primary current lineation
   Shrinkage cracks

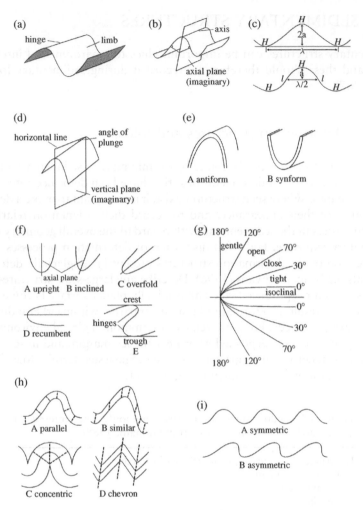

Figure 8.11 Field description of folds, assembled from information in Park (1997): (a) The hinge and limbs of a fold; (b) How to define the axis and axial plane of a fold; (c) How to measure the amplitude (*a*) and wavelength (λ) of a fold with respect to its hinge (*H*) and inflexion point (*I*); (d) How to measure the plunge of a fold. The angle of plunge is measured from the horizontal, in a vertical plane; (e) How to describe the closing direction of a fold, showing an antiform (upward-closing fold) and a synform (downward-closing fold); (f) How to describe the attitude of a fold. A–D are upright, inclined, overfold and recumbent folds, while E shows the distinction between crest and trough lines and hinge lines in an inclined fold; (g) Classification of folds into gentle, open, close, tight and isoclinal based on the inter-limb angle; (h) Types of fold profile. Parallel fold: dashed lines show constant layer thickness measured perpendicular to fold surface. Similar fold: dashed lines show constant layer thickness measured perpendicular to axial surface. Concentric fold: dashed lines are radii of a circle. Chevron fold: dashed lines are kink planes separating straight fold limbs; and (i) Fold symmetry: a set of folds are regarded as symmetric if the limbs are of equal length and asymmetric if limbs are of unequal length

8.7.2 Faults

Faults are the planar discontinuities often present within sediments. Where such faults exist, their style and orientation should be recorded (Figure 8.12). If there are distinctive marker horizons in the sediment it may be possible to measure displacement along faults as well as their orientation. Faults can be grouped into sets according to their orientation to determine if there are regional trends in orientation that can be related to process. Consistent orientations are typical of processes such as ice-overriding or ice-push where external forces are from a consistent direction. Random faults are more indicative of processes such as the removal of ice support from a sediment-ary body or the melting of included ice masses.

8.8 PALAEOCURRENT DATA

As the name suggests, these are data that provide information on the direction of sediment transport in the past. The techniques of palaeocurrent

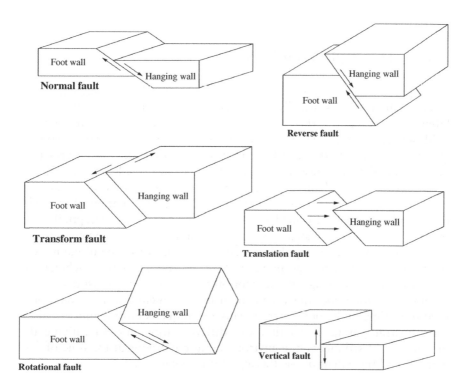

Figure 8.12 The terminology used in describing different types of faults

analysis can be applied to any sediment that has been transported by water, for example in glaciofluvial environments, where former currents can be inferred from the sedimentary structures present. Graham (1988) considers that palaeocurrent data can be grouped into two classes:

1. Properties that acquire directional significance only when mapped regionally, including attributes such as the presence/absence of some distinctive feature, for example boulder type or mineral assemblage and scalars (the magnitude of a property such as particle size, roundness or unit thickness). These data are recorded routinely in the field as part of most sedimentary descriptions.

2. Properties that provide directional information when measured at the point of observation. These properties may indicate a line of flow (e.g. primary current lineation, groove marks, symmetrical ripple crests) or a unique direction of flow (e.g. imbrication, cross-stratification, flute marks). In deposits with well-developed sedimentary structures, such as cross-stratification, it is often worth collecting additional data on the direction of foreset dips to obtain the average local flow direction.

8.9   OTHER PROPERTIES

Investigations of most other sedimentary properties (e.g. geotechnical and mechanical properties) usually require equipment that is only available in a laboratory setting and therefore require samples to be removed from the field (see section 8.10). However, there are several mechanical properties of glacigenic sediments that can be readily tested in the field using pressuremeter tests, plate load tests and *in situ* direct shear tests. Klohn (1965) and Radhakrishna and Klym (1974) provide examples of tests for measuring both the shear strength and the elastic moduli of deformation of the sediment and describe their field operation. The magnetic susceptibility of sediments can also be tested in the field using portable equipment such as the Bartington MS2 susceptibility meter, although this type of equipment cannot be used close to high-power radio transmitters, heavy machinery or large iron objects. Magnetic susceptibility can also be measured in the laboratory if time does not permit field measurement and it is commonly used in dealing with core material. This technique is most commonly used for lithostratigraphic correlation and differentiation, for determining the provenance of sediments, the environmental conditions at the time of formation, and the extent of weathering and pedogenesis.

8.10 FIELD SAMPLING TECHNIQUES

Field sampling procedures are guided firstly by the aim of the research and secondly by the laboratory techniques to be used on return from the field. For this reason, it is imperative to think through your research design before embarking on sampling in the field. There is little point in sampling material that is not directly relevant to the research question and for which you do not have access to the correct laboratory equipment for further analysis. The key to successful sampling is to follow set procedures and to ensure that detailed records are kept in a notebook of sample locations, sample numbers and their context. Always number and label samples in a systematic and logical order with a waterproof or permanent marker and give each sample a unique number or sample code (Box 8.3). Once collected, samples

---

Box 8.3   Sample coding in the field

Many glaciological and geomorphological field projects require the collection of samples in the field, which are then transported back to the laboratory for subsequent analyses. When the time comes to analyse these samples in the laboratory, it is essential that you know exactly from where and when each sample was collected. To do this, samples are usually given unique numbers or sample codes in the field. Each sample is then labelled with that unique code so that it can be identified at a later date by comparison with a field notebook. Different fieldworkers use different methods of numbering or coding their samples, but here is one recommended approach. Each sample should be given a code relating to the name of the person who collected it, the year of collection, the location and a unique number. The unique number should increase incrementally every time a new sample is collected. For example, the first sample collected by James Bond (JB) in the year 2004 at Trapridge Glacier (TRG) would be coded as JB/04/TRG/01. Subsequent samples would be coded JB/04/TRG/02, JB/04/TRG/03, JB/04/TRG/04 and so on. If this fieldworker then moved to a second glacier, say Black Rapids Glacier (BRG), in the same year, numbering would change to JB/04/BRG/01, JB/04/BRG/02 and so on. No two samples should ever be given the same sample code, even if they are from the same locality, as this inevitably leads to confusion. However, letters can be appended to the sample number if required to quickly identify the nature of individual samples. For example, all

Box 8.3 (Continued)

sediment matrix samples might be coded as **a**, whilst samples of clasts for morphological analysis might be coded as **b**, giving JB/04/TRG/01a and JB/04/TRG/01b, and so on. The fieldworker would record in his/her notebook relevant information about the sample, including details of its context and possible significance. Each sample should be carefully bagged (preferably into a sealed plastic sample bag) and labelled with the sample code using a waterproof marker pen. Never be tempted to write a sample code on a piece of paper and put it into the bag along with the sample – moisture from the sample invariably causes the ink on the paper to run and the writing quickly becomes illegible. A quick means of remembering how to use this coding system is shown below.

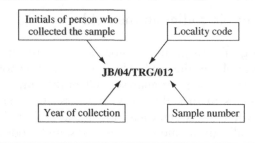

must be carefully transported and stored to avoid damage. When you return from fieldwork you must be able to identify precisely where each sample came from and what it represents. If you cannot ensure this is the case, it is not worth collecting samples in the first place.

8.10.1 Dry-sieving for particle-size determination

Glacigenic sediments typically comprise particles from the smallest fraction (clay) to the largest (boulders). Bulk samples of several kilograms or more are therefore required to accurately reflect the full particle-size distribution of the sediment. Since the collection and transport of such large samples is seldom possible, many workers approach this task as a two-stage process by (1) estimating the proportion of mud, sand and gravel in the field and (2) collecting ~100-g samples of the matrix (normally the sand fraction and below) for laboratory dry-sieving purposes (Figure 8.13). Before taking a sample, clean

(a)

(b)

Figure 8.13   Collecting bulk sediment samples for laboratory analysis: (a) Sediment sampling from ice debris at the terminus of a contemporary glacier in Svalbard; and (b) Sediment sampling from exposures on the glacier forefield. The sediment is collected from a cleaned part of the face to avoid surface weathering and put directly into a clean, labelled sample bag for transport back to the laboratory (Photographs: N.F. Glasser)

the outcrop face with a spade or trowel to ensure that the sample is not contaminated by slumped or weathered material (Gale and Hoare, 1992). Samples should consist only of the matrix, and must not be biased by the inclusion of outsize clasts. Carefully collect the matrix by scraping material into an open sample bag held beneath the desired deposit. Be sure to write the sample number and its location on the outside of the sample bag using a waterproof pen. You should also note the context from which the sample was collected on the bag as well as in a notebook. If the sample is collected from a sedimentary facies described within a logged section, mark the sample numbers on the log for extra security.

8.10.2   Sampling for clast mesofabrics, microfabrics and microstructural description

These techniques require the collection of undisturbed, orientated samples of diamicton and for this reason sampling is described for all three techniques together. Samples should not be taken from the top 2 metres of a diamicton unit, or within 1 m of the bottom of the unit because clasts at this point may not have been transported far from their source and therefore have not acquired a representative fabric. Gale and Hoare (1991), Carr *et al.* (2000) and Lachniet *et al.* (2001) all offer advice on methods of sampling. Samples should be collected from an exposure only once all weathered, slumped, or otherwise disturbed material has been removed. This normally requires the outer 30 cm of the face to be removed with a spade or trowel. Larger orientated samples for mesofabrics (normally around 150 mm × 105 mm × 80 mm) can be sampled directly into a metal sampling frame (effectively a large cutter) and transferred within it for safe transport to the laboratory. The sample must always be marked clearly with a North arrow, and the upper surface marked clearly with the word TOP. For clast microfabrics and microstructural descriptions, sediments are sampled using smaller metal tins (approximately 100 mm × 80 mm × 40 mm in size) inserted over a carved-out sediment block. These tins are sometimes known as Kubiena tins. The sediment-filled tins are carefully removed from the face of the exposure, and excess sediment is removed until lids can be fitted over both ends of the tin. Care must be taken not to disturb the structural integrity of the sediment block during sampling. Be sure to note the sample number and orientation. Samples are then stored in airtight plastic bags until ready for thin section preparation. This involves impregnating the sample with a polyester resin and cutting thin sections from the hardened block (Carr and Lee, 1998).

8.10.3 Sampling sediments for geotechnical properties

Most laboratory tests of the geotechnical properties of sediment (e.g. oedo-meter tests, direct shear and triaxial compression tests) require undisturbed block or cored samples to be removed from the field (Loiselle and Hurtubise, 1976). The general rule of thumb is that the sample size must be sufficient to accurately reflect the properties that are to be tested in the laboratory. For example, oedometer consolidation tests require samples with a diameter of at least 150 mm and a thickness of at least 50 mm. Sample size also depends on the capability of the laboratory equipment to be used. For example, most triaxial consolidation cells can only accommodate specimens up to 150 mm in diameter, although some larger cells can take specimens up to 1 m in diameter. More specialized tests, such as the hydraulic conductivity of fractured till, may require samples sizes as large as 400 kg (Grisak *et al.*, 1976).

8.10.4 Sampling sediments for chemical properties

Most analyses of chemical properties are performed in the laboratory on the <2-mm fraction. The guiding principle for sampling is that the value obtained in the laboratory for a particular chemical component should be an accurate representation of its actual concentration of that material in the field. This requires samples of at least 1 kg of a material even if these are later subdivided in the laboratory into the smaller (1–20 g) samples that most chemical analyses require (see Gale and Hoare, 1991; pp. 249–250 for further guidance on this topic).

8.11 FABRIC ANALYSIS: GENERAL CONSIDERATIONS

Fabric refers to the orientation of particles in a given sediment body. In glacial environments, techniques of fabric analysis are applied only to diamictons (poorly sorted, unconsolidated sediments with a wide range of particle sizes). Sometimes fabric is evident as a visible preferred orientation of gravel particles in the field, whilst sometimes fabrics can only be detected at a microstructural scale. Fabric is determined by measuring the declination and inclination of the long axis (*a*-axis) of elongate clasts. Data can be collected at three spatial scales: as a clast macrofabric, mesofabric or microfabric. Clast macrofabrics are recorded *in situ* in the field using gravel-sized clasts (size range normally

10–100 mm). Both mesofabrics (size range variable but normally clasts in the range 2–10 mm) and microfabrics (size range <2 mm) require orientated samples to be collected and returned to laboratory conditions. The techniques for recording these fabrics are outlined in sections 8.12–8.14.

8.12   CLAST MACROFABRICS

8.12.1   General considerations

Clast macrofabrics and clast microfabrics involve the field measurement of the declination (compass orientation) and inclination (dip angle) of clasts in diamictons. Both measurements are made using a hand-held compass-clinometer. Clast macrofabrics have been widely used by glacial geologists and geomorphologists because of the belief that these techniques can:

1. provide useful information about directions of ice movement;
2. differentiate between modes of deposition;
3. differentiate between different sedimentary facies;
4. provide an indication of the relative strain undergone by a sediment (Box 8.4).

Box 8.4   Can clast macrofabrics be used to distinguish sedimentary facies?

Bennett *et al.* (1999) set out to test the long-held assertion that clast macrofabrics can be used to distinguish between different glacigenic sediments. They compiled a data set of 111 clast fabrics measured in and around the margins of contemporary glaciers in Svalbard, Alaska, Iceland and Switzerland. Of these fabrics, 88 were measured in exposures of glacigenic sediments, including facies interpreted as meltout tills, glacigenic debris flows, deformation tills, lodgement till and glaciomarine diamictons. The remaining 23 fabrics were measured in debris-rich basal ice exposed at contemporary glacier margins. All debris-rich basal ice samples were based on the measurement of 25 clasts, and those from the sedimentary facies were based on the measurement of 50 clasts. In both cases, all clasts measured had *a:b*-axis ratios of >3:2. The authors plotted the fabrics on both triangular eigenvalue plots and bivariate eigenvalue plots in order to compare

---

their results with other published studies. They found little discrimination between the sedimentary facies sampled and that all the populations overlap to some extent. They concluded that, because of this overlap, clast macrofabrics alone do not provide a basis for distinguishing between the different sedimentary facies in the glacial environments studied. Bennett *et al.* (1999) therefore advocated caution in using clast macrofabric data as a genetic fingerprint in the glacial environment, whilst recognizing that these fabrics may still be useful as indicators of ice flow directions and relative strain in sediments.

Bennett, M.R., Waller, R.I., Glasser, N.F., Hambrey, M.J. and Huddart, D. 1999. Glacigenic clast fabrics: Genetic fingerprint or wishful thinking? *Journal of Quaternary Science*, 14(2), 125–135.

Clast macrofabrics can be recorded as 2D (declination only) or 3D (declination and inclination) data. Suitable sites are those where possible topographic influences on former ice flow are minimal and where fabrics can be recorded at a point far enough down in a unit to minimize the danger of disturbance by soil development or cryoturbation. The chosen site must also be safe, that is not below overhanging cliffs or under loose boulders. Remove all magnetic equipment from the immediate vicinity of the compass-clinometer, as these may affect readings. In some glacierized areas it may prove necessary to excavate sections from which to collect data. For example, Benn (1995) described how fabric data were collected from pits dug on the forefield of Breidamerkurjökull, Iceland, to investigate the fabric signature of different tills as well as in the flank of a flute (a streamlined landform). It can take a considerable amount of time to collect a clast fabric, so you should ensure that the site chosen is amenable and offers some protection from the weather. Experienced fieldworkers can often collect a set of 50 measurements in around an hour. This time can be reduced if there are two people, so that one can measure and one can record. In some instances, for example when collecting basal ice fabrics, it can even pay to have three people collecting and recording (Figure 8.14).

8.12.2 Three-dimensional clast macrofabrics

There is little agreement in the literature about how many individual clasts need to be measured to make a clast fabric meaningful. Some workers have

(a)

(b)

Figure 8.14 Collecting clast macrofabric data in the contemporary glacial environment, Svalbard: (a) In this situation two people are collecting fabric measurements while a third person is recording the data. This speeds up data collection considerably (Photograph: J.L. Etienne); and (b) Close-up view of clast macrofabric data collection. In this example, the compass is being used to measure the inclination of the *a*-axis of the clast (Photograph: N.F. Glasser)

based their studies on sets of 25 measurements (e.g. Domack and Lawson, 1985), whilst most workers advocate 50–100 clasts (see Bennett *et al.*, 1999 for further discussion). Three-dimensional clast macrofabric data are normally collected for the *a*-axis of samples of 50 prolate clasts. This means selecting only those clasts with an *a*-axis to *b*-axis ratio greater than 3–2. Analyses of clast macrofabric data sets show that there are systematic variations in fabric strength with clast size. Clasts with an *a*-axis longer than 20–30 mm show a much stronger preference for parallel orientation than do shorter clasts (Kjaer and Krüger, 1998). For this reason, it is important to be consistent in the selection of clasts for measurement, ensuring that they are all within a given size range. Some workers (e.g. Benn, 1995) have therefore selected only those clasts in the size range of 30–125 mm. There are also systematic differences in till fabrics over short distances (e.g. Young, 1969), so clasts should be chosen from within a single, well-defined area (say 0.5 m^2) rather than from across a wide area.

Clast macrofabrics are collected as follows:

Step 1   Using a compass, measure the declination (the compass orientation) of the long axis (*a*-axis) of the clast to the nearest 1° (a number between 0 and 360°). Ensure that the compass is held horizontally so that the needle does not stick on its housing. Declination should be recorded in the direction of inclination (dip angle) of the clast. For example, a clast that is aligned exactly east–west and whose *a*-axis is inclined to the west (i.e. dips downwards towards the west) has a declination of 270° not 90°. Some workers determine the declination by first removing individual clasts from the sediment, inserting a wooden rod or pencil into the hollow that is left in the matrix, and then measuring the orientation of the rod or pencil.

Step 2   The inclination (dip angle) of the long axis of the clast from the horizontal (a number between 0 and 90°) is then determined using the clinometer on a compass-clinometer. Again, this instrument should be read to the nearest 1°. The figure recorded should be the average of the inclination of the upper and lower surfaces of the clast in the direction parallel to the declination of that axis (Kirby, 1969). Once these two measurements have been made on a clast, the clast should be discarded. It is good practice to be consistent in the order in which measurements are recorded. By convention the compass orientation is followed by the dip angle. For example, a clast with its long axis oriented at 70° and dipping at an angle of 14° would be recorded as 70°/14°. Note that measurements are recorded as a three-digit number (compass orientation) and a two-digit number (dip angle) to avoid later confusion of the two measurements.

Step 3   Repeat steps (1) and (2) until a representative sample (normally 50 clasts) has been collected.

Step 4   On return from the field, the data can be processed into a format that allows comparison with other published studies (see section 8.12.4).

Other measurements can also be collected in conjunction with simple 3D clast macrofabrics to increase the statistical significance and reliability of the data set. Benn (1995) described how data were collected on the forefield of Breidamerkurjökull, Iceland, for not only the declination and inclination of clast *a*-axis, but also for: (1) poles to *a–b* planes, (2) poles to erosional facets, and (3) the orientation of clasts with stoss-and-lee morphologies.

8.12.3   Two-dimensional clast macrofabrics

For 2D clast macrofabrics, the declination (but not the inclination) of the *a*-axis is recorded. The procedure is outlined in section 8.12.2 (step 1). Again, it is best practice to remove each clast from the section and discard it after it has been measured to avoid repeat measurements. The advantage of using 2D clast macrofabrics over 3D fabrics is that strong fabrics can be established quickly, since they are much less time-consuming to perform in the field. The disadvantage is that any consistent dip orientations in otherwise apparently random samples will be missed.

8.12.4   Clast macrofabric data presentation

The data collected can be plotted as 2D fabrics on **rose diagrams** (declination only) and 3D fabrics (declination and inclination) on a **stereographic projection** (a **stereonet**) (Figure 8.15). Rose diagrams are effectively circular histograms, which summarize clast orientation in groups or intervals (normally 5 or 10° intervals). No information on dip is presented on a 2D rose diagram. Stereonets are a more powerful method of displaying 3D data. The orientations of lines and planes are plotted relative to the centre of a sphere, called the projection sphere. The projection sphere is divided in half by a horizontal plane called the projection plane. Compass directions are indicated on the projection plane. All measurements are made relative to horizontal, so only the lower hemisphere of the projection sphere is used. The two most common stereographic projections are the **equal-angular** or **Wulff stereonet**, and the **equal-area** or **Schmidt stereonet**. Of the two, the Schmidt equal-area stereonet is more commonly used for plotting clast fabric data because this method preserves size and is therefore visually more pleasing than the equal-angular stereonet.

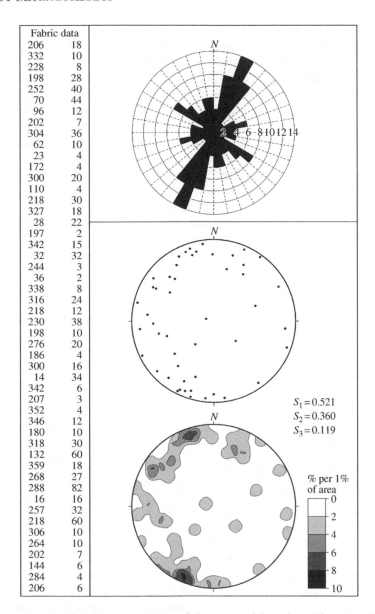

Fabric data	
206	18
332	10
228	8
198	28
252	40
70	44
96	12
202	7
304	36
62	10
23	4
172	4
300	20
110	4
218	30
327	18
28	22
197	2
342	15
32	32
244	3
36	2
338	8
316	24
218	12
230	38
198	10
276	20
186	4
300	16
14	34
342	6
207	3
352	4
346	12
180	10
318	30
132	60
359	18
268	27
288	82
16	16
257	32
218	60
306	10
264	10
202	7
144	6
284	4
206	6

$S_1 = 0.521$
$S_2 = 0.360$
$S_3 = 0.119$

% per 1% of area
0
2
4
6
8
10

Figure 8.15 Graphical representation of clast macrofabric data. The original data, collected from englacial debris in Soler Glacier in Patagonia, are listed in the left-hand column and also shown plotted as a 2D rose diagram (upper stereonet) and as 3D point and contour data (bottom two stereonets)

The procedure for plotting lines and planes on the Schmidt and Wulff stereonets is identical. The *a*-axis orientation of the clast (declination) is recorded by measuring clockwise around the outer edge of the sphere, whilst the dip angle of the clast (inclination) is recorded on the concentric circles that make up the stereonet. A vertical clast plots as a single point at the exact centre of the stereonet. A plunging clast plots as a single point at a location determined by its angle of plunge: a steeply plunging clast plots close to the centre of the stereonet and a shallower plunging clast plots closer to the edge of the stereonet.

Stereonet plots are constructed either by hand (plotting the information on tracing paper laid over a stereonet template) or using existing software packages (entering the data into a computer program, which automatically plots the data onto a stereonet). A number of commercially available software programs exist for this purpose, such as the 'Rose' and 'Stereo' facilities provided as part of the RockWare Utilities package (http://www.mathstat.com.au/products/science/rockware_utilities.htm). In addition, the GEOrient (http://www.earthsciences.uq.edu.au/~rodh/software/) and StereoNett (http://homepage.ruhr-uni-bochum.de/Johannes.P.Duyster/stereo/stereoload.htm) programs can be downloaded from the Internet and used to plot rose diagrams and stereonet projections.

Using a software package to analyse fabric data has two major advantages. The first advantage is that most packages allow you to contour the data on the stereonet in addition to simply displaying the data as points, which makes it easier to visualize the strength or weakness of the fabric. A second advantage is that some of these packages automatically calculate statistical parameters based on the data. The most commonly calculated statistical parameters are the eigenvalues and the eigenvectors. These are essentially measures of the strength and orientation of the directional properties of the data (Benn, 1994; Benn and Ringrose, 2001). Sets of observations are resolved into three mutually orthogonal eigenvectors (V_1, V_2 and V_3). The principal eigenvector, V_1, is parallel to the axis of maximum clustering in the data and V_3 is normal to the preferred plane of the fabric. The degree of clustering about the respective eigenvectors is given by the eigenvalues S_1, S_2 and S_3. The relative magnitudes of S_1, S_2 and S_3 reflect the shape of the fabric data. These eigenvalues can be used to calculate an isotropy index $(I = S_3/S_1)$ and an elongation index $(E = 1 - (S_2/S_1))$. The isotropy index measures the similarity of a fabric to a uniform distribution, and varies between 0 (all observations confined to a single plane) and 1 (perfect isotropy). The elongation index measures the preferred orientation of a fabric in the V_1/V_2 plane, and varies between 0 (no preferred orientation) and 1 (a perfect preferred orientation with all observations parallel) (Benn, 1994). Interpretations of clast macrofabric data are then made possible by comparing the stereonets and calculated statistical parameters with other published studies (see Table 8.1 for further details).

8.13   CLAST MICROFABRICS AND MICROSTRUCTURAL DESCRIPTION

These two techniques can be used in the laboratory to analyse thin sections from samples originally collected in the field or cross-sections of cores. Sample collection and transportation for making thin sections is described in section 8.10.2. Clast microfabric analysis involves measuring the orientation of elongated sand-sized grains under the microscope. The measurement procedure and plotting of data are similar to those described above for clast macrofabrics. Microstructural description (sometimes also referred to as micromorphology) is the systematic description of features identified at low magnification (between 10x and 100x). In this method, thin sections are viewed under a petrological microscope using plane and cross-polarized light. Van der Meer (1996) has provided a good description of this technique, and Van der Meer (1993) and Carr *et al.* (2000) have defined the terms used in the description of microstructures. One of the main applications of this technique in recent years has been discriminating between subglacial and glaciomarine sediments (Van der Meer, 1993; Hart and Roberts, 1994; Carr, 2001; Table 8.8).

Table 8.8   Suggested criteria for discriminating between subglacial and glaciomarine sediments using microstructural analysis (modified from Carr, 2001)

	Subglacial sediments	Glaciomarine sediments
General features	Crushed quartz grains	Coarse, winnowed plasma texture
	Edge rounding on grains	Dropstone structures
	'Non-winnowed' matrix	*in situ* marine microfossils
		Graded bedding structures
		Absence of edge rounding on grains
		No plasmic fabric developed
Planar 'brittle' deformation	Grain lineations	
	Symmetrical pressure shadows	
	Augen-shaped intraclasts	
	Unistrial and masepic fabrics	
Rotational 'ductile' deformation	Rotation features (e.g. circular alignments, galaxy structures, asymmetric pressure shadows)	
	Rounded soft sediment pebbles	
	Pervasive masepic fabrics	
	Non-associated planar features	
Microfabric characteristics	Unidirectional vertical microfabric	Distinct vertical or near-vertical microfabrics in the vertical plane

8.14   CLAST MESOFABRICS

Clast mesofabrics are determined in the laboratory on fine gravel clasts (normally in the size range of 2–10 mm) in orientated blocks or cores of sediment using the same basic procedure as clast macrofabrics. This scale of fabric analysis is adopted when time in the field is at a premium, when sites are only temporarily exposed, or when a sample is only available at this scale, for example, from a borehole. Sample collection and transportation are described in section 8.10.2.

8.15   LABORATORY ANALYSIS

A large number of laboratory tests are possible on samples collected in the field (Table 8.1). The main aim of these tests is to provide additional information on the depositional environment (if unknown) of a sedimentary facies, or to provide extra information on the properties of the sediment in order to gain insight into its behaviour under certain conditions. Laboratory analyses commonly undertaken by glacial geomorphologists divide the behaviours conveniently into those concerned with **physical** properties and those concerned with **chemical** properties (Goudie, 1990).

8.15.1   The physical properties of sediment

This includes particle-size analysis, the surface textures of particles in a sediment using SEM, measurements of the mass, density, porosity and void ratio, the permeability and infiltration capacity, moisture content, bulk fabric and structure, X-ray radiography and microstructural analyses (Table 8.1). Particle-size distributions of matrix samples are determined using a combination of wet- or dry- (mechanical) sieving techniques to isolate the gravel and sand fractions, together with sedimentation techniques such as pipette or measurement by laser granulometer, X-ray SediGraph and Coulter counter for the finer silt and clay fractions (McManus, 1988). Particle-size results are generally expressed either as a histogram showing weight or number in size class, or as a cumulative frequency *versus* size or as double logarithmic plots following the method outlined by Hooke and Iverson (1995). The advantage of cumulative frequency plots is that grain-size envelopes can be identified for groups of samples and that more than one facies can be displayed on a single graph. Ternary diagrams can also be used to display sand–silt–clay ratios, which allow facies envelopes to be drawn around individual data points (Figure 8.16). Recent studies have also used

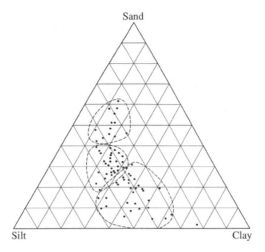

Figure 8.16 Particle-size data for 76 till samples from three tills in northeastern Ohio plotted on a sand–silt–clay ternary diagram (modified from Karrow (1976) with permission from the Royal Society of Canada). The dashed lines enclose three different populations, inferred to represent three different tills

mineral magnetic analyses for lithostratigraphic correlation and differentiation of glacigenic sediments (Walden *et al.*, 1987, 1996, 1999). A further series of laboratory tests are designed to examine the strength of materials (their geotechnical properties). These are determined by their particle-size distribution and mineralogy, the nature of the succession within which they occur, the stress history that they have undergone, and the presence, frequency and orientation of joint planes within them (Boulton, 1976). Laboratory tests of these properties include shear strength measurements, penetrometer tests and oedometer tests (Table 8.1). In some cases these tests can be used to differentiate between glacigenic sediments (Figure 8.17).

8.15.2   The chemical properties of sediment

These are clues to the chemistry of the rock from which the sediment is derived or the processes acting in the substrate upon which that sediment is developed. Laboratory techniques exist to determine the concentration of a number of chemical components including sodium, potassium, calcium, calcium carbonate-equivalent and sulphate, together with the complex chemical grouping that comes under the heading of organic matter. Each of these properties provides information on the source of the material (e.g. its position within the lithosphere or atmosphere, its organic content), the

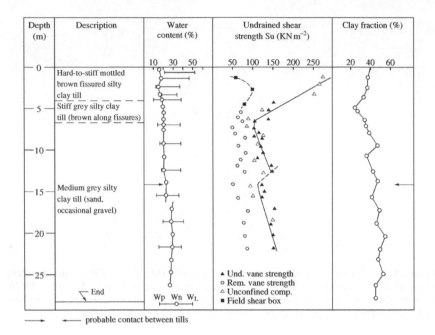

Figure 8.17 Summary diagram, showing the variation of engineering properties with depth in a 29-m core from Sarnia, Ontario, Canada. Horizontal arrows mark the inferred contact between two different tills (modified from Quigley and Ogunbadejo (1976) with permission from The Royal Society of Canada)

extent of weathering and pedogenesis, and the movement of chemicals through the material. These techniques can be grouped into studies of provenance, environmental conditions at the time of formation, and extent of weathering and pedogenesis. Laboratory tests include the electrometric determination of pH and conductivity, the determination of calcium carbonate and sulphate content, the determination of acid-extractable sodium, potassium and calcium content by flame photometry, atomic absorption spectrophotometry and X-ray diffraction of clay minerals for provenance studies (e.g. Ehrmann *et al.*, 2003).

8.16   INTERPRETING THE ENVIRONMENT OF DEPOSITION OF SEDIMENTS

By integrating your results it is possible to deduce the most likely mode of origin of the sediments examined. It is good practice to list the reasons (i.e. diagnostic characteristics) for your selection (Figure 8.18). Inference is

Evidence for terrestrial glaciation

Abraded surfaces
> Grooved, striated and/or polished rock surfaces
> Crescentic gouges, friction cracks, chattermarks
> Striated boulder pavements and bullet-shaped clasts
> Roches moutonnées
> Nye channels, 'p-forms' and 's-forms'

Diamict and muddy boulder gravel beds with:
> Little or no stratification
> Irregular thickness, typically several metres (but may be up to tens of metres)
> Preferred clast orientation
> Lenses of sand/gravel (glaciofluvial)
> Shear structures parallel to depositional surface

Depositional landforms
Sediments occur in close association with landforms such as moraines or eskers

Association with extensive sheets of sand and gravel
Inferred glaciofluvial origin

Evidence of glaciomarine/glaciolacustrine deposition

Massive or stratified diamicts, often tens to hundreds of metres thick, with gradational boundaries
Dropstones in stratified units
Random clast fabric
Slight sorting or winnowing at top of beds, leaving lag deposits
Association with *in situ* fossils
Association with rhythmites (varves in lakes; tidal rhythmites in fjords)
Association with re-sedimented deposits (subaquatic gravity flows)

Evidence common to both environments

Clasts
Range in size up to a few metres
Range in lithologies, reflecting the terrain over which ice flowed
Constant mix of clasts over a wide area
Range in shape from very angular to rounded (e.g. sub-angular to sub-rounded predominant in
    subglacial debris), sometimes influenced by prior transport history
Surface features include facets and striations from transport within zone of basal traction, and
    flat-iron or bullet-nosed shapes from persistent glacial transport
Fragile clasts survive well
Calcareous crusts in some deposits

Grains
Surface of quartz grains show typical fracture characteristics
Surface of quartz grains show chattermarks
Unstable (ferromagnesian) minerals remain relatively unweathered

Other evidence of cold climate

Ice wedge casts and cryoturbation structures
Sorted stone circles, polygons and stripes
Solifluction lobes
Association with loess

Figure 8.18 Principal criteria that can be used to support a glacial origin for diamictons. Diagram modified from Hambrey (1999), reproduced by permission of Terra Antartica Publications

based on interpretation of the sediment characteristics (Box 8.5). You need to make inferences about process (e.g. energy regime, distance of transport) in order to work out the environment(s) of deposition. Remember that some of the characteristics are the result of processes that may have occurred prior to deposition (e.g. transport). The best way to arrive at interpretations is by comparison with previous studies in the published literature, particularly process-related or modern analogue studies (Table 8.1). You may not always be able to find comparable studies in the literature and should not be afraid to make rational interpretations based upon the sedimentology,

---

Box 8.5   Interpreting debris transport pathways in an Icelandic glacier from till properties

Kjaer (1999) presented the results of a study of the sediments on the forefield of Sléttjökull, a temperate outlet glacier draining the Mýrdalsjökull ice cap in Iceland. His study was designed to examine the effect of subglacial transport on sediment properties at a site where ice-flow direction and velocity, glacier thermal regime, sediment transport distance and source are all known. Recent recession of the glacier front has left a well-developed series of ice-marginal moraines, fluted ground moraine and minor outwash plains on the forefield of the glacier. Immediately in front of the glacier, revealed only by this frontal recession, is a large bedrock knob composed of palagonite (a basaltic tuff) whilst the surrounding bedrock is basalt. The different lithologies mean that the evolution of sedimentary facies can be established by examining sediment properties at various intervals down-ice of the palagonite bedrock knob. Kjaer sampled sediments at increasing distances from the bedrock knob and undertook analyses of grain-size distribution, geochemistry, mineral magnetic properties and clast morphology (clast shape, roundness and striations on clasts) (Figure B8.5). The distinctive lithologies were used to gain insight into how these properties varied with distance from the glacier (a surrogate for former transport distances). One of the conclusions of his study is that very short transport distances (only around 250 m) are required for the different sedimentological properties to reach a 'mature' state. Thus within 250 m of transport distance significant changes occur in the grain-size distribution, geochemistry, mineral magnetic properties and clast morphology. Beyond this distance, variations suggest that till properties are a function of parameters other than transport distance. This study illustrates neatly how a multi-methodological study can offer new insight into the interactions between a glacier and its substratum.

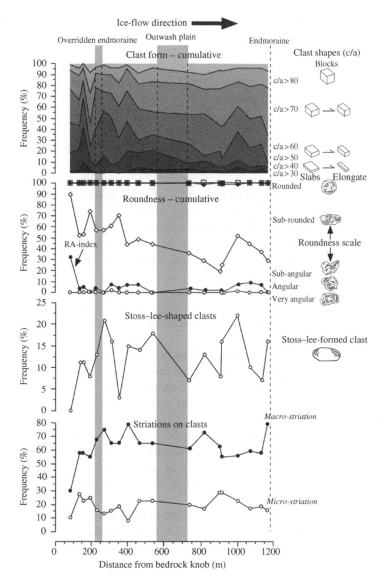

Figure B8.5 Variations in clast morphology (clast shape, clast roundness, stoss–lee-shaped clasts and striations on clast surfaces) with distance along a transport path on the forefield of Sléttjökull, Iceland, according to Kjaer (1999). Shaded areas indicate major geomorphological features serving as guidelines. Reprinted from Kjaer (1999) with permission from Elsevier

Kjaer, K.H. 1999. Mode of subglacial transport deduced from till properties: Mýrdalsjökull, Iceland. *Sedimentary Geology*, **128**, 271–292.

structure and other attributes of the sediments examined. Some useful environmental indicators are as follows:

- The particle-size distribution is arguably the most important feature. Tills commonly have coarse particle-size distributions and are poorly sorted. A well-sorted sediment often represents selective separation of fractions due to either (i) deposition of coarse material in rivers because the fall in flow velocity means it is unable to transport large particles (thus fine material is transported downstream and deposited in low-energy environments, such as lakes or floodplains) or (ii) preferential sorting by waves and longshore drift on a beach (very poorly sorted sediments are often indicative of basal glacial transport, which tends to mix material from a variety of sources).
- *Texture*: Predominantly coarse materials are indicative of beach or (glacio)fluvial gravels (high-energy environments). Predominantly fine materials are indicative of floodplain alluvium or lake sediments (low-energy environments).
- *Clast roundness*: Beach gravels are commonly well rounded; river gravels are less well rounded; basal glacial material is generally sub-angular to sub-rounded; scree material and supraglacial debris are typically angular or very angular.
- *Surface textures on clasts*: The widespread occurrence of features such as striations, grooves, chips and facets on clasts almost certainly indicates basal glacial transport. Bullet-shaped clasts are also indicative of basal glacial transport. Percussion marks can occur on beach and fluvial gravels.
- *Colour*: Different colours may reflect different sources of material (i.e. from different lithologies). Colour may also be the result of weathering processes, the development of a soil, or the deposition of iron (red, brown) or manganese (black) in response to changing water table conditions.
- *Organic content*: A high organic content is usually indicative of floodplain alluvium or lake sediments. Glacigenic sediments are normally accompanied by little or no organic material.

8.17 PRESENTATION OF SEDIMENTOLOGICAL DATA

The appropriate presentation of sedimentological data is very important. Data must be presented in a manner that enables an impartial reader to visualize the sediments present and to understand how interpretations were drawn from the data collected. Krüger and Kjaer (1999) have suggested one method of doing this, applicable to both contemporary and ancient glacial environments. Their **data chart** is essentially a field log combined with post-fieldwork data analysis (e.g. clast macrofabric data, information on clast shape and roundness, percentage of striated clasts), together with an overall genetic interpretation of the sediments present. This data chart provides a neat method of displaying the data collected in a standard format and it can be extended or modified to suit individual circumstances (i.e. columns can be removed or added). Another effective method for displaying sedimentological data is to place field logs alongside digital photographs of the sediment. This allows the reader to see both a photograph of the original sediment, and its description and interpretation.

8.18   STUDENT PROJECTS

There are a number of possibilities for research projects that use the techniques of glacial sedimentology. The techniques outlined in this chapter can be applied equally to both contemporary glacial environments and previously glaciated areas. Some possibilities are as follows:

1. Most sedimentological studies require an overall aim or purpose. In former glacial environments (glaciated environments) the purpose is normally to use a sedimentary succession to attempt a reconstruction of events or to interpret the former depositional environment. A study might therefore use sedimentological techniques to examine the internal composition of a section in a single glacial landform (e.g. a drumlin or an esker) in order to elucidate its mode of formation, or a study might involve the description of several sedimentary exposures in close proximity to attempt a 'landsystems', or regional reconstruction of events.

2. The most powerful studies are those that link sediments to landforms to produce landform–sediment associations. For example, in the contemporary glacial environment, one might study the internal composition of flutes in front of a receding glacier and attempt to link the sediment types and their properties to styles of glacier motion.

3. Some studies use a variety of techniques to test the origin of sedimentary successions (e.g. field description of sediments combined with field sample collection and post-fieldwork laboratory analysis of micro-structures or some other chemical or physical property of the sediment).

4. Field studies might involve combining techniques to look at the interaction of variables, for example combining clast macrofabrics with studies of surface striation on those clasts to study styles of deformation.

5. Using clast lithological studies to determine the origin of clasts in glacigenic sediments (e.g. their source areas and dilution). A typical question that might be answered in this way is whether or not clast lithology varies horizontally or vertically through a particular section.

9

Mapping glaciers and glacial landforms

9.1 AIM

The aim of this chapter is to describe the techniques that can be employed in the mapping of glaciers and glacial landforms in both terrestrial and aquatic (marine and lacustrine) settings. The motivation for such studies is that former glaciers and glacial landforms provide insight into the operation of glaciological processes that cannot be studied at modern glaciers. We describe the techniques that are commonly used in both modern (*glacierized*) and former (*glaciated*) glacial environments. Techniques covered include the extraction of data from topographic maps, the analysis of remotely sensed images for surface landform identification and mapping, geomorphological mapping and techniques used in structural glaciological mapping. We also describe the most commonly used techniques for 3D mapping of sediments and landforms including GPR and electrical resistivity surveys, the techniques employed in marine geophysics and seismic profiling, and the mapping and measurement of landform change over time. This chapter outlines the key stages in a research project on the assumption that most workers will follow a progression from the initial choice of a field area, through the mapping exercise itself, to completion of a final map.

9.2 GENERAL CONSIDERATIONS

Planning and preparation are essential for the successful mapping of glaciers and glacial geomorphological features. Schytt (1966) described the overall

Field Techniques in Glaciology and Glacial Geomorphology Bryn Hubbard and Neil Glasser
© 2005 John Wiley & Sons, Ltd

purpose of glacier mapping, and Blachut and Müller (1966) discussed some of the practical problems and solutions in constructing glacier maps using a variety of examples, including the White Glacier in the Canadian Arctic, the Salmon Glacier in the Rocky Mountains and the Khumbu Glacier in the Himalayas. In addition, an entire issue of the *Canadian Journal of Earth Sciences* (Volume 3, Number 6, published 1966) is devoted to this theme, reporting the results of discussions at a symposium on glacier mapping. For those embarking on glacial geomorphological studies, Demek and Embleton (1978) and Gardiner and Dackombe (1983) provided discussions of the various approaches used in geomorphological mapping. Waters (1958) described the techniques used in morphological mapping.

9.3   AIMS OF THE MAPPING AND THE AREAL EXTENT OF THE MAP

It is important to establish the **aim of the mapping**, as this will determine the areal extent of the map, the scale at which it is mapped and the techniques used in the mapping. In glacial geomorphological studies, for example, mapping efforts are normally restricted to areas only where there are well-developed glacial landforms and sediments. Mapping will concentrate on these landforms and their relationship with periglacial, nival and slope forms rather than all aspects of geomorphology. However, many of the recent advances in glacial geomorphology and palaeoglaciology have come from mapping initiatives in areas where the evidence of glaciation is minimal (Kleman, 1994; Kleman and Hattestrand, 1999). In these areas, the objective may be to map not only glacial landforms, but also their association with non-glacial landforms such as tors, blockfields and plateau remnants (Kleman and Stroeven, 1997; Stroeven *et al.*, 2002).

The **scale of investigation** determines the mapping techniques to be used. The two most common mapping techniques used by glaciologists and glacial geomorphologists are as follows:

- *Remote-sensing techniques* (e.g. the interpretation of aerial photographs and satellite images), which are normally applied to areas of $10-1000 \, km^2$ or greater. The term 'remote sensing' includes both satellite imagery and aerial photography. Satellite images are normally used for the mapping of very large areas such as outlet glaciers from ice sheets, whereas aerial photographs are normally used for smaller valley glaciers.
- *Field surveying and field mapping* (e.g. geomorphological mapping, geophysical surveys), which are generally applied to areas less than $10 \, km^2$. This category also includes investigations at the scale of individual landforms, which are applied to areas of $1-100 \, m^2$.

9.4   DESK-BASED STUDIES

Once the aim of the project has been established, most investigations start with a **desk-based study**. At a bare minimum this normally includes a literature search on the area and the analysis of any existing topographic or thematic maps. A preliminary glacier map or geomorphological map based on the interpretation of aerial photographs and satellite imagery may also be produced at this stage.

9.4.1   Literature searches

Literature searches may be carried out quickly and efficiently using the Internet and other electronic literature databases. These allow keyword searches to be made using title or subject keywords, the names of specific geographical areas, authors' names and journal titles.

9.4.2   Analysis of published maps

Large amounts of useful information can often be obtained from **published maps** and **atlases,** in both paper and digital form. These documents show the current extent of glacier ice, and historical maps and atlases can be used to document changes in spatial extent over time. Maps can be used to determine the elevation, orientation, dimension and volume of contemporary glaciers and associated landforms, as well as to estimate basic parameters for glaciers such as the equilibrium line altitude (ELA) using a variety of different methods (Leonard and Fountain, 2003). One popular method is to calculate the **accumulation-area ratio** (AAR), which is the ratio of the accumulation area to the total area of the glacier (e.g. Aa, 1996). AAR values for contemporary glaciers are commonly between 0.5 and 0.8, and a ratio of 0.6 ± 0.05 is considered a characteristic value for valley glaciers in steady state. The ELA is calculated by estimating from contour lines on maps the total area of a glacier in the accumulation and ablation areas. The approximate position of the ELA can also be estimated by finding the 'inflection point' in the contours on a glacier surface, since glacier surfaces are commonly convex below the ELA and concave above it.

9.4.3   Morphometric analysis

Morphometric analysis is a method of extracting information from published maps. King (1966) provides an overview of the types of information that can be extracted in this way. Contour maps are widely used in morphometric studies, where the aim is to gain an understanding of the evolution of landforms through quantitative description. Studies of the morphometry of glacial valleys (Box 9.1) provide a good example of this technique. Contour maps can be used to construct cross-sectional or long profiles

---

Box 9.1   The morphometry of glacial valleys in the Tien Shan, China

Over the last two decades a number of authors have studied the cross-sectional morphology of glacial valleys, dispelling the myth that glacial valleys are 'U'-shaped and suggesting that the cross-profile of glacial valleys is better described by power laws ($y = ax^b$) and quadratic equations ($y = a + bx + cx^2$), depending on the glaciological and geological setting of the valleys in question. However, these studies have traditionally relied on measurements from single cross-sections within individual valleys, rather than exploring the longitudinal variation in cross-profiles along valleys. Yingkui *et al.* (2001) attempted to quantify these along-valley changes using the morphology of glacial valleys in the Tien Shan mountains of China. Fifty-six cross-sections of troughs were compiled from 1:50 000 topographic maps with 20 m contour intervals. On each cross-section 10–15 data points were obtained. The cross-sections were located so as to follow systematically changes in the morphology of valleys both with and without tributaries. The authors used a new model, the variable width/depth ratio (VWDR) model, to analyse the data collected. In the VWDR model, the cross-sectional shape of a valley is expressed as a function of the width/depth ratio at various heights above the valley floor. Two derived parameters, m (a measure of the breadth of the valley floor) and n (a measure of the steepness of the valley sides), are required for this model. Yingkui *et al.* (2001) concluded that variations in both m and n at cross-profiles along individual valleys are related to longitudinal changes in glacier erosion potential along the valleys. These changes in glacier erosion potential are, in turn, a function of location relative to former glacier equilibrium line altitude and location relative to confluences. Yingkui *et al.* (2001) make the important point that their conclusions are based on a field study in only one mountain range, and that investigations from other areas are required to test the universal applicability of their findings.

Yingkui, L., Gengnian, L. and Zhijiu, C. 2001. Longitudinal variations in cross-section morphology along a glacial valley: A case-study from the Tien Shan, China. *Journal of Glaciology*, **47**, 243–250.

of landforms, and to calculate slope angles and gradients from contour line spacing. Other landscape features that have been described quantitatively from maps include the length and gradient of slopes, maximum relief and drainage density. Contour maps and digital elevation models can also be used for area-altitude calculations or **hypsometry** (Strahler, 1952; Brocklehurst and Whipple, 2004). The most useful scales for studies of glaciers and glacial landforms are likely to be 1:50 000, 1:25 000 or 1:10 000 scale maps (or even smaller scale maps in some specialist cases), since these maps contain the most detailed information. King (1966) provided information on how glacial features such as hanging valleys, valley steps and diffluent channels can be recognized on contour maps. One of the classic studies based on map information is that of Chorley (1959), who studied the shapes of drumlins from contour maps. Rose and Letzer (1975) later compared map-derived measurements from contours on 1:25 000 maps with field mapping of drumlin distribution. Map-derived values such as drumlin elongation ratios, tapering values and stoss-point ratios were measured reasonably accurately from maps but the size, spacing and densities of drumlins were found to be consistently inaccurate. The number of individual drumlins identified from map analysis was found at best to be 50% of all drumlins identified during fieldwork. This shows that, although maps can yield substantial amounts of information, there is sometimes no substitute for fieldwork.

Troeh (1965) attempted to describe the **shape of landforms** on maps with 3D equations. The formula that best describes landforms is

$$Z = P + SR + LR^2$$

Where,
Z is elevation at any point;
P is the elevation at the centre of the feature described by the equation;
S is the slope gradient of the surface at that centre;
L is half the rate of change of the slope gradient with radial distance;
R is radial distance from the centre of the feature.

Although Troeh (1965) used the examples of alluvial cones and pediments, this type of analysis could easily be applied to glacial landforms.

Information derived from contour maps can also provide the basis of regional studies. For example, Peterson and Robinson (1969) and Robinson *et al.* (1971) used information derived from topographic maps to produce trend surface maps of cirque altitudes across Tasmania and Scotland.

9.4.4 Digital map data

Many of the tasks involved in extracting this information are made much easier using digital information in a **Geographical Information System (GIS)**.

A GIS can store and manipulate large quantities of data, and allows the automation of simple tasks such as the calculation of areas and slope angles. A GIS also allows data to be input as individual layers, making the task of synthesizing data from numerous sources much easier. Recent advances in GIS make quite sophisticated analyses possible. For example, Etzelmüller and Björnsson (2000) described the potential for combining map-based data with glaciological theory in a GIS. Using the Vatnajökull ice cap (Iceland) as an example, they showed how digital photogrammetry can be combined with GPR and GPS data to create **Digital Elevation Models (DEMs)** of the subglacial relief and the surface topography of the ice cap. They then used a GIS to compute basal shear stress and flow velocity distributions for Vatnajökull, to predict the distribution of major subglacial meltwater channels and to calculate the potential annual runoff from the ice cap. Brown *et al.* (1998) used geomorphological measures such as elevation, relative relief, roughness and slope gradient derived from DEMs to distinguish between different types of **glacial landscape** in southeastern Michigan, USA. This approach rests on the assumption that a set of measurements describing topographic form can be used to distinguish between geomorphologically distinct landscapes. In this case, a supervised classification scheme was successfully used to distinguish between four planar relief assemblages (lake plain, outwash plain, outwash spillway and till plain) and two types of hilly landscape (moraine and stagnation moraine).

9.5 REMOTELY SENSED DATA

9.5.1 Mapping from remotely sensed images: General principles

Because glacial environments are remote, it is difficult to monitor glaciers on a regular basis, and field-based measurements are sometimes impossible. On remote glaciers, glaciologists and glacial geomorphologists rely on **remotely sensed images** to derive information about glacier characteristics. Remotely sensed images are those taken from above the Earth's surface, from either airborne platforms (e.g. cameras mounted on aeroplanes or helicopters) or spaceborne platforms (e.g. satellites). Hall and Martinec (1985), Jensen (1996), Lillesand and Kiefer (2000) and Rees (2001) give further information on the physical principles of remote sensing, and Gao and Liu (2001) and König *et al.* (2001a) provide recent reviews of the applications of remote sensing to glaciology and glacial geomorphology. Many field-based projects start with the production of a map from aerial photography or a satellite image, which is later refined by fieldwork on the ground.

Table 9.1 Optical remote-sensing techniques commonly used in glaciological research (modified from Gao and Liu, 2001)

Platform	Instrument	Sensor/band	Spatial resolution	Swath width	Revisit period
Airplane	Camera	Photographs	Variable	Variable	Variable
Airplane	Scanner	AVIRIS			
Satellite	Landsat	RBV	79 m	98 km	16 days
		MSS	79 m		
		TM	30 m		
		ETM	15 m		
Satellite	SPOT	PAN	10 m	60 km	26 days
		XL	20 m		
Satellite	NOAA	AVHRR	1.1 km	2400 km	12 hours

A number of airborne and spaceborne sensors can provide useful information on the glacial environments. This section is concerned principally with **optical sensors** (as opposed to microwave/radar or laser sensors) because the images that they collect can be interpreted visually (Table 9.1). The choice of sensor depends upon the requirement for spatial (in both 2D and 3D), temporal and spectral information. The **spatial resolution** of an observing sensor refers to the distance between the nearest objects that can be resolved and is given in units of length (e.g. metres). Typically, airborne sensors have high spatial resolution at the expense of a wide image swath, whereas satellite-acquired images tend to provide a broader overview, but in less detail. The moderate (10–100 m) spatial resolution provided by the Landsat TM sensor (25 m), Landsat ETM (15 m) and SPOT HRV and HRVIR (10 m Pan, 20 m multispectral) are typically used for monitoring regional changes in glacial extent. The finer spatial resolution provided by airborne sensors (such as the Airborne Thematic Mapper) or very high resolution satellite sensors like IKONOS are more appropriate for detailed mapping of glacier surfaces and to detect the variability in reflectance characteristics across a glacier terminus.

9.5.2   Aerial photography

Aerial photography is a particularly powerful tool for landform identification, landform mapping, for mapping ice surface structures and for making glacier restitutions (reconstruction of a previous state or position). The advantage of aerial photographs is that they are normally recorded in visible light, and therefore show glaciers and landforms in a form that is familiar to the human eye (Figure 9.1). Despite recent advances in satellite remote sensing, aerial photography remains the highest spatial resolution data set

(a)

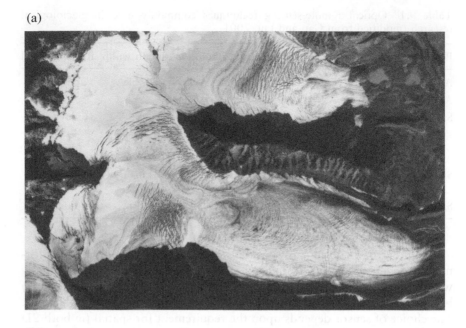

(b)

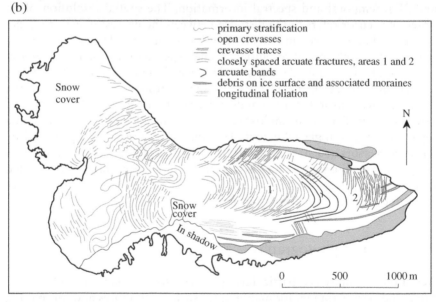

Figure 9.1 (a) Vertical black and white aerial photograph of Storglaciären, northern Sweden, taken in late summer (18 August 1980) when the snow line has receded up-glacier to reveal the bare ice surface; and (b) structural glaciological interpretation of the surface features visible on the aerial photograph. Diagram modified from Glasser *et al.* (2003) with permission of Blackwell Publishing Ltd

available. Repeat surveys in many glacierized regions mean that multi-temporal coverage exists, representing an important archive of glacier extent and variation. The spatial resolution of aerial photography is sufficient to provide detail on land surface properties that cannot be obtained by satellite imagery, for example detail such as glacier surface texture, glacier structures and interactions with the surrounding environment. Aerial photographs are particularly useful for mapping in remote or inaccessible areas (Arnold, 1966; Brandenberger and Bull, 1966; Helk, 1966; Lindner *et al.*, 1985). Skilled interpretation of an aerial photograph is required if it is to be used for landform mapping or for identifying particular structures on a glacier surface. Landforms or features identified on aerial photographs are normally then transferred onto base maps for presentation purposes. Aerial photograph interpretation involves the description of glaciers or landforms from either single photographs (**binocular**) or pairs of overlapping photographs (**stereoscopic**).

Aerial photographs fall into two categories: vertical and oblique (Dickinson, 1979). **Vertical aerial photographs** are taken with the optical axis of the camera perpendicular to the horizontal plane of the ground. **Oblique aerial photographs** are taken with the optical axis of the camera forming an angle of less than 90° with the ground. Aerial photographs are normally taken in a series of parallel strips along predetermined flight lines. These flight lines are organized such that adjacent strips overlap, usually by around 40%. Along each flight line the photographs are taken so that there is also a forward overlap, normally 60%. Overlapping photographs can be used to view the ground stereoscopically (i.e. in three dimensions) using a **stereoscope** or stereoscopic glasses. Stereoscopic vision works on the principle that pairs of pictures of the same area, but acquired from different angles, can be arranged such that the human brain interprets the image as a 3D view of the ground. There are two basic rules that should be followed when looking at aerial photographs under a stereoscope:

1. Each of the photographs should be directly under one of the eyes.
2. Shadows cast by objects in the photographs should always be towards the viewer. If this rule is not followed, a 'pseudoscopic' effect of inverted relief may be produced.

Vertical dimensions are considerably stretched in aerial photographs (a process known as **relief stretching**). This occurs because, although the average distance between the pupils is ∼6 cm, stereopairs are normally exposed from tens of metres apart in the sky.

Notional scales as a representative fraction (RF) of individual photographs may be calculated from the formula:

$$RF = \frac{f}{H}$$

where f is the focal length of the camera and H is the height of the camera above the ground. It is important to remember that the scale of an individual photograph is not uniform as a result of variations in the path of the aircraft, which mean that objects can appear distorted. This is especially the case around the edges of the photograph where the distance from the camera to the ground is greater than at the centre of the photograph.

Until the development of digital soft-copy photogrammetry, elevation data was generated using a stereoplotter, an instrument that supplicates the exact relative position and orientation of the airborne camera at the moment the two images are acquired. By analysing relative displacements brought about by image parallax (the apparent change in relative position of stationary objects caused by a change in viewing position) it is possible to construct a miniature landscape model from the overlap on a stereo pair and compute elevation of a number of photograph points. Stereo pairs of overlapping photographs can be used to construct contour maps using digital photogrammetry (Wolfe, 1983). To do this, pairs of digital images are scanned and analysed using a computer with image-processing capabilities. The images may be from satellite or airborne scanners or conventional aerial photography. Stereo pairs of aerial photographs (showing an overlap of 60% or greater) provide the highest-resolution digital elevation data (usually 5–50 m in the horizontal). The images are displayed on the screen for operator interpretation, enhanced by image processing, or subjected to image correlation in order to form a DEM.

One of the acknowledged difficulties of photogrammetric contouring lies in obtaining stereoscopic vision in featureless areas, such as the accumulation areas of glaciers, where continuous snow cover means that shadows, structural features, debris and dust are absent from the surface. Østrem (1966) presented a solution to this problem, which used 'dye bombs' on the glacier surface as markers before aerial photographs are taken. Two hundred bags of powdered dye, each weighing 3–5 kg, were packed in paper bags and dropped as 'dye bombs' from an aircraft onto the glacier surface. Bags had to be dropped from between 50 and 100 m above the ice surface to achieve the best spread of dye during explosion on impact. The coloured markers left by the dye on the ice surface were clearly visible in the air photographs taken later in the same day. In this way, features were added to the monotonous areas of snow cover to enable stereoscopic representation. Østrem (1966) calculated the total cost of the surface colouring operations to be 25% of the cost of the aerial photography.

Unfortunately there are few set rules for aerial photograph interpretation, although both Anderson *et al.* (1998) and Kirkbride *et al.* (2001) offered some practical guidance on the procedures that ought to be used when mapping glacial landforms. Essentially, there are two methods for mapping: (1) drawing directly onto transparent draughting film overlain on unfolded topographic maps (e.g. 1:50 000 scale contour maps) or (2) mapping onto overlays on individual aerial photographs, which are later geo-referenced and warped to the appropriate National Grid. The first method is most commonly used because it is more time-efficient and generally least expensive. Mapping directly onto topographic maps, however, introduces some loss of spatial precision in locating landform boundaries on maps. Complex, often gradational, boundaries must also be generalizing into pen lines; a 0.3-mm pen line on a 1:50 000 scale map is equivalent to 15 m on the ground.

Aerial photographs are especially useful in mapping glacier fluctuations where a number of photographs of the same location are available at time intervals (Field, 1966; Meier, 1966; Petrie and Price, 1966). Multi-temporal aerial photographs can also be used in glaciology in the mapping of ice-surface structures (see section 9.5.4.4), and mapping of surface morphology for measuring flow velocities (Haefeli, 1966). Aniya *et al.* (2002) described the practicalities of undertaking such an exercise using vertical aerial photographs acquired from a light aircraft with a 6 × 6-cm format camera. Images were captured over the ablation area of Glaciar Soler, an outlet glacier of the Northern Patagonian Icefield, in the summers of 1984, 1986 and 1999. The photographs were taken so that there was more than 50% forward overlap to ensure stereoscopic coverage. Manual interpretation of the aerial photographs included the identification of features on the ice surface such as debris-covered and debris-free ice, medial moraines, band ogives and wave ogives, individual crevasses and crevasse patterns, supraglacial streams and ponds and ice-cored moraines. These authors also measured the spacing of ogives on the glacier surface and used these to calculate average flow velocities of the glacier to compare with their field measurements. Other more complex calculations are also possible from analysis of the movement of surface features such as crevasses. Harper *et al.* (1998) examined the distribution and change of crevasse patterns over time on Worthington Glacier, Alaska. They found that splaying and transverse crevasses are translated by flow from several tens of metres to 100 m, and that they last no more than 1–2 years.

9.5.3   Satellite remote sensing

Satellite remote sensing is most useful for those studies concerned with large spatial areas. The number of satellites orbiting the earth has increased dramatically in recent years and a variety of commercially available imagery now exist. The most commonly used images are Landsat TM (Thematic Mapper) and ETM (Enhanced Thematic Mapper), SPOT (Systeme Probatoire pour l'Observation de la Terre), IKONOS, ERS-SAR (European Remote-sensing Satellite Synthetic Aperture Radar), ASTER (Advanced Spaceborne Thermal Emission and Reflector Radiometer) and NOAA AVHRR (Advanced Very High Resolution Radiometer) scenes. One of the advantages of satellite images is that they provide large swath widths of the earth – a single Landsat TM scene with a spatial resolution of 30 m captures an area equivalent to hundreds of individual aerial photographs. Paul (2002) used Landsat TM imagery to calculate areal change of 235 Austrian glaciers between 1969 and 1992, a task that would have required interpretation of hundreds of aerial photographs. Mapping of landforms from satellite remote sensing is also possible and is now a widely used tool in ice sheet reconstructions. The most common satellite imagery for this purpose has been Landsat, although other sensors such as AVHRR are becoming increasingly important. For example, Fahnestock *et al.* (2000) used a composite AVHRR image to map flow stripes and rifts on the Ross Ice Shelf, Antarctica.

9.5.4   Applications of remote-sensing techniques in glaciology and glacial geomorphology

The applications of remote sensing in glaciology and glacial geomorphology are as follows:

1. creating inventories and mapping glaciers, ice shelves and sea ice;
2. differentiating between snow and ice for the identification of accumulation and ablation areas;
3. measuring mass balance and snowmelt runoff;
4. mapping glacier structures and surface morphology;
5. determining glacier velocities;
6. mapping the distribution of landforms and sediments.

9.5.4.1   Creating inventories and mapping glaciers, ice shelves and sea ice

This category includes the determination of 2D variables such as the spatial extent of glaciers, ice shelves and sea ice as well as the monitoring of glacier evolution and decay (Box 9.2). This may involve the creation of contour maps from either terrestrial photogrammetry or vertical aerial photography. For example, Konecny (1966) tested various methods of volumetric calculations during surveys of a number of glaciers in the Rockies, Ellesmere Island, Alaska and the Yukon and found terrestrial photogrammetric surveys to be particularly valuable in this respect. Andreassen *et al.* (2002) used **geodetic methods**, where the cumulative net balance is calculated from glacier surface elevation changes measured in different year, to study volume changes for seven glaciers in Norway. These authors used digital photogrammetric methods to produce digital terrain models (DTMs) for the

---

Box 9.2 Mapping changes in glacier extent using field and remote-sensing techniques

Ramírez *et al.* (2001) demonstrated how a number of different techniques were combined to monitor changes in the physical extent of Glaciar Chacaltaya in the Cordillera Real, Bolivia. These authors collected information about the evolution of the glacier over the last 60 years using direct (monthly) mass balance measurements across a network of stakes on the glacier surface since 1991, surface topographic measurements of the glacier collected since 1992, stereophotogrammetric restitutions from air photographs taken in 1940, 1963 and 1983, and reconstructions of former glacier extent since its Little Ice Age maximum from the distribution of moraines on the glacier forefield. A GPR survey at a frequency of 50 MHz was also conducted in 1997 in order to map the subglacial topography and to estimate the present ice thickness. The modern topographic and GPR surveys were used to create a reference map of the glacier, from which a digital terrain model was produced. The authors were then able to reconstruct the area and volume of the glacier at various times in the past using the digital terrain model as a base. The overall conclusion of this study is that the glacier has experienced a major recession since the Little Ice Age, losing some 89% of its surface area. If the current rate of shrinkage is maintained, Ramírez *et al.* (2001) estimated that the glacier would completely disappear within 10–15 years.

Ramírez, E., Francou, B., Ribstein, P., Descloitres, M., Guérin, R., Mendoza, J., Gallaire, R., Pouyaud, B. and Jordan, E. 2001. Small glacier disappearing in the tropical Andes: A case study in Bolivia, Glaciar Chacaltaya (16°S). *Journal of Glaciology*, **47**, 187–194.

---

glaciers from pairs of vertical aerial photographs. DTMs were produced from time-separated sets of photographs of the glaciers to monitor changes in surface elevation over periods of between 20 and 30 years. Surface-elevation changes were calculated by subtracting the DTMs on a cell-by-cell basis for each of the glaciers. The change in water equivalent over this time was then obtained by multiplying the difference grid with the density of ice ($900 \, \text{kg m}^{-3}$). These changes in water equivalent were found to be in general agreement with traditional field-based measurements of mass balance on the seven glaciers. Andreassen *et al.* (2002) concluded that the geodetic methods that they employed were equally as reliable as field-based measurements, but considerably less time-consuming and expensive.

Remotely sensed images are now also widely used to create glacier inventories for remote or inaccessible areas (Box 9.3), and to create regional scale inventories (Field, 1966; Kääb *et al.*, 2002; Paul *et al.*, 2002). Dwyer (1995) provides details of how Landsat MSS and TM images were used to map the dynamics of tidewater glaciers in East Greenland. Medium-scale resolution satellite imagery such as this is often sufficient to detect glacier-margin fluctuations while providing the wider spatial overview that

---

Box 9.3 Producing glacier inventories from satellite imagery and aerial photography

Data concerning the extent of glacier cover and glacier variations from the icefields of Patagonia (southern South America) are scarce compared with records from the Northern Hemisphere. This is partly because the region is sparsely inhabited and partly because the icefields are remote, inaccessible and plagued by persistent bad weather. Map coverage of the icefields and surrounding area is also poor. As a result, as recently as 1996 no detailed inventory of the Southern Patagonian Icefield (13 000 km^2) and its outlet glaciers existed. Aniya *et al.* (1996) published the first such inventory based on interpretation of the extent of ice cover on satellite images and aerial photography. They created a mosaic of the Southern Patagonian Icefield from Landsat TM (Thematic Mapper) scenes taken on 14 January 1986. Using the satellite images in combination with the aerial photographs, Aniya *et al.* (1996) were able to identify the locations of the glacier divides on the Southern Patagonian Icefield and to demarcate individual glacier drainage basins. As a result they recognized a total of 48 outlet glaciers draining the Icefield. A supervised classification of the Landsat TM images, using bands 1 (0.45–0.52 μm), 4 (0.76–0.90 μm) and 5 (1.55–1.75 μm), allowed them to discriminate between areas of snow and ice across the Icefield and therefore find the location of the transient snowline on each glacier. The transient snowline was used as a surrogate for the equilibrium line altitude in order to calculate the extent of the accumulation and ablation areas of each glacier. The combination of satellite imagery and aerial photography, together with published topographic maps, allowed Aniya *et al.* (1996) to compile information for each outlet glacier on:

1. the location of the snout in latitude and longitude;
2. overall length in km;
3. total area in square km;
4. accumulation area (km^2) and its aspect (compass orientation);

5. ablation area (km²);
6. AAR, the ratio of the accumulation area to the total area;
7. ELA in m;
8. presence or absence of iceberg calving at the terminus;
9. highest elevation in m;
10. lowest elevation in m;
11. overall relief (highest elevation–lowest elevation).

The use of satellite imagery therefore allowed Aniya *et al.* (1996) to complete the first inventory for an otherwise inaccessible glacier area. This type of inventory is important because it is a benchmark against which future glacier variations can be measured.

Aniya, M., Sato, H., Naruse, R., Skvarca, P. and Casassa, G. 1996. The use of satellite and airborne imagery to inventory outlet glaciers of the Southern Patagonia Icefield, South America. *Photogrammetric Engineering and Remote Sensing*, 62, 1361–1369.

finer resolution photogrammetry cannot. Using multi-temporal Landsat imagery and basic manual digitizing of glacier boundaries, Williams *et al.* (1997) tracked changes in the margin of Vatnajökull, Iceland, over a 19-year time period, revealing recession of 1413 m for one glacier and 1975 m for another. These measurements were within 33 m of actual ground observations and 121 m of extrapolated ground observations, respectively. The authors highlighted the advantage of using the satellite sensor data with respect to monitoring fluctuations along the whole of the glacier boundary as opposed to ground observations, which were restricted to the 23 stations located along the margin. Li *et al.* (1998) documented similar work, focusing on glacier-margin fluctuations over a 21-year time span in the Tibetan Plateau using both TM and MSS imagery. They detected advance in the northern glaciers and recession in the southern glaciers, ranging from 50 to 100 m a^{-1}. Given the fact that no other method could reasonably make a comparable analysis over this time period and the strong correlation of satellite with ground-based measurements, both reports confirmed the value of Landsat TM data for monitoring margin fluctuations.

9.5.4.2 Differentiating between snow and ice for the identification of accumulation and ablation areas

Remotely sensed data is widely used in mass balance studies, where the researcher needs to be able to differentiate between the accumulation and ablation areas on a glacier. Grain size and the liquid water content of snow and ice can be detected by optical, microwave and radar sensors. These spectral classification techniques are outlined in section 7.3.2.1 in relation to investigations of glacier mass balance.

9.5.4.3   Measuring mass balance and snowmelt runoff

Volumetric changes can be calculated from contour line shifts on maps derived from terrestrial photogrammetry and field surveys (Kick, 1966), although there is the need to ensure the accuracy of the photogrammetric maps if these are to be used as the basis for volumetric calculations (Paterson, 1966). Remote sensing can also be used to calibrate against ground measurements of, for example, surface albedo (Knap *et al.*, 1999). König *et al.* (2001b) used SAR to determine the firn-line altitude on Austre Okstindbreen, Norway, and suggested that monitoring of the firn-line altitude rather than the ELA is the better method for remotely sensed mass balance studies. König *et al.* (2002) used synthetic aperture radar (SAR) to study superimposed ice formation on Kongsvegen, Svalbard, and compared the satellite-derived data with fieldwork observations.

For smaller-scale alpine glaciers, the fine resolution offered by photogrammetric techniques is ideal. Kääb and Funk (1999) presented a method of modelling mass balance on Griesgletscher, Swiss Alps, based on combining a number of key data sets, three of which were derived from photogrammetry: glacier surface elevation, its change with time and surface velocities. Glacier surface elevations were generated using stereo models, and change over time was analysed by the comparison of the resulting DEMs, while surface velocities were derived by following the movement of distinct features on the glacier tongue and measuring their displacement in three dimensions over a given time period. The authors found a high correlation between photogrammetric results and field surveys, commenting on the appropriateness of this technique for debris-covered glaciers because of the distinct contrast of the surface materials. Finally, since the pioneering work of Robin (1966), research has focused on direct measurement of ice surface elevation changes using satellite altimetry (e.g. Bamber *et al.*, 2001).

9.5.4.4   Mapping glacier structures and surface morphology

Glacier structures and surface morphology are readily mapped from both satellite imagery and aerial photography. Hambrey and Dowdeswell (1994) mapped the structural attributes of the surface of the Lambert Glacier–Amery Ice Shelf system, Antarctica (the world's largest glacier system), using Landsat TM imagery and used these structures to provide insight into the flow regime of the glacier system. Dowdeswell and Williams (1997) also used Landsat MSS and TM digital images to identify and map the distribution of surge-type glaciers in the Russian High Arctic, an inaccessible area for which there is little aerial photography. They used late-summer imagery, when there is minimum snow cover, to map features indicating surge-type behaviour such as looped medial moraines, potholes formed on the glacier surface during the quiescent phase and heavily crevassed surfaces indicative of glaciers in the active phase of a surge cycle. Bindschadler *et al.* (2002) used a combination of satellite images from Landsat, IKONOS and ERS1, together with aerial photography, to study the evolution of ice dolines on the surface of the Larsen Ice Shelf, Antarctica. Wessels *et al.* (2002) used ASTER imagery to monitor and measure the growth of supraglacial lakes on the surface of glaciers in the Himalaya. Kirkbride (1995) illustrated how time-separated aerial photographs can be used to track the movement of large supraglacial boulders on the surface of the Tasman Glacier, New Zealand. By tracking the displacement of these boulders on successive aerial photographs from 1957, 1971 and 1986, he was able to produce maps of surface ice flow vectors for the glacier. A final example is the study of Fahnestock *et al.* (2000), who used an enhanced composite AVHRR image, consisting of 15 separate images, to map the distribution of flow stripes and rifts across the Ross Ice Shelf, Antarctica. This composite image allowed them to map the development, deformation and rotation of individual rifts and crevasses, as well as the provenance of individual flow stripes, on the surface of the ice shelf. From this, the authors were able to develop a four-stage flow history that accounts for the structural features preserved in the ice shelf.

Structural maps of glaciers at the valley scale and above can easily be produced from single or pairs of overlapping vertical aerial photographs. Structures on the ice surface can be recorded from aerial photography by producing simple tracings of visible features onto acetate sheets, which can then be scanned as digital files and manipulated in a suitable graphics package. Packages such as Adobe Illustrator are particularly suitable for this purpose because related features can be grouped as layers for publication purposes. The general characteristics of structures, dimensions and cross-cutting relationships can be recorded, together with compass orientations. An example of an aerial photograph of a glacier (Storglaciären, northern Sweden) and a structural interpretation based on this photograph shows the type of interpretation that is possible (Figure 9.1). Enlargements of high-resolution aerial photographs can be used to map individual structures, such as those present on the surface of Griesgletscher (Hambrey and Müller, 1978). Aerial photographs can also be used to make general observations about surface morphological features of glaciers, for example mapping and monitoring the drainage of ice-dammed lakes and calculations of the volume of water involved from repeat aerial photographs (Helk, 1966).

This is seldom straightforward; especially where problems arise in delineating the margins of active glaciers with thick snow or debris cover (Avsiuk *et al.*, 1966; Kasser and Roethlisberger, 1966). Some structures, especially those containing liquid water such as meltwater ponds, can also be detected by infrared mapping techniques (Poulin and Harwood, 1966).

9.5.4.5 Determining glacier velocities

Simple 2D velocity measurements can be made from aerial photographs by monitoring the movement of surface structures through time and from repeat terrestrial photogrammetric measurements. Time-lapse terrestrial photography can also be used to determine glacier velocities. These techniques are described in section 7.4 in relation to field investigations of glacier velocities.

9.5.4.6 Mapping the distribution of landforms and sediments

The distribution of landforms and sediments is normally determined by the manual interpretation of satellite images or aerial photographs, although direct measurements of landforms are possible using stereo pairs of aerial photographs and digital photogrammetry (Welch and Howarth, 1968; Aniya and Welch, 1981). Holmlund *et al.* (1996) used photogrammetric methods to produce estimates of sediment volume changes in the proglacial area of Storglaciären, northern Sweden, and concluded that the present erosion and sedimentation are in balance. Woodruff and Evenden (1962) tested the accuracy of making measurements of geomorphological features from aerial photographs against those taken directly from maps. They concluded that aerial photographs provide a good source of data for the measurement of linear features (such as rivers), areal parameters (such as basin shape) and gradient parameters (such as hypsometric data), especially in areas for which high-quality topographic maps do not exist.

The interpretation of remotely sensed imagery requires greater skill than map reading. A satellite or camera records every detail of the Earth's surface with none of the selectivity of a map, whereas a cartographer has chosen which details to show. As Dickinson (1979; p. 308) pointed out: 'Not only is nothing labelled but most things are presented from a totally unfamiliar viewpoint and the totality of detail may confuse as much as it helps, at least until some basic skills are acquired.' Avery (1977) and Dickinson (1979) provided some general guidance on the principles of aerial photograph interpretation. Objects are normally recognized on air photos by their size and shape, their colour or tone and their relationship with other features with which they are associated. Since most aerial photographs are black and white, landform mapping generally involves the identification of features based on changes in tone or contrast.

Ford (1984) discussed some of the relative merits of landform mapping using Seasat SAR and Landsat MSS images. Because SAR uses microwave energy, it is able to penetrate clouds and is not affected by seasonal differences in solar radiation. There are no problems with shadows (often cast by neighbouring mountains). Seasat SAR has a spatial resolution of 25 m. Ford demonstrated the use of SAR in mapping glacial landforms (drumlins in Ireland) and modern glaciers (Alaska Range). Features such as drumlin width and length could be mapped from SAR but not from Landsat MSS because of the lower spatial resolution of Landsat MSS and problems with shadowing effects. In the modern glacial environment Ford demonstrated the potential for mapping features such as medial and lateral moraines, stagnant ice and fluted ground.

9.6   GEOMORPHOLOGICAL MAPPING

Although Goudie (1990) expressed surprise at the assertion of Demek and Embleton (1978; p. 313) that geomorphological mapping is 'the main research method of geomorphology in numerous countries', this type of mapping remains one of the most important techniques in glacial geomorphological research. Geomorphological maps are cartographic representations of the Earth's landscape that contain information about the morphology, genesis and age of its relief. Geomorphological maps should show the spatial distribution of landforms, account for their genesis and, in some cases, provide the age of the landforms. In contrast to topographic maps, geomorphological maps are essentially interpretative and so geomorphological mapping requires an appreciation of the complex nature of landforms and the processes of formation (Lowe and Walker, 1997).

Two types of geomorphological map are commonly distinguished. **General geomorphological maps** cover all aspects of the relief and geomorphology of an area. **Themed geomorphological maps** cover only one particular aspect of the relief, for example its glacial, periglacial or nival geomorphology. Here, we are concerned only with themed maps, specifically glacial geomorphological maps. If properly constructed, geomorphological maps can provide information on landform morphology, relief genesis, relief age, and the spatial arrangement of relief elements and their interrelationships. All geomorphological maps rely on some form of **landform classification;** the grouping or ordering of landforms by dimensions, age, origin or evolution. This type of classification rests on the assumption that all landforms have a characteristic shape (or form) that can be used for their identification. Landforms are often composed of characteristic sediments or sediment associations that in turn are the product of specific processes operating at present or having operated in the past (Box 9.4). Unfortunately, the principles of geomorphological classification are not generally agreed and different workers have therefore advanced their own classification schemes (Demek and Embleton, 1978; Goldthwaite, 1988).

Box 9.4   Geomorphogical mapping in upland glaciated areas

Anderson *et al.* (1998) presented the results of geomorphological mapping of 44 km^2 of upland glaciated terrain in a remote mountain range called Macgillycuddy's Reeks in south-west Ireland. Their geomorphological maps were compiled from both aerial photography and field mapping. The aim of their mapping was to arrive at a morphostratigraphic classification of the landscape (morphostratigraphy is the subdivision of a landscape on the basis of relative age). Field mapping was carried out using enlarged base maps reproduced from 1:25 000 scale Irish Ordnance Survey maps. The accuracy of the field mapping was checked by inspection of computer-enhanced images of aerial photographs that were enlarged to the same scale as the topographic base maps (Figure B9.4). To overcome the problems of distortion between topographic maps and aerial photographs a system of grid corrections was used to realign the images until an acceptable fit was achieved. The geomorphological maps were then redrawn, scanned, and digitized using Freehand software. On their geomorphological maps these authors identify five main landform assemblages: (1) rockwalls, (2) drift, (3) talus slopes, (4) mountain-top detritus and (5) alluvial fans and debris cone complexes. Rockwalls are *in situ* outcrops of rock, either with smoothed and polished surfaces (glaciated) or showing intense frost shattering (periglacial). 'Drift' is the collective term used to describe a range of glacigenic sediments and sedimentary landforms. Talus slopes are accumulations of coarse, angular debris below rockwalls. They take a variety of forms but are usually found in cones or sheets. Associated landforms are rock avalanches, protalus ramparts and protalus rock glaciers. Mountain-top detritus is the term used to describe the frost-weathered regolith which often mantles the summits and slopes of mountains in cold climate regions. Finally, alluvial fans and debris cone complexes are fan-shaped accumulations which develop at stream tributaries and at the base of incised slopes. From the landforms present on their geomorphological maps, Anderson *et al.* (1998) were able to identify three morphostratigraphic zones in their study area. Each of these zones is related to a distinct glacial

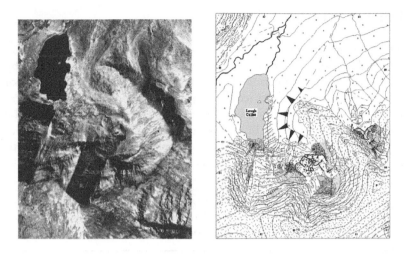

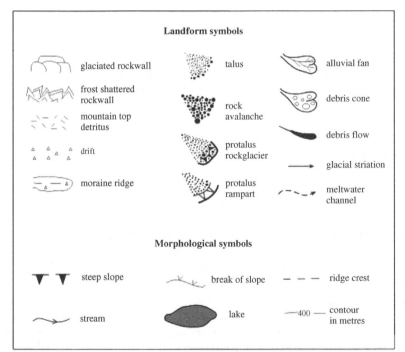

Figure B9.4 (a) Vertical aerial photograph and geomorphological map of glaciated terrain, Lough Callee, Macgillycuddy's Reeks, south west Ireland. The geomorphological map was constructed from the aerial photograph. (b) The symbols used in the mapping are also shown. Reproduced from Anderson *et al.* (1998) by permission of the Society of Cartographers

---

Box 9.4   (Continued)

phase. These are: (1) glacial landforms associated with a period of maximum glaciation (22–18 ka BP) when the mountains were inundated by
a large ice sheet, (2) a period of glacial readvance between 18 and 13 ka
BP during the decay of this ice sheet, and (3) a period of local glaciation
when only small glaciers formed in cirque basins, believed to represent the
Younger Dryas (11–10 ka BP). This work neatly illustrates the principles
of geomorphological mapping and the application of this technique to
unravelling the history of glaciation in a mountainous terrain.

Anderson, E., Harrison, S. and Passmore, D.G. 1998. Geomorphological mapping and
    morphostratigraphic relationships: Reconstructing the glacial history of the Macgil-
    lycuddy's Reeks, south west Ireland. *Society of Cartographers Bulletin*, 30, 9–20.

9.7   FIELD MAPPING

The collection of data in the field is often a major component of all glacial
geomorphological research projects. Field research can involve the mapping, measuring and monitoring of both landforms and processes. Here we
describe the main techniques in relation to landform recognition, studies of
bedrock morphology, field surveying, GPR in sediments, electrical resistivity measurements and marine geophysical surveys.

9.7.1   Landform recognition

Glaciers and ice sheets produce a huge variety of **landforms** and **sediments**. One
of the major challenges of glacial geomorphological mapping is therefore the
recognition and delineation of these landforms and sediments, and their genetic
interpretation. This problem arises because most landforms are not clearly
spatially defined by a typical feature or set of features and there is therefore no
uniform method of landform recognition that can be applied to geomorphological mapping. Landform recognition depends to a large extent on the
observer and their experience of glacial environments. Box 9.5 describes some
of the most commonly encountered glacial depositional landforms, together
with guidance on the delineation of these landforms (i.e. how to actually mark
these features onto a map). If you intend to undertake a research project that
involves landform recognition, it pays to be familiar with the full range of
landforms likely to be encountered. The classification scheme for the mapping
of glacial landforms proposed by Demek and Embleton (1978; pp. 296–302) is

Box 9.5   The identification and mapping of glacigenic landforms

Many different glacigenic landforms will be encountered during a typical field mapping project. Table B9.5 provides some guidance on how to identify these landforms and how to define their position on a geomorphological map, together with an indication of their overall glaciological significance (Source: information contained in Hambrey (1994), Bennett and Glasser (1996), Benn and Evans (1998), Kirkbride *et al.* (2001) and Lukas (2002)).

Table B9.5 Identification criteria for commonly encountered glacigenic landforms and their glaciological significance

Landform	Identification criteria	Boundary defined by	Significance
Ice-scoured bedrock	Widespread exposures of bare bedrock with smoothed, striated or plucked upper surfaces	Outermost extent of bare bedrock	Evidence of glacier ice at its pressure-melting point
Drift limit	The edge of the cover of a glacigenic deposit that is not also marked by a moraine ridge. Drift limit may be identified by a change in vegetation type or by change in the density of glacially deposited boulders	Outermost extent of glacigenic deposit	Drift limit marks the extent of a glacier advance
Moraine ridge	A single ridge or collection of ridges composed of glacigenic material. Typically 1–10 m (but exceptionally 100 m) in height. Sharp or rounded crests and either linear, curved, sinuous or sawtooth in plan	Lower concave break of slope. The crestline orientation can also be recorded if this is well-defined	Moraine ridge marks the lateral or terminal extent of a glacier advance (except in the case of 'hummocky moraine'; see below)
'Hummocky moraine'	A seemingly chaotic assemblage of irregular hummocks and hollows. 'Hummocky moraine' often shows order when viewed on aerial photographs, or when mapped in detail in the field	Outermost extent of 'hummocky moraine'. The crestline orientations of individual hummocks can also be recorded if the scale of the map permits	Formed by the slow melting of stagnant, debris-covered ice, by deposition at receding glacier margins, or the release of material from proglacial or englacial thrusts

Box 9.5   (Continued)

Table B9.5   (Continued)

Landform	Identification criteria	Boundary defined by	Significance
Fluted moraine	Groups of straight, elongated ridges of glacigenic material. Typically 10–100 m in length and 1-m wide	Upper convex and lower concave breaks of slope of individual features (where scale permits)	Subglacial modification of sediment. Flutes are formed parallel to direction of ice flow
Till sheet	A cover of glacigenic material with no surface form or expression. Often associated with isolated boulders and bounded by a drift limit or moraine ridge	Outermost extent of glacigenic material	Extent of till sheet marks the extent of a glacier advance
Drumlins	Smooth, streamlined, oval-shaped or elliptical hills with steep or blunt stoss (up-ice) faces. Composed of a variety of glacigenic sediments. Typically 5–50-m high and 10–3000-m long, with length-to-width ratios of less than 50. Occur in swarms	Upper convex and lower concave breaks of slope (where scale permits)	Widespread subglacial modification of sediment. The steeper, blunt end usually indicates the up-ice face
Kame terrace	Flat-topped terrace along valley sides. Typically up to 10 m in height. Usually composed of sand and gravel. Steep slopes commonly represent former ice-contact slopes. Upper surface may be pock-marked by kettle holes	Upper convex break of slope. Similar terrace heights can be indicated by distinctive shadings and patterns	Formed when sediment is deposited by meltwater flowing laterally along an ice margin, commonly against a valley side
Kame	Sub-conical hills or segments of flat-topped terraces typically up to 10 m in height. Usually composed of sand and gravel. Steep slopes commonly represent former ice-contact slopes	Upper convex break of slope. Crestline orientations of individual kame segments can also be recorded if the feature is discontinuous	Fragmentary features, formed in a similar manner to kame terraces, but often in ice-walled tunnels, against steep valley sides or in front of a glacier
Esker	Usually sinuous in plan and composed of sand and gravel. Some eskers are single-crested, whilst others are braided in plan. Eskers may climb uphill or cross low cols	Upper convex and lower concave breaks of slope. Crestline orientations of individual esker segments can also be recorded if the feature is discontinuous	Glaciofluvial landforms created by the flow of meltwater in subglacial, englacial or supraglacial channels

Meltwater channels	Channels cut in rock or sediment, often with abrupt inception and termination and lack of modern catchment. Subglacial meltwater channels may breach cols, displaying convex-up long profile	Thalweg of channel, location of channel inception and termination. Arrow to indicate direction of former drainage can also be recorded if known	Evidence of former meltwater discharge routes
Trimline	Line separating areas of solifluction from extensive gullying or areas of mountain-top detritus/ weathered material from upper limit of ice-scoured bedrock	Lower limit of solifluction or weathering and upper limit of extensive gullying or ice-scoured bedrock	Former vertical dimensions of a glacier or englacial thermal boundary

Benn, D.I. and Evans, D.J.A. 1998. *Glaciers and Glaciation.* London, Arnold.

Bennett, M.R. and Glasser, N.F. 1996. *Glacial Geology: Ice Sheets and Landforms.* London, John Wiley, 364pp.

Hambrey, M.J. 1994. *Glacial Environments.* London, UCL Press.

Kirkbride, M.P., Duck, R.W., Dunlop, A., Drummond, J., Mason, M., Rowan, J.S. and Taylor, D. 2001. Development of a geomorphological database and geographical information system for the North West Seaboard: Pilot study. *Scottish Natural Heritage Commissioned Report BAT/98/99/137.* Scottish Natural Heritage Edinburgh.

Lukas, S. 2002. Geomorphological evidence for the pattern of deglaciation around the Drumochter Pass, Central Grampian Highlands, Scotland. Unpublished MSc Thesis (Diplomarbeit), Faculty Geosciences, Ruhr University of Bochum, Germany, 115pp.

a good starting point for this, as is the classification and description of glacigenic landforms presented by Goldthwaite (1988; pp. 273–277).

It is unlikely that any two geomorphological maps compiled independently by different workers of the same area will be identical, especially in glaciated areas. This is because landform recognition and interpretation is often a subjective process. Thus, features identified as moraine ridges by one worker might be interpreted as periglacial landforms (e.g. fossil rock glaciers) by another. One solution to this problem is to use glaciological theory to test the validity of the mapping. For example, this can be achieved by using one of a number of methods to calculate former ELAs for the mapped palaeoglaciers (Figure 9.2). Another means of extending this type of analysis where landforms are of an ambiguous origin is to use the formulae provided by Carr (2001) to estimate former glacier ablation gradient and the temperature at the former equilibrium line, to calculate glacier mass balance, total velocity and basal sliding rates. These data can then be compared with contemporary glaciers to determine if the reconstructed glaciers are within

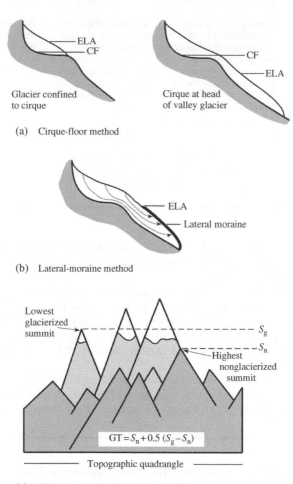

(a)   Cirque-floor method

(b)   Lateral-moraine method

(c)   Glaciation-threshold method

Figure 9.2   Common methods used to derive past ELA for glaciers. (a) *Cirque-floor method*: The ELA is inferred to lie above the altitude of the cirque floor (CF). (b) *Lateral moraine method*: The up-glacier limit of a lateral moraine approximates the ELA of the glacier that produced the moraine. (c) *Glaciation-threshold method*: The average altitude between the highest nonglacierized summit (S_n) and the lowest glacierized summit (S_g) defines the glaciation threshold (GT) in an area. (d) *Altitude-ratio method*: In the median-altitude variant of this method, the ELA lies midway in altitude between the head of the glacier (A_h) and the terminus (A_t). In the terminus-head altitude ratio (THAR) approach, the THAR equals the ratio of the altitude difference between the terminus and the ELA divided by the total altitude range of the glacier. The ELA can be estimated by adding the altitude of the terminus to the product of the total altitude range and an assumed THAR. (e) *Accumulation-area ratio (AAR) method*: The AAR is the ratio of the accumulation area (S_c) to the total area of the glacier (where S_a is the ablation area). Empirical studies suggest that for glaciers in steady-state (SS), an AAR of 0.65 ± 0.5 is appropriate. Diagram reprinted from Porter (2001), with permission from Elsevier

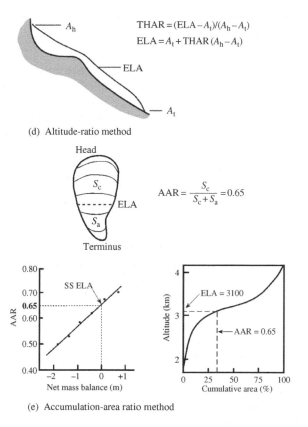

$$THAR = (ELA - A_t)/(A_h - A_t)$$
$$ELA = A_t + THAR\,(A_h - A_t)$$

(d) Altitude-ratio method

$$AAR = \frac{S_c}{S_c + S_a} = 0.65$$

(e) Accumulation-area ratio method

Figure 9.2 (Continued)

acceptable limits. Carr (2001) applied this method to five former glaciers in the mountains of North Wales and, using this approach, was able to discriminate between those glaciers of Younger Dryas age and those relating to an earlier episode of glaciation. Finally, it is also possible to combine landform mapping with sedimentological studies to produce maps of landform–sediment assemblages for glacial environments (e.g. Etienne *et al.*, 2003) and to use maps of sediment types to delimit sediment transport pathways through a glacier (Box 9.6).

Box 9.6   Mapping the distribution of landforms and sediment types for former glaciers

Benn (1989) and Evans (1999) provide good examples of how the techniques of landform mapping, field surveying and sediment analysis can be combined in studies of glacial debris transport and deposition.

Box 9.6　(Continued)

Both authors were interested in understanding the nature of the debris inputs to a glacier, how this debris is modified during transport, and how these factors combine to shape the resulting landforms. Evans (1999) studied the moraines deposited by a formerly coalescent glacier in the Jardalen cirque complex, western Norway. Field techniques used included geomorphological mapping to establish the location of free rock faces within the cirque complex, as well as the distribution of moraines, drift limits and periglacial trimlines in the cirques. Data were obtained from contours on topographic maps to calculate average slope gradients, and these measurements were combined in an equation to calculate the free rock face areas lying above the former glacier and the extent of basin asymmetry (effectively the 'inputs' to the system). The moraines of the cirque complex were mapped in the field and assigned latero-frontal, lateral or medial status with respect to the former glacier. Cross-profiles were obtained by surveying the moraines with surveying equipment, and profiles were drawn up and cross-sectional areas calculated (effectively the 'outputs' from the system). Clast-form analysis was undertaken on the surveyed moraines, including measurement of the relative dimensions of the three principal axes (the a-, b- and c-axes; see section 8.5.1) of clasts and degree of edge rounding for samples of 50 clasts. The percentage of the sample with $c{:}a$ axial ratios of ≤ 0.4 (the C_{40} index) was calculated, together with the RA index (the percentage of angular and very angular clasts). Finally, the distribution of the C_{40} and RA data was mapped within the cirque complex and contoured in order to determine whether any spatial trends existed. These data were then compared with probable debris transport paths through the former glacier. Evans (1999) concluded that moraine asymmetry is linked to basin asymmetry and the free rock face area above the glacier, although the relationship is complicated by the reworking of existing debris by the former glaciers. The distribution of clast form indices supports the widely held belief that moraine debris includes proportionately more actively transported debris in a down-glacier direction.

Benn, D.I. 1989. Debris transport by Loch Lomond Readvance glaciers in northern Scotland: Basin form and the within-valley asymmetry of lateral moraines. *Journal of Quaternary Science*, 4, 243–254.

Evans, D.J.A. 1999. Glacial debris transport and moraine deposition: A case study of the Jardalen cirque complex, Sogn-og-Fjordane, western Norway. *Zeitschrift für Geomorphologie N.F.*, 43, 203–234.

9.7.2 Mapping bedrock morphology

It is also possible to produce maps of the distribution of landforms relating to glacial erosion. This approach rests on the assumption that there is a discrete family of bedrock surface features that indicate the presence of former glaciers and the nature of ice/bedrock contact conditions (Box 9.7). This type

---

Box 9.7 Mapping bedrock morphology in the forefield of receding glaciers

Many contemporary glaciers are now receding across bedrock surfaces that formerly lay directly beneath the glacier. By mapping the distribution of landforms and sediments on these bedrock surfaces it is possible to make inferences about the former subglacial conditions beneath the glaciers. Walder and Hallet (1979) reported the results of a mapping project on the limestone bedrock exposed by recession of Blackfoot Glacier, a small cirque glacier in Glacier National Park, Montana, USA. Plane table and adelade (survey instruments used in map-making) were used to produce a detailed topographic map with elevation contours over an area of c.5000 m² of bedrock. Primary and secondary control points were located on this map, against which the position of landforms on the bedrock were to be recorded. Landforms and sediments on the former glacier bed were then identified and mapped directly onto this base map. The landforms and sediments mapped were those deemed to reflect different degrees of modification by glacial abrasion, water erosion, chemical dissolution and chemical precipitation. Walder and Hallet (1979) used the following criteria to identify features:

Chemically altered areas: Distinguished by the presence of abundant, furrowed, subglacially precipitated calcite deposits on the down-glacier sides of bedrock protuberances. They indicate close ice/rock contact beneath the former glacier, separated by a thin water film.
Abraded areas: Distinguished by the scarcity of both subglacially precipitated calcite deposits and solutional furrows, with abundant striations. They represent areas of intimate ice/rock contact.
Cavities: Areas of steeply inclined bedrock, usually down-glacier of bedrock highs or breaks in slope, which exhibit extensive solutional features and lack of evidence of ice contact. Some of these features also have a thin, discontinuous coating of subglacial precipitate around their peripheries. These areas are interpreted as the sites of former water-filled subglacial cavities.

Box 9.7   (Continued)

Precipitate-filled depressions: Shallow (<50 mm) bedrock depressions nearly filled with subglacial precipitate. These features are interpreted as representing former shallow subglacial cavities.

Channels: Narrow, elongated depressions, typically 50–250-mm deep, 100–200-mm wide and 2–5-m long, with well-developed solutional features. Many have a thin discontinuous coating of subglacial precipitate. They are interpreted as subglacial meltwater channels ('Nye channels').

Karst: Narrow, deep, linear to sublinear bedrock depressions formed by dissolution along two joint sets. They tend to interconnect with underground passages, and are therefore interpreted as glacially modified sub-aerial karst features.

Walder and Hallet (1979) estimated their detailed map to be accurate to within 300–400 mm, with elevations accurate to 100 mm. From the landform/sediment relationships on their map, Walder and Hallet (1979) concluded that a nearly continuous, non-arborecsent network of cavities acted as the primary meltwater drainage system beneath the glacier. They estimated that nearly 20% of the former glacier sole was separated from the bed by water-filled cavities and that the remainder of the glacier–rock interface was characterized by a thin water film. Although the ideas presented by Walder and Hallet (1979) were developed numerically by a second study at Glacier de Tsanfleuron in Switzerland (Sharp *et al.*, 1989), these types of bedrock maps are still relatively rare. Whilst originally time-consuming to produce, the advent of new technology such as EDMs and dGPS now make these maps relatively fast to produce.

Sharp, M., Gemmell, J.C. and Tison, J.-L. 1989b. Structure and stability of the former subglacial drainage system of the Glacier de Tsanfleuron, Switzerland. *Earth Surface Processes and Landforms*, **14**, 119–134.

Walder, J.S. and Hallet, B. 1979. Geometry of former subglacial water channels and cavities. *Journal of Glaciology*, **23**, 335–346.

---

of exercise can be achieved at a variety of scales: at the micro-scale of individual bedrock outcrops (Glasser *et al.*, 1998b; Figure 9.3), at a valley scale (Thorp, 1981; Sharp *et al.*, 1989a; Lawson, 1996) and at the scale of an entire glacier (Walder and Hallet, 1979; Sharp *et al.*, 1989b). In many ways, this type of exercise is more straightforward than mapping glacial depositional landforms because landforms of glacial erosion are generally easier to recognize, classify and map than their depositional counterparts (Laverdiere *et al.*, 1979; Glasser and Bennett, 2004).

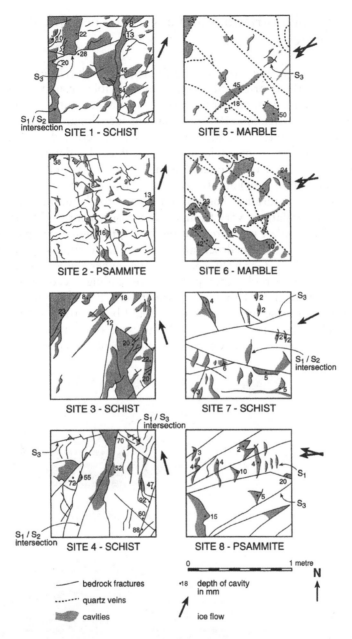

Figure 9.3 Examples of small-scale maps ('micro-maps') showing the effects of glacial erosion on Proterozoic metamorphic bedrock surfaces in front of a receding glacier in Svalbard. Each map is exactly 1 m² (1 m × 1 m). Individual maps were produced by placing a 1 m × 1 m quadrat (graduated at 10-cm intervals) onto the bedrock surface and transferring the position of bedrock fractures, quartz veins, former subglacial cavities and measurements of cavity depth onto a scale map. Diagram reproduced from Glasser *et al.* (1998) with permission from the University of Chicago

9.7.3   Mapping striations and associated bedrock surface features

Striations are one of the most commonly mapped bedrock features, principally because their orientation provides valuable insight into former ice-movement direction, former ice/bedrock contact conditions and ice sheet thermal regime (Kleman, 1990). Striations are long, narrow, parallel millimetre-scale scratches in the substrate that are created by rock debris embedded in the base of a glacier (Figure 9.4). To determine striation orientation, a number of individual striations are measured in close proximity (e.g. all within 1 m^2) and the results plotted either directly on to base maps or as rose diagrams. There is no agreed total number of measurements: recommendations vary from single measurement to 50 striations at each site. The simplest measurement method is to align a compass parallel to each striation on the bedrock and read off the compass orientation in degrees (Figure 9.4). Generally speaking, the higher the variability in striation direction on a rock outcrop, the more readings need to be taken to produce a statistically reliable mean orientation. Note that when plotting striation data on base maps, striations should be recorded as lines, not as arrows. This is because striations only give the orientation, and not the direction, of ice movement. Overall ice movement directions can only be

Figure 9.4 Measuring the orientation of striations on a bedrock outcrop in the field. The compass housing is placed alongside a striation and the orientation is read from the compass to the nearest degree (Photograph: N.F. Glasser)

inferred from relationships with other glacial landforms such as rat-tails, medium-scale stoss-and-lee features, crescentic gouges and glacial smoothing of outcrop irregularities (Jansson *et al.*, 2002) and from the topographic setting (Lawson, 1996). Where more than one set of striations exist (i.e. striations are **cross-cutting**) it is possible to determine the relative age of ice flow trends (Figure 9.5). Jansson *et al.* (2002) suggest that the following criteria can be used to achieve this:

(a) Striae preserved in lee-side positions relative to other striae sets are older.
(b) Striae that occur on crests between grooves and coarse striae are younger.
(c) Striae cut into other striae or grooves are younger.

It is also possible to map and classify individual striae into one of three morphological categories (Iverson, 1991). Type 1 striae become progressively wider and deeper down-glacier until they end abruptly, often as deep steep-walled gouges. They are considered by Iverson (1991) to form as a striating clast ploughs forward and downward, before either the striator point breaks off the clast or the torque on the clast is sufficiently large so that it rotates out of the groove. Type 2 striae start and terminate as faint, thin traces. They steadily broaden and deepen until they reach a maximum width and depth near their centre point. They are explained by sharp striator points that are rotating as they slide. The point initially has a large ploughing angle, causing progressive incision of the striation. Deeper ploughing causes more rapid clast rotation as the torque on the clast increases. Rotation, together with comminution of the clast point, reduces the ploughing angle so that there is a steady reduction in striation depth. Thus, Type 2 striae indicate clasts that slow down until the maximum striation depth is reached and then steadily accelerate, until at the striation terminus the clast has the same velocity as the ice. Type 3 striae begin abruptly as deep gouges and then become progressively narrower and shallower down-glacier. They form where a striator point contacts and indents the bed. Clast rotation with little displacement along the bed produces a low ploughing angle, so that a gradual reduction in indentation depth occurs as sliding proceeds.

Measurement of the orientation of bedrock gouges and cracks (a family comprising chattermarks, crescentic gouges and crescentic cracks) is slightly different. A sense of ice movement direction can sometimes be gained from measurement of the orientation of the ends (or horns) of individual fractures, although it is unclear whether the horns point up-ice or down-ice. Where this is the case, overall ice movement directions can only be inferred from relationships with other glacial landforms and sediments and from the topographic setting. Where these features occur as bedrock fractures, Harris (1943) found that the direction of forward dip of the fracture provided

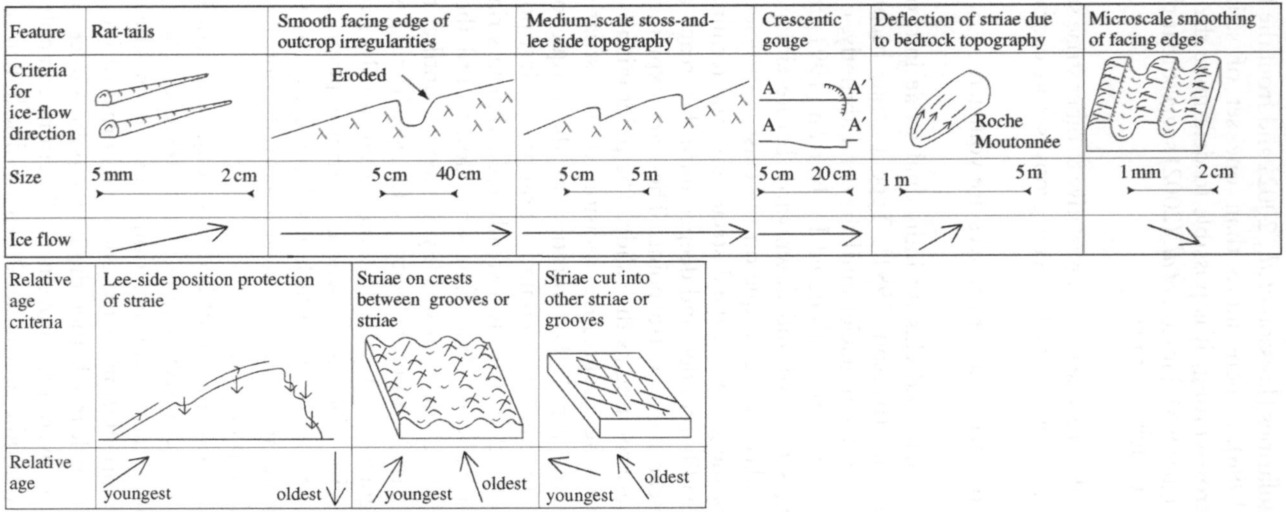

Feature	Rat-tails	Smooth facing edge of outcrop irregularities	Medium-scale stoss-and-lee side topography	Crescentic gouge	Deflection of striae due to bedrock topography	Microscale smoothing of facing edges
Criteria for ice-flow direction		Eroded		A A'	Roche Moutonnée	
Size	5 mm — 2 cm	5 cm — 40 cm	5 cm — 5 m	5 cm — 20 cm	1 m — 5 m	1 mm — 2 cm
Ice flow	↗	↗	↗	↗	↗	↗

	Lee-side position protection of striae	Striae on crests between grooves or striae	Striae cut into other striae or grooves
Relative age criteria			
Relative age	youngest ↘ oldest	youngest ↗ oldest	youngest ↙ oldest

Figure 9.5 Criteria used in the interpretation of ice-flow direction and for establishing cross-cutting relationships between sets of striae of different ages. Reprinted from Jansson et al. (2002), with permission from Elsevier

a good indication of ice movement direction. In all of the features examined by Harris, the forward dip of the fracture was orientated parallel to ice flow. One potentially fruitful method of mapping small-scale bedrock features is to use close-range stereo photogrammetry to record the detail of bedrock surfaces so that digital images can be used for more complex analysis, for example analysis using small-scale DEMs (Wingtes, 1985).

9.7.4   The micro-roughness of bedrock surfaces

Roughness elements at the millimetre-to-centimetre scale can be readily measured in the field using a micro-roughness meter (MRM) and profilometer. The MRM has been used successfully (McCarroll, 1991, 1992; Hubbard *et al.*, 2000), as has the portable device described by Shoshany (1989) to record the surface microtopography on a variety of different bedrock surfaces in the field. The MRM is made up of a track, graduated in millimetres, mounted on a tripod along which a perpendicularly orientated stylus can run freely. The stylus is attached to a strain dial that registers displacement of the stylus. To record a roughness profile with an MRM the following steps are followed:

1. The MRM is laid down in the desired orientation on the surface to be measured. Ensure that the MRM is stable on its surface before you begin data collection.
2. The stylus is located at $X = 1$ mm and lowered to the ground surface.
3. While the stylus tip is in contact with the ground surface, the stylus-to-dial attachment is adjusted to give a dial reading of some hundreds (anywhere between 400 and 600 is ideal – this procedure is simply to ensure the dial can register displacements both up and down from this first point).
4. The exact Z reading is recorded at $X = 1$.
5. The stylus is allowed to rise off the bedrock and is moved on 1 mm (by hand) and lowered to the bedrock surface. A second reading is taken.
6. The process is repeated every 1–100 mm (i.e. 10 cm).

A standard carpet fitter's gauge, or 'profilometer', is also commonly used to measure bedrock roughness at the centimetre scale. The profilometer is a comb equipped with fine, closely spaced teeth that advance or retreat through the housing when depressed against a rough surface. To record a roughness profile with a profilometer the following steps are followed:

1. The profilometer is depressed against the bedrock surface until the imprint of the rock surface is registered in the pattern of the profilometer's teeth.
2. The roughness profile is then set into the pattern of the teeth upon removing the comb from the surface.

3. The roughness imprint is transferred to graph paper by tracing over the outline of the teeth, held steadily and as flat as possible against a sheet of graph paper. Use a sharp pencil to trace the profiles on to the graph paper, taking care not to force or bend the teeth.
4. Following fieldwork, the outline is 'digitized' by reading points every 1 mm along the graph paper and expressing their Z co-ordinates to 0.1 mm.

McCarroll (1997) has published details of a spreadsheet that can be used to calculate the roughness index of rock surfaces measured in the field with an MRM or profilometer.

Bedrock roughness can also be investigated at the metre-scale and above using standard surveying equipment (a tripod, engineer's level and survey staff) or a hand-held GPS. Finally, it is worth remembering that rock-surface roughness and weathering are intimately related, so that you may wish to combine studies of surface roughness with studies designed to test the depth of rock weathering such as the Schmidt hammer (section 10.6.1) and weathering rind thickness (section 10.6.2).

9.8   FIELD SURVEYING TECHNIQUES

Field surveying is used for a number of purposes in both glaciology and glacial geomorphology. In field glaciology, surveys are used to measure rates of surface glacier motion, to fix in space the location of measurement points on a glacier surface such as the location of boreholes, to define glacier boundaries and to record rates of recession through repeat annual surveys. Field surveying is commonly used in geomorphological research to make contour maps of a feature or group of features at very large scale (e.g. maps at scales of 1:500 are possible, showing features that would never be visible on say 1:10 000 maps), to construct cross-sectional profiles across landforms and to locate their position accurately in space. This type of survey can also be used to create themed maps that include only features selected by the operator, for example to plot the occurrence of linear features such as flutes or the boundaries of different sediment types that are not marked on ordinary maps.

A number of survey instruments can be used to determine slope angles, construct profiles and make simple maps (Table 9.2).

9.8.1   Compass clinometer

Sighting onto an object with a compass clinometer is a quick and easy method of determining slope angles, but is generally only accurate to ±5°

Table 9.2 Survey instruments commonly used in field glaciology

Instrument	Operating scales	Main uses	Accuracy	Examples of usage in glaciology
Compass clinometer	1–10 m	Determining slope angles Constructing crude slope profiles	±5°	Quick estimates of glacier snout, moraine or rockfall slope angles
Abney level	1–100 m	Determining slope angles Constructing slope profiles	±1°	Comparing proximal/distal slopes on moraines Establishing horizontal levels, for example former lake shorelines, correlating terrace fragments
Engineer's level and total station with electronic distance measurement (EDM) device	1 m–1 km	Determining slope angles Constructing detailed slope profiles Making simple maps and locating objects in relative space	±1° Millimetre accuracy possible using an electronic distance measurement (EDM) device	Glacier velocity surveys Mapping the position of ice margins and features on a glacier surface Profiles of complex terrain such as bedrock roughness Moraine profiles
Global positioning system (GPS)	1 m–10 km	Determining the height of objects Making maps and locating objects in absolute space	Millimetre accuracy possible if used in differential mode; otherwise metre accuracy	Glacier velocity surveys Mapping the position of ice margins and features on a glacier surface Recording one-off positions such as sample locations

and should therefore only be used where no other methods are possible, for example in very steep or rocky terrain.

9.8.2 Abney Level

An Abney Level is a surveying instrument consisting of a spirit level, protractor and a sighting tube, used to measure the angle of inclination of a line from the observer to a distant point, for example a slope measurement. Abney Levels are relatively cheap, portable and easy to use. The easiest means of collecting slope readings with an Abney Level is to have two operators, one sighting with the level and one operating as a 'target'.

Using an Abney Level to make a slope reading:

- The person sighting with the Abney Level determines his or her eye height on the target person, for example shoulder, chin, mouth, eyes, etc. (the ideal is to have a field partner who is the same height so that measurements can be made from eye height to eye height).
- The 'target' person moves away to a point at the desired distance from the person sighting with the Abney Level.
- The person sighting with the Abney Level views the 'target' through the sighting tube and levels the instrument by turning the spirit level until the bubble appears on the centre line at the appropriate point on the 'target'.
- The person sighting with the Abney Level removes the instrument from the eye and reads off the slope angle from the scale engraved on the side plate. Angles are normally recorded to the nearest degree.

Using an Abney Level to construct a topographic profile: Abney Levels can also be used to construct topographic profiles across a landscape (e.g. to record changes in slope angle) or across individual landforms (e.g. to characterize moraine morphology or moraine asymmetry, to calculate moraine volumes). The instrument is used in exactly the same way as outlined above except that, when constructing profiles, it is also necessary to measure the horizontal distance between the two operators with a tape measure. When measuring such a profile, the two operators move across the landform in tandem, measuring and recording slope angles with the Abney Level and horizontal distance with the tape measure. It is important to mark and record all major breaks in slope on the profile so that the profile accurately reflects the true shape of the landform. Missing out breaks of slope will 'smooth out' the shape of the landform. Remember also to record up-slope measurements as positive numbers and down-slope measurements as negative numbers. Once completed, the measurements can be used to construct a scale profile of the landform for presentation purposes. Other data, for example the distribution of sedimentary facies or sample locations, can be added to the profile if desired.

9.8.3 Engineer's level/total station

The apparatus required for this type of survey comprises a tripod, Engineer's level (that needs initial levelling) and a survey staff. By sighting from a tripod-mounted Engineer's level to the survey staff, it is possible to construct topographic profiles in a similar manner to that outlined in section 9.8.2. An Engineer's level can also be used to take repeat readings to a distant object or marker, for example to undertake a glacier velocity survey. To use an Engineer's level, or any other tripod-mounted optical instrument, the following steps must be followed.

- Set the tripod up on firm ground in good light and in full view of the area being surveyed.
- Spread the tripod feet well apart and place them so that the tripod plate is approximately level.
- Press the tripod shoes firmly into the ground, or set them into cracks or chipped depressions if you are forced to set the tripod up on bedrock.
- Level the instrument with particular care.
- Carry out all checks in the order prescribed for the instrument.

Many fieldworkers use a **total station** (an Engineer's level with a built-in Electronic Distance Measurement (**EDM**) device), which is excellent for rapid and accurate surveying. The EDM measures the distance from the total station to a pole-mounted reflective prism controlled by a second operator. Distance and angle measurements can be stored in the total station for later download. Most total stations are provided with software that allows these data to be used in digital mapping, for example to make contour maps and DEMs.

If there is no access to these instruments, a simple device for measuring slope angles and slope profiles is described by Blong (1972). This is simply constructed using a standard builder's level and a staff, but lacks the accuracy of more sophisticated equipment.

9.8.4 Global positioning systems (GPS)

Global positioning systems was developed originally for military purposes and is funded by and controlled by the US Department of Defense. GPS provides specially coded satellite signals that can be processed in a GPS receiver, enabling the receiver to compute position, velocity and time.

Normally, a minimum of four GPS satellite signals are used by the receiver to compute positions in three dimensions. Using a single hand-held receiver, 10 m accuracy is possible, but this can be increased to millimetre accuracy if the GPS is used in differential mode with a hand-held receiver and a fixed control receiver.

9.9 GROUND-PENETRATING RADAR AND SHALLOW SEISMIC REFLECTION INVESTIGATIONS OF SEDIMENT BODIES

Ground-penetrating radar has been used successfully in both glacierized and glaciated environments to investigate the 3D properties of glacigenic sediments (Box 9.8). The principles of GPR and the use of GPR equipment in the field are outlined in Chapter 6. For a full review of the application of GPR to sedimentary products, see Neal (2004). Radar frequencies of between 50 and 200 MHz provide the best penetration and resolution in the clastic material typical of most glacigenic sediments. Frequencies of

---

Box 9.8   Ground-penetrating radar investigations of glacial sediments

Ground-penetrating radar, sometimes also known as georadar, is an inexpensive method of determining sedimentary facies and structures in situations where widespread surface exposures do not exist. The geophysical properties underlying the use of GPR are described in Chapter 6 in relation to investigations on glaciers, but GPR can also provide continuous profiles across unconsolidated clastic sediments in a fast and non-destructive manner. GPR has been successfully applied to a wide variety of Quaternary and glacial sediments, including detecting the extent of buried ice in push moraines (Lønne and Lauritsen, 1996), mapping the architecture of glaciofluvial sediments (Beres *et al.*, 1995), identifying deformation structures in glacial sediments (Jakobsen and Overgaard, 2002) and calculating the volume of talus slopes (Sass and Wollny, 2001). Variations in the radar wave permittivity of unconsolidated sediments cause reflections when radar waves are passed into sediment. Lithological properties, particularly particle-size variations and changes in water content, are the primary causes of these variations. The different reflection patterns caused by these variations can be grouped into distinct radar facies that can be used to characterize and interpret sedimentary facies. Radar frequencies of between 50 and

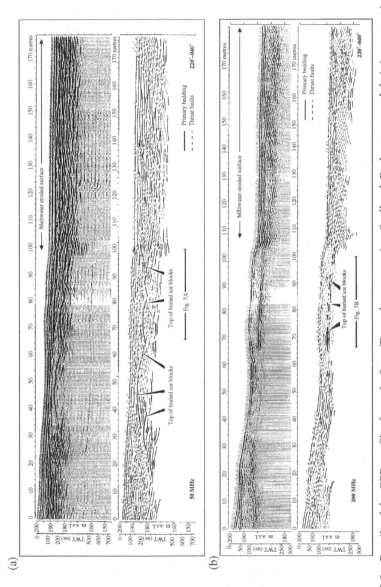

Figure B9.8 Details of the GPR profiles from the Scott Turnerbreen moraine, Svalbard. Radargrams and their interpretation are based on: (a) 50-MHz antennae; and (b) 200-MHz antennae. Former ice flow was from left to right (Lønne and Lauritsen (1996))

Box 9.8   (Continued)

200 MHz are most commonly used in clastic sediments, although frequencies as low as 25 MHz may be used in loose debris such as talus that contains large void spaces (Sass and Wollny, 2001). Typical antennae separations are 0.5, 1.0 or 1.5 m with step sizes of 0.5 m. Used in this way, GPR has proved capable of detecting both sedimentological and glaciotectonic structures in sand and gravel bodies. The sedimentary structures that can be recognized include sub-horizontal bed boundaries, channels and large-scale cross-bedding. Glaciotectonic structures appear on GPR profiles as inclined bed boundaries, thrust planes and folds.

Beres, M., Green, A., Huggenberger, P. and Horstmeyer, H. 1995. Mapping the architecture of glaciofluvial sediments with three-dimensional georadar. *Geology*, 23(12), 1087–1090.

Jakobsen, P.R. and Overgaard, T. 2002. Georadar facies and glaciotectonic structures in ice marginal deposits, northwest Zealand, Denmark. *Quaternary Science Reviews*, 21, 917–927.

Lønne, I. and Lauritsen, T. 1996. The architecture of a modern push-moraine at Svalbard as inferred from ground-penetrating radar measurements. *Arctic and Alpine Research*, 28(4), 488–495.

Sass, O. and Wollny, K. 2001. Investigations regarding Alpine talus slopes using ground-penetrating radar (GPR) in the Bavarian Alps, Germany. *Earth Surface Processes and Landforms*, 26, 1071–1086.

25 MHz or lower may be used in loose debris such as talus slopes or rock glaciers that contain large void spaces.

Onshore shallow seismic reflection surveys employ a seismic source at ground surface level to investigate the sedimentary properties of the material beneath the ground surface. A full explanation of seismic survey methods is provided in section 9.11. The general principle is that an acoustic source, or 'shot', often a sledgehammer and a metal strikeplate on the ground surface, is used to generate a short pulse of sound that penetrates the ground beneath. Reflection of energy takes place at boundaries between sediment layers of different acoustic impedance, with the reflection strength depending on the impedance contrast. The return signals from each sound pulse are collected by geophones on the ground surface to build up a picture of the subsurface sediments and their relationship to the ground surface. Shallow seismic reflection techniques have been used successfully by Harris *et al.* (1997) to investigate the structure of glacigenic sediments underlying a beach at Dinas Dinlle in North Wales, UK. These authors used five stacked sledgehammer blows on a steel plate as an acoustic source, detected by 100-Hz geophones buried 1.5 m apart beneath beach level. The shot-to-geophone offsets ranged from 1.5 to 39 m. Using this method, the authors reported that 350 24-channel records were recorded over a three-day period. Data processing is required to remove low-frequency ground roll and high-frequency electrical noise from the records, after which the vertical resolution of the final section is approximately 1 m.

9.10 ELECTRICAL RESISTIVITY SURVEYS

The basic principle behind **electrical resistivity** (sometimes also known as **electrical resistivity tomography** or **ERT**) is that layers below the surface can be identified and mapped because different geological materials have different electrical properties. The methods used to measure the properties of geological materials can be divided into two types (Burger, 1992), of which we cover only the first type:

1. methods using **applied currents**, including electrical resistivity, induced polarization and electromagnetic surveying;
2. methods using **naturally occurring currents**, including telluric surveying, magnetotelluric surveying and self-potential.

In all electrical resistivity methods an electric current is introduced into the surface of the earth with two electrodes, and the voltage drop across the surface of the ground is measured using another two electrodes. Because electrical flow disperses through the ground, these surface measurements provide information about the electrical character of materials below the earth's surface. This is termed the **resistivity** of the material, which provides an indication of the type of material being sampled (Table 9.3).

Table 9.3 Typical resistivities of some common geological materials (based partly on data in Milsom, 2002)

Substrate	Typical resistivities (Ω)
Wet silt/clay	1–100
Topsoil	50–100
Gravel	100–600
Weathered bedrock	100–1000
Loose sand	500–5000
Well-jointed bedrock	Low 1000s
Unjointed bedrock	High 1000s

Electrical resistivity is used in glaciological and glacial geomorphological studies, as well as in mapping the water table, locating buried stream channels, mapping contaminant plumes and mapping karst topography. For example, Evin *et al.* (1997) used electrical resistivity measurements to investigate the geometry of the permafrost and the amount of ice in sediments on rock glaciers in Grizzly Creek, southwest Yukon. Using this technique, they were able to infer that there was no extensive massive ice in the rock glaciers. Salem (2001) demonstrated how electrical resistivity measurements were used to map the distribution of sediments and water in a glacial aquifer in northern Germany based on the distribution of different sediment types (silts, sands and gravels) and their water content. Baines *et al.* (2002) have described a new technique called electrical resistivity ground imaging (ERGI), which is essentially a geophysical technique for producing high-resolution profiles of the shallow subsurface (<200 m). The advantage of ERGI is that it can be used to detect subsurface changes in lithology below clay and silt layers, a situation where other techniques such as GPR do not. The main difference between ERGI and ERT is that ERGI uses measurements from the ground surface, whereas ERT commonly involves measurements from borehole to borehole or from borehole to surface. Using examples from a variety of environmental settings, Baines *et al.* (2002) demonstrated that it was possible to map the geometry and lithology of sand and gravel channel fills and valley fills, even when buried beneath clay and silt.

The equipment used for measuring electrical resistivity in the field typically includes an ammeter, a voltmeter, four electrodes (two for source/sink, two for voltmeter/potential), connecting wire and a power source (DC or AC, batteries in series or motor-driven generator). The battery is connected through the ammeter to a resistor in a simple electric current. The battery supplies a potential difference from terminal to terminal, and the charge moving through the system is called current ($i = q/t$, where i is current, q is charge and t is time). Electrode spacing is an important consideration, and the arrangement for current electrodes and potential electrodes depends on the field objectives. If the goal of the project is to map a lateral variation in a geological material, then a constant-spacing spread of electrodes is used. Conversely, if the goal is to map the depth variation in a geological material, then an expanding spread is used. Over the years, many variations have been tried but only three are now in popular usage: the *Wenner Method*, the *Schlumberger Method* and the *dipole–dipole array* (Milsom, 2002).

Care should be taken in choosing the precise location and topographic setting of sites for electrical resistivity surveys, especially in and around built-up areas. Potential sources of errors in measurements include:

- buried pipelines, which change flow paths because of high conductivity;
- buried cables, which change flow paths because of high conductivity;

- topographical variations, which distort the potential field;
- anisotropic material, for example a bedrock joint that is filled with water that causes the electrical current to have a preferred pathway.

9.11   AQUATIC (MARINE AND LACUSTRINE) GEOPHYSICAL TECHNIQUES

The geophysical techniques outlined in this section can be applied equally to both ocean (**marine**) and lake (**lacustrine**) environments, and are only part of a suite of geophysical and sedimentological data collection techniques that can be used to map the sub-aquatic distribution of sediment bodies. Sediment characteristics beneath water are most commonly mapped using seismic techniques, by obtaining sediment samples and by collecting underwater photographs and video. The equipment used for these tasks is controlled from a vessel, normally a ship, on the water surface. Stoker *et al.* (1997) provided an excellent practical review of the methods used in these fields and the volume edited by Davies *et al.* (1997) provides detailed guidance on how to interpret the data collected. For further information on the geophysical principles behind these techniques see Urick (1983) and for information on the sedimentary processes operating in these environments see Kennett (1982), Dowdeswell and Scourse (1990) and Dowdeswell and O'Cofaigh (2002).

9.11.1   Seismic Techniques

Seismic techniques include **shallow seismic reflection profiling** and **sidescansonar systems** (Figure 9.6). Both techniques employ acoustic methods to investigate the sea-bed morphology.

9.11.1.1   Seismic reflection methods

These methods depend on the generation and detection of acoustic waves (Stoker *et al.*, 1997). An acoustic source is used to generate a short pulse of sound from a ship, which passes through the water and penetrates the sea bed. Reflection of energy takes place at boundaries between sediment layers of different acoustic impedance, with the reflection strength depending on the impedance contrast. The return signals from each sound pulse are collected by hydrophones on the ship. In this way it is possible to 'see through' layers in the sea bed, and as the ship moves across the water a profile of the subsurface structure of the sea bed emerges (Box 9.9). The

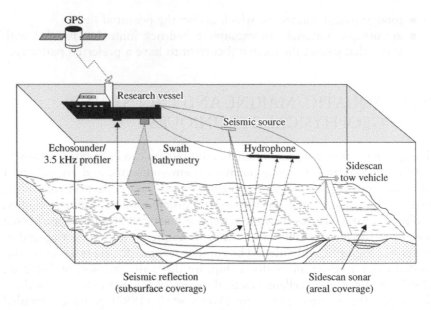

Figure 9.6 Mapping the seafloor using ship-borne equipment. Echosounders and swath bathymetry can be used to build up a picture of the topography of the seafloor, whilst high-resolution seismic-reflection profiling methods can provide a picture not only of the seafloor topography but also of its subsurface structure. Sidescan sonar, operated from a tow vehicle, can provide wide areal coverage of the seafloor topography. A satellite GPS is used to fix the geographical position of the profiles

---

Box 9.9   Seismic mapping of trough-mouth fans

Large continental-scale ice sheets have reached the edge of the continental shelf at repeated intervals during the Quaternary. When they do so, these ice sheets deposit large amounts of sediment on and immediately in front of the shelf-break. In some cases these sediment bodies form fan- or delta-like protrusions at the mouths of glacial troughs and channels crossing the continental shelf and ending on the shelf-break. These sediment accumulations have been termed 'trough mouth fans' (TMFs) by Vorren and Laberg (1997). TMFs are important palaeoclimatic indicators because they represent times when palaeo-ice sheets advanced as far as the shelf-break. These authors identified eight TMFs on the northwest European glaciated continental margin from sidescan sonar investigations, varying in size between 2700 and 215 000 km². The largest TMFs are those on the eastern margin of the Norwegian-Greenland Sea, namely the North Sea TMF, the Bear Island TMF and the Storfjorden TMF. The largest of the

TMFs, the Bear Island TMF, covers an area about the size of Iceland. Detailed investigation of TMF internal stratigraphy is difficult because these features are now submerged beneath the ocean. The only practical method of investigating their internal composition is to combine seismic studies with sediment cores. The most detailed stratigraphy is shown by 3.5-kHz seismic profiles, which enabled Vorren and Laberg (1997) to determine that many TMFs are composed of stacked units of debris flows. As a result, they inferred that the sediments deposited at the shelf-break by the ice sheets were remobilized and transported downslope by debris flows.

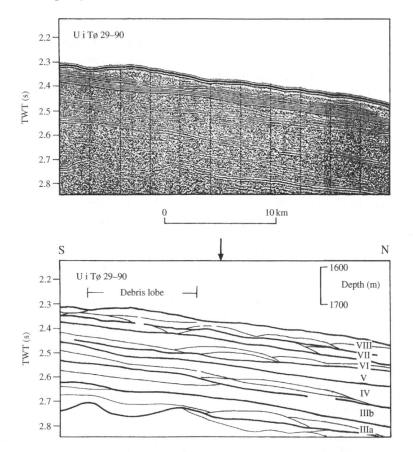

Figure B9.9   Cross-profile of stacked debris-flow sets on the Bear Island TMF (Figure 8 of Vorren and Laberg, 1996)

Vorren, T.O. and Laberg, J.S. 1997. Trough mouth fans: Palaeoclimate and ice-sheet monitors. *Quaternary Science Reviews*, **16**, 865–881.

resolution of the seismic profile is dependent on the frequency of the acoustic energy; the higher the frequency the better the resolution. High-frequency (1–10 kHz) systems such as **boomer, pinger** and **parasound** systems provide excellent near-surface information, penetrating up to 100 m into the sea bed. **Airgun, watergun** and **sleevegun** pulses operate at lower frequencies (10–100 Hz) and can penetrate 1–2 km, but with a corresponding reduction in resolution.

9.11.1.2   Sidescan sonar systems (**swath bathymetry**)

These techniques involve the use of sound to produce **sonograms**, described by Stoker *et al.* (1997) as the marine equivalent of aerial photographs. Sidescan sonar systems do not provide detail of the subsurface structure of the sea bed, rather they use the detected echoes of a transmitted pulse to give an indication of the surface relief (bathymetry) and surface texture of the sea floor. The principle is simple: an acoustic pulse is sent via a transducer. The acoustic pulses are reflected at the sea floor and the reflected echoes are received at the transducer. The elapsed time between the outgoing pulse and the return echo is a measure of the water depth. Sidescan systems can be either hull-mounted or towed along behind a ship. Towed systems produce better data because they are free from the background noise and roll associated with a ship on open water. The frequencies used determine the width ('the swath') of the seafloor that is illuminated. Thus, low-frequency (6.5 kHz) systems can illuminate a swath up to 30 km in width, whilst high-frequency (500 kHz) systems only illuminate a swath of 150 m in width. Strong reflections (high backscatter) from boulders, gravel and vertical features facing the sonar transducers are white; weak reflections (low backscatter) from finer sediments or shadows behind positive topographic features are black. The seafloor is typically surveyed in swaths 100–500-m wide; the swaths are then collated into a mosaic to form a composite image of the survey area.

9.11.2   Glacial geomorphological applications of seismic techniques

The application of swath-mapping technology during the 1990s has provided a wealth of new information about the morphology of the seafloor, and this technology has been applied particularly to the continental shelf and seafloor adjacent to the Antarctic Peninsula (e.g. Canals *et al.*, 2002). These authors used a ship-mounted Simrad EM12S system to produce profiles of the seafloor, demonstrating the presence of mega-scale ridges and grooves, as well as drumlinized topography. These data were combined with 175 in.^{3} airgun high-resolution seismic reflection profiles to investigate the internal structure of the shelf margin. Wellner *et al.* (2001) demonstrated how these techniques can be combined, collecting seismic records, swath bathymetry data and sediment cores for the Antarctic continental shelf. Swath data were collected with a SeaBeam 2100 hull-mounted system and consist of 120 beams of 12-kHz data.

9.11.3   Obtaining sediment samples

9.11.3.1   Sediment traps

Sediment traps are a useful means of quantifying rates of contemporary sediment transport in rivers and for assessing the relative importance of different inputs to marine and lacustrine systems. For example, Lewis *et al.* (2002) used sediment traps to investigate spatial and temporal changes in sedimentary processes in Bear Lake, Devon Island, Nunavut, Canada Lake, a glacier-fed lake at the downstream end of a 5.5-km-long sandur. They deployed 33 sediment traps on 12 fixed moorings early in the summer when the lake was ice-covered. Each of the moorings supported traps 15 m below the water surface and 1.5 m above the lake floor. The traps consisted of a metal cone attached to a 500-ml bottle. The area of the opening of the cone was 249 cm^2 and the cone had an aspect ratio (height to diameter) of 0.9. A plastic baffle with 1.3-cm^2 openings was placed over the cone to reduce current-generated resuspension, which would otherwise result in under-trapping of sediment (Bloesch and Burns, 1980). Lewis *et al.* (2002) deployed their arrays of moorings for periods of between 1 and 3 weeks. At the end of each deployment period, the moorings were slowly raised, supernatant water was siphoned from the cone and the bottles were removed and replaced. They also undertook recordings of changes in the lake level, bathymetric conditions and lake currents over the duration of the survey, as well as recording meteorological conditions around the lake such as air temperature and rainfall.

   Similar sediment transport studies are also possible in stream channels in proglacial areas using **bedload traps** (e.g. Warburton and Beecroft, 1993) and monitoring of **suspended sediment concentrations** (e.g. Hodgkins, 1996; Hodson *et al.*, 1997; Hodson and Ferguson, 1999). Suspended sediment concentrations are normally obtained by repeated sampling of a stable reach of the proglacial stream. Water samples can be obtained by hand or automatically using a sampler such as the Epic Products 1011 sampler (Hodgkins, 1996). Sampling interval is variable, but most studies have sampled at intervals between 1 and 3 hours. After collection, samples are filtered in a Perspex pressure-filtration apparatus through pre-weighed Whatman grade 40 papers (pore size 8 μm). The filter papers are then dried and reweighed to determine the suspended sediment concentration. Concurrent measurements of river discharge are also required for meaningful interpretation of suspended sediment concentration data sets.

9.11.3.2   Sediment coring techniques

Sediment cores are used where the aim is to recover a sample of sediment from beneath a water body or where no surface exposures are available. This is especially the case in the marine and lacustrine environments and much of the technology has developed in order to satisfy the requirements for deep-sea drilling from ships (Kennett, 1982). A variety of different coring methods exist: each approach is different and must be considered in terms of cost and the type of information sought. In sedimentological investigations, the primary concerns are the length and continuity of the sequence, and the amount of mechanical disturbance of the sediments that can be tolerated.

Conventional rotary drilling techniques have the largest penetration depth (up to thousands of metres in some cases) but have the major disadvantage that they disturb sediments during drilling and therefore do not always preserve sedimentary structures in soft sediments (although they may be used successfully in older, lithified glacigenic sediments). Rotary drilling techniques were used extensively by the Deep-Sea Drilling Program and the Offshore Drilling Program on glaciated continental shelves. The problem of drilling disturbance is overcome by using **piston coring** and **hydraulic piston coring techniques**, which are rapidly punched into sediment without rotary action. Conventional piston coring techniques (e.g. **gravity corers**) employ a tight-fitting piston inside a core barrel, with penetration energy supplied by weights fitted to the top of a drilling rig. They are capable of taking individual cores around 20 m in length, which can be increased if successive cores can be taken from the same hole. Hydraulic piston coring can obtain cores up to 200 m. Both are particularly effective in glaciomarine and glaciolacustrine environments, which are generally dominated by unconsolidated material. **Powered corers** (e.g. vibro-corers and percussion corers) can be used to penetrate relatively hard sediment types because the corers are physically driven into the sediment, normally by compressed air or electricity, rather than falling free into the sediment layer.

9.11.3.3   Grab samplers and dredges

Other systems, such as **grab samplers** and **dredges**, are capable only of obtaining near-surface samples. Grab samplers are buckets or segments that drive into the sediment layer and enclose and retain a layer. One of the most commonly used is the Shipek grab, capable of obtaining up to 2 litres of sediment. This device is lowered overboard on a rope. Once in contact with the sediment layer it automatically snaps shut, trapping the sediment inside. Dredges consist of a solid metal frame opening linked to a collecting bag, which is tied behind a ship on a wire rope. The dredge is weighted so that it remains in contact with the seafloor. Note that grab and dredge samples are almost invariably disturbed and do not always accurately reflect the composition of the seafloor, for example if fine material is lost during recovery. Dredge samples are particularly difficult to interpret because they are not derived from a single point but from transects across the seafloor.

9.11.4   Collecting underwater photographs and video

The most common method of obtaining underwater photographs and video footage is to use a **remotely operated vehicle (ROV)** to explore the sea or lake floor. ROVs are fully submersible mini-submarines that can be deployed from a vessel on the water surface and controlled by an umbilical cord. They can be used to investigate the glacier–water–sediment interface and to gain access to otherwise inaccessible underwater areas such as the underside of ice shelves and the termini of calving glaciers. The design, installation and operating procedures for the use of these machines are described in detail by Dowdeswell and Powell (1996). Most ROVs are equipped with a video or photographic camera, a light source, devices to measure water salinity, turbidity and temperature, imaging systems such as sidescan sonar, and bucket dredges for sediment sampling. ROVs can also be customized with other items of equipment as required to suit the requirements of individual research projects.

9.11.5   Field observations of icebergs and iceberg calving

9.11.5.1   Field observations of icebergs

Observations and tracking of very large icebergs and their oceanic effects are generally made from remote-sensing platforms (e.g. Arrigo *et al.*, 2002). Field observations are made primarily from aboard ships, in either fjordal settings or the open ocean. Hotzel and Miller (1983) examined records of iceberg observations in the Labrador Sea in order to investigate the functional relationships between the dimensions of icebergs and the ratios between their linear dimensions. These authors presented data for the length (defined as the longest horizontal dimension at the water plane), width, height (the distance between the water line and the highest point of an iceberg), draft (calculated using the power curve formula [draft $= 3.781 \times$ length$^{0.63}$]) and mass of icebergs. They also outlined how individual icebergs can be classified according to their morphology into five groups: pinnacle, drydock, domed, blocky and tabular. Dowdeswell (1989) used oblique and vertical aerial photographs, together with data collected with airplane-mounted 60 MHz radio-echo sounding equipment, to investigate the nature of icebergs calved from Svalbard glaciers and relate these to glacier morphology and dynamics. Dowdeswell and Forsberg (1992) described observations of icebergs in the inner part of Kongsfjorden, Svalbard, made directly from a launch along transects across the fjord. These authors recorded the maximum width and freeboard of 295 icebergs and presented the data as size–frequency distributions, concluding that iceberg calving in the area was dominated by the production of relatively large numbers of small icebergs. Dowdeswell and Forsberg (1992) also collected data for a number of environmental variables that enabled them to calculate iceberg melt rates and therefore to estimate the 'life expectancy' of icebergs calved into the fjord. The environmental variables collected were measurement of salinity and temperature with depth (CTD) using a Sensordata-200 CTD meter, measurement of iceberg velocities through survey from the fjord shore in order to estimate drift speeds under calm and stormy conditions and a sidescan sonar survey (200 m) track of seafloor morphology to look for evidence of scouring by iceberg keels. The size, frequencies and free-boards of icebergs have also been recorded using ship's radar and sextant outside fjord mouths in Svalbard (Dowdeswell and Forsberg, 1992) and Greenland (Dowdeswell *et al.*, 1992).

9.11.5.2   Field observations of iceberg calving

Observations of large-scale calving events can be made from remote-sensing platforms such as RADARSAT imagery (Fricker *et al.*, 2002), but field observations are generally confined to the scale of individual glaciers. Warren *et al.* (1995) described the magnitude and frequency of iceberg calving events from Glaciar San Rafael, an outlet of the North Patagonian Icefield, by recording the size, location and characteristics of iceberg calving during daylight hours over the space of 32 days in 1991 and 1992. For each individual calving event, the time, location and type of calving (submarine or subaerial) were recorded. A simple iceberg classification scheme was used to record the size of each calving event based on: (1) the proportion of the ice-cliff height involved, (2) the size of the resulting iceberg, and (3) the size and number of waves in the proglacial area generated by each event. The size categories were then given a mean volume in order to calculate the calving flux from the glacier: small

$(25 \, m^3)$, medium $(2500 \, m^3)$, large $(25\,000 \, m^3)$ and extreme $(250\,000 \, m^3)$. This method provides a semi-quantitative means of estimating daily calving flux, although there are acknowledged errors in assigning a volume to event types and in the fact that nocturnal observations were not possible. A further method of mapping and monitoring iceberg calving events is by repeat photography of ice cliffs, a method used successfully by Kirkbride and Warren (1997) at Maud Glacier, New Zealand. These authors took photographs at intervals of 0.25–7 days from the same position on an adjacent moraine to study changes in the ice cliff at the glacier. The repeat photography was supplemented by measurements of the ice cliff using an EDM theodolite and from velocity measurements from aerial photogrammetry.

9.11.5.3 Sampling from icebergs

Icebergs also provide an opportunity to sample the ice and sediment of calving glaciers. Dowdeswell and Dowdeswell (1989) obtained samples of debris from Svalbard icebergs and used these, in conjunction with calculated iceberg melt rates, to estimate sedimentation rates from icebergs. Sedimentation rates calculated from icebergs range from 1 to $7 \, mma^{-1}$. Samples obtained in this way can also be used to provide information on the particle-size distribution of iceberg debris, and to calculate sediment concentrations by weight within icebergs by melting iceberg samples, filtering the water and weighing the residual debris. Hunter *et al.* (1996) used the sediment in icebergs and in glacier ice at contemporary tidewater glaciers to calculate the debris content of different ice types (basal, englacial and supraglacial ice samples) and ultimately to calculate debris fluxes from three tidewater glaciers in Alaska. Finally, Glasser and Hambrey (2001) have reported how sediment samples can be obtained from debris-rich upturned icebergs frozen into sea ice. The sea ice provides a safe platform for skidoo access to icebergs, which are extremely difficult to access when floating in open water.

9.12 MAPPING GLACIER STRUCTURES (STRUCTURAL GLACIOLOGY)

The principles of mapping and measuring structures on glaciers (**structural glaciology**) are similar to those of structural geology, involving the description of structures in ice and snow such as folds, faults, crevasse traces and foliation, together with the assignment of relative ages and interpretations to these structures. For a recent review of progress in this field, see Hambrey and Lawson (2000). Structural glaciological studies normally involve four stages:

1. observations from remotely sensed images (see section 9.5.4.4);
2. field observations;
3. laboratory studies;
4. modelling studies.

9.12.1   Field observations

Structural glaciological field observations require access to the glacier surface, so are only possible where the surface of a glacier can be crossed safely. Heavily crevassed and snow- or debris-covered glaciers are not suitable for such studies. A rope, crampons and an ice axe are essential safety requirements for structural investigations. It is easiest to make structural glaciological observations when the ice surface is bare (i.e. snow-free). Field visits are therefore best planned for late in the ablation season; ideally when the transient snowline has receded to its maximum up-glacier position but before the first new snowfall. This allows for the maximum measurement area on the glacier. Before undertaking fieldwork it is essential that the observer gain prior knowledge of structural glaciology from the reading of descriptions of structures on other glaciers (see for example Allen *et al.*, 1960; Ragan, 1969; Hambrey, 1975, 1977, 1994; Hambrey and Milnes, 1975; Hambrey and Müller, 1978; Hooke and Hudleston, 1978; Casassa, 1992; Lawson *et al.*, 1994; Glasser *et al.*, 1998b; Hambrey *et al.*, 1999; Hambrey and Lawson, 2000). This will aid the identification of features in the field. Features that can be recorded include sedimentary stratification, foliation, crevasses, crevasse traces, folds, transposed foliations, axial planar foliation and thrusts. Guidance on the field description of folds is given in section 8.7.1.

Since the surface of a typical glacier may be crossed by thousands of such structures, it is clearly impractical to map or measure all features present. One way around this is to build up a picture of the structures present by mapping and measuring the orientation of structures systematically at a number of sample points along a series of transect lines. By convention, transect lines are usually marked out across the ice surface in a direction orientated perpendicular to glacier flow. The start and finish of transects can be marked with cairns constructed from boulders, and the locations of transects can then be recorded using standard surveying equipment (see section 9.8). For a valley glacier or outlet glacier, structures will typically be measured at 50-m intervals near the snout and at 100-m intervals in the upper basin (Figure 9.7). A 10–25 m spacing is commonly employed near the glacier margins, where structural orientations tend to change rapidly

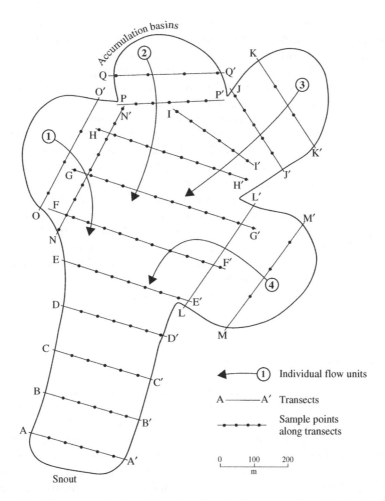

Figure 9.7 Plan view of a hypothetical valley glacier, consisting of a single glacier tongue fed from four separate accumulation basins or tributary cirques. Lines from A–A′ to Q–Q′ represent transects across the glacier surface, along which structural glaciological measurements may be made. In this example, transects are arranged systematically from the glacier snout to the upper accumulation basins so that individual transects are aligned perpendicular to ice flow. In the lower parts of the glacier sample points are located at 50-m intervals along each transect, increasing to 100-m intervals in the upper basins. The number and spacing of measurements can be varied to account for the distribution of different structures on the ice surface (for example by increasing the number of measurements at the confluence of flow units). The number of sample points and overall coverage will also be determined by the position of the transient snowline on the glacier

with distance. The spacing of measurement points and distances along transects can be checked with a 30-m tape or simply by pacing between sample points. The orientation of transects can be determined with a compass or hand-held GPS, if available. Data concerning the orientation of individual structures can be collected using conventional strike-and-dip measurements or recorded as dip and dip direction. At each sample point you should also make observations about the nature of the structures observed, including the type of bedding (e.g. stratified, veined, lenticular), the size of features observed and the relationships between structures (e.g. cross-cutting relationships, offset relationships, relationships between ice structures and debris). Visual observations, sketches and photographs of ice characteristics can also supplement these structural measurements. Features to look for include the shape, size and distribution of inclusions such as air pores, bubbles, cavities, solid impurities (debris), channels and cells containing liquid and any structural inhomogeneities created by them. Observations of ice crystals, including crystal shapes and the average crystal size, should also be recorded. Simple crystallographic fabric analysis can also be performed in the field if the ice surface is weathered and crystals have etched out sufficiently.

9.12.2   Laboratory analysis

This type of analysis is available only if: (a) intact ice blocks can be transported back to a laboratory without the blocks melting during transit, and (b) a suitable laboratory (e.g. walk-in freezer or cold-lab) is available in which to conduct laboratory analyses. Microscopic measurements and descriptions can be made in these laboratories, including the shape and size of crystals, the spatial relationships of the crystals, the relationships between crystals and inclusions, and crystal microfabric measurements (Shumskii, 1964; Hansen and Wilen, 2002).

9.12.3   Modelling studies

Theoretical and numerical modelling studies provide useful tests of the development of structural sequences. For more information on these models see Hubbard and Hubbard (2000) and Lliboutry (2002).

9.12.4   Presentation of structural glaciological data

Structural glaciological data can be presented in both **map format** or plotted on **equal-area stereographic projections** to show the orientation of structures in three dimensions. Structural maps derived from aerial photograph interpretation alone usually show only the distribution of different structural features across the glacier surface (Figure 9.1a,b). However, maps derived from field measurements and field observations show the location of sample points on the glacier surface as well as the trend of structural attributes (Figure 9.8). Data obtained in the field are presented on equal-area stereographic projections (a projection of points from the surface of a sphere onto a plane), which are a useful means of graphically representing 3D structural data. There are two commonly used types of spherical projection: the stereographic (equal-angle) and the Lambert or Schmidt (equal-area) projections (Suppe, 1985). Of the two, the Lambert equal-area projection is probably the most widely used. Data are plotted either by hand, using a piece of tracing paper placed over a co-ordinate net, or by entering the dip angle (plunge) and dip direction (azimuth) of groups of features into a program such as 'Stereo' (one of the utilities in 'Rockware', a geological software package). This program automatically plots the data points onto a spherical projection, as well as contouring the data (if required). Azimuths are measured clockwise from the north and plunges from the outside in. Each spherical projection should be clearly labelled with a title to show the structural attributes that it represents, as well as indicating the number of points plotted (e.g. $n = 50$ to indicate a sample size of 50) so that a reader is able to tell the sample size upon which it is based.

9.13 FINAL MAP COMPILATION

9.13.1 Map scales

Map scale is particularly important because as it is reduced, increasing generalization and grouping of forms is necessary. The most detailed and useful glacier maps are therefore those presented at scales of between 1:5000 and 1:10 000, such as the 1:10 000 map of the White Glacier, Axel Heiberg Island (Blachut and Müller, 1966). Ewing and Marcus (1966) examined several hundred glacier maps and identified three scale categories: large scale (<1:40 000), medium scale (1:40 000–1:300 000) and small scale (>1:300 000). These authors pointed out that as the scale of the map increases, the number of specific elements that can be shown on the map also increases. Large-scale maps are therefore desirable for the presentation of detailed features such as glacier surface structures, but small-scale maps are useful for the presentation of larger portions of glacierized landscapes. Similar principles of scale apply to geomorphological maps, with the scale of presentation determined by the aim of the map.

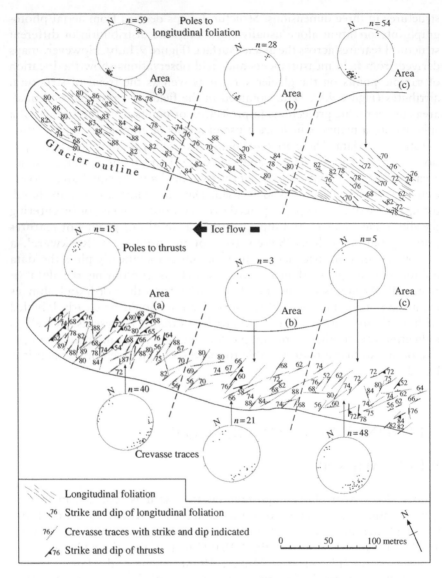

Figure 9.8 Map of the snout of the glacier Kongsvegen in Svalbard showing the structural attributes mapped on the surface of the glacier during fieldwork (modified from Glasser *et al.*, 1998b). The glacier surface has been sub-divided into three areas (a, b and c) for convenience. The upper diagram shows the trend of longitudinal foliation on the glacier surface, together with measured strike and dip. The poles to longitudinal foliation in areas a, b and c are shown in the stereograms, with the sample size indicated in each case by $n = 59$, $n = 28$, etc. The lower diagram shows poles to thrusts and poles to crevasse traces. All stereograms are Lambert equal-area projections (lower hemisphere). Reprinted from the *Journal of Glaciology* with permission of the International Glaciological Society

9.13.2 Cartographic design

Blachut and Müller (1966) provided an excellent overview of the design and presentation of glacier maps, including recommendations concerning scale, contour interval, plotting, symbols and shading, the delineation of the glacierized area and other units, and the use of colour. Ewing and Marcus (1966) also listed some of the possibilities for improving the cartographic design of glacier maps, including the need for colour to highlight glacier contours on the ice surface and to distinguish glacier contours from topographic contours. These authors also highlighted the growing need for the standardization of symbols in glacier cartography, although Ewing and Marcus (1966) stopped short of providing a set of such symbols. Blachut and Müller (1966) provided a key for the symbols used on their White Glacier map, which can be readily used in most circumstances (Figure 9.9a). Similarly, in an attempt to standardize the mapping symbols and legends used in the cartographic representation of geomorphology, Demek and Embleton (1978), Cooke and Doornkamp (1990) and Gardiner and Dackombe (1983) all provided suggested symbols for use when compiling geomorphological maps (Figure 9.9b). Although these schemes may in some cases need to be adapted to account for special circumstances, they provide a useful framework for geomorphological mapping. Finally, a key to the symbols used, a scale bar and a North arrow must be included on all maps.

9.14 MAPPING AND MEASUREMENT OF LANDFORMS CHANGE OVER TIME

Previous studies of mapping and measurement of landform change over time include sedimentary descriptions of the processes involved in resedimentation of glacigenic sediments in proglacial areas (e.g. Boulton, 1967, 1968; Lawson, 1979) but provide little or no quantitative data concerning rates of these processes (cf. Bennett *et al.*, 2000). One recent study that has attempted to redress this balance is that of Kjaer and Krüger (2001), who studied the ablation of ice-cored moraines (areas with sediment-covered stagnant glacier ice) and the formation of dead-ice moraines (ice-free areas created as a result of ablation of ice-cored moraines) over a 4-year period at the terminus of the Kötlujökull Glacier in Iceland. These authors used repeat precision-levelling of ice-cored moraines (accuracy ±1 mm using an EDM theodolite) from a fixed point to create 3D maps of the ice-cored topography and contour diagrams of their surface features. From these measurements, Kjaer and Krüger (2001) were able to calculate rates of surface lowering of the moraines as the ice-cores degraded. They also combined their measurements with sedimentological observations in order

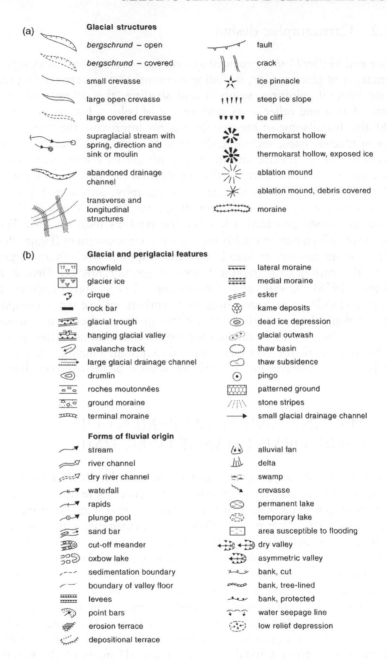

Figure 9.9 Suggested symbols for use (a) in glacier mapping (from Blachut and Müller, 1966); and (b) in geomorphological mapping (from Cooke and Doornkamp, 1974). Note that neither of these schemes is exhaustive, so that although they include the most commonly encountered landforms, additional symbols may need to be added to cover those landforms not included here

to make inferences about the process–form relationships in these rapidly changing environments.

9.15 STUDENT PROJECTS

The mapping techniques outlined in this chapter support a range of research possibilities at both contemporary and former glaciers, as well as opportunities for mapping glacial landforms.

- Morphometric analyses of glacial landforms from map and fieldwork measurements, including glacial valley cross-profiles, long profiles and cirque morphometry.
- Reconstructions of former ELAs from cirque altitudes, together with glaciological reconstructions of former cirque glaciers.
- Glacial geomorphological mapping projects using remotely sensed data and field investigations. These types of mapping projects are suitable to both former and contemporary glacial environments. Glacial geomorphological maps can be used as the basis for more detailed glaciological reconstructions.
- Producing structural glaciological maps from either remotely sensed data or field investigations. These might consist of maps of relatively obvious surface structures such as crevasse patterns or ogives from remotely sensed images or else more complex field observations and measurements on glaciers.
- Investigations of the relationships between ice surface structures and sediments, structural glaciological controls on debris transport and deposition, and the relationship between ice surface structures and surface drainage on glaciers.
- Investigations of the 3D structure of glaciers, which combine surface structural measurements and ice radar.
- Producing maps of bedrock features such as striations and other glacial erosional landforms and relating them to former ice flow directions and ice dynamics. Allied to this are studies of bedrock roughness on proglacial areas and producing maps of the distribution of landforms and sediments on bedrock surfaces to make inferences about the former subglacial conditions.
- Field investigations that involve mapping the distribution of different debris types across a glacier and determining debris transport pathways through glaciers.
- Observations of iceberg calving at modern glacier margins to investigate the relationships between calving rates, water depths and meteorological conditions.

- Quantifying rates of sedimentation in proglacial environments such as lakes and rivers using lake sediment traps, river bedload sampling and suspended sediment sampling.
- Investigations of the 3D structure of sediment bodies using GPR systems and shallow seismic reflection methods.
- Mapping and measurement of landform change over time, for example the repeat survey of changes in ice-cored landforms and the effects of different debris concentrations on surface glacier ablation.

10

Monitoring and reconstructing glacier fluctuations

10.1   AIM

The aim of this chapter is to describe the techniques that can be employed to monitor and reconstruct glacier fluctuations over time. These techniques fall into three main groups: (i) techniques applied to contemporary glaciers, including remote sensing (satellite imagery, aerial photograph interpretation and digital photogrammetry); (ii) field techniques such as mapping and the use of historical documents, which can be used to extend this record back in time; and (iii) numerical-age and relative-age estimates obtained from samples originally collected in the field, which can be used to elucidate the timing and duration of former glacial events. Although we have included discussion of some of the field sampling and laboratory techniques used to obtain numerical-age estimates for periods of former glacier extent, including radiocarbon (^{14}C) dating, cosmogenic isotope dating and luminescence dating, this chapter is by no means a complete list of all geochronological methods. Rather we have selected only those most commonly used in monitoring and reconstructing glacier fluctuations. Equally we have chosen not to cover techniques used mainly in correlating Quaternary glacial events (e.g. tephrochronology and palaeomagnetism). Excellent accounts of these techniques and their application to Quaternary science can be found in Lowe and Walker (1997) and Brigham-Grette (1996).

Field Techniques in Glaciology and Glacial Geomorphology   Bryn Hubbard and Neil Glasser
© 2005 John Wiley & Sons, Ltd

10.2   REMOTELY SENSED IMAGES

10.2.1   Basic principles

The principles of **aerial photography** and **satellite remote sensing** are covered in section 9.5. The most common method of measuring glacier fluctuations is by the analysis of multi-temporal (i.e. time-separated) aerial or terrestrial photographs and high-resolution satellite images, which can be used to reconstruct glacier front positions and changes in glacier morphology over the time sequence available (Box 10.1). Interest in this field of study is inspired chiefly by the desire to understand the relationship between glacier fluctuations, climate and potential sea-level change (Dyurgerov and Meier, 1997a,b). In this context, the most powerful studies are perhaps those that have linked observed glacier fluctuations to numerical models of glacier behaviour (Oerlemans, 1988, 1994, 1997a,b; Oerlemans and Fortuin, 1992; Raper *et al.*, 1996; Schmeits and Oerlemans, 1997). Remotely

---

Box 10.1   Mapping glacier fluctuations from repeat aerial photography

Despite their size, relatively little is known about the dynamics and recent behaviour of the two Patagonian icefields. In one of the first studies of its kind, Aniya and Enomoto (1986) demonstrated how the variations of outlet glaciers of the North Patagonian Icefield could be analysed through time using aerial photographs taken in 1944–1945, 1974–1975 and 1983–1984. The aerial photographs used were a combination of oblique and near-vertical images, either commercially available or taken by the authors from an aircraft with a hand-held camera. From their analysis, Aniya and Enomoto (1986) concluded that the behaviour of the six outlet glaciers studied can be summarized in one of three ways: (1) fairly consistent recession over the 40-year period (two glaciers), (2) rapid recession during the period 1974–1984 (two glaciers), (3) rapid recession between 1944 and 1958. Simple photogrammetric measurements were also made on the photographs, demonstrating surface lowering on the glaciers of between 40 and 120 m between 1944–1945 and 1983–1984. Finally, these authors compared the pattern of glacier recession with meteorological observations (precipitation and temperature) made over the same time period at a nearby meteorological station, but found that the meteorological data showed little relationship with the observed patterns of glacier behaviour. This study was later extended to cover the larger South Patagonian Icefield using a variety of remotely sensed data, including oblique and vertical aerial photographs, Landsat MSS, Landsat TM and SPOT scenes (Aniya *et al.*, 1997).

Aniya, M. and Enomoto, H. 1986. Glacier variations and their causes in the Northern Patagonian Icefield, Chile, since 1944. *Arctic and Alpine Research*, 18, 307–316.

Aniya, M., Sato, H., Naruse, R., Skvarca, P. and Casassa, G. 1997. Recent glacier variations in the Southern Patagonian Icefield, South America. *Arctic and Alpine Research*, 29, 1–12.

sensed images can also be used to map and monitor short-lived geomorphological events such as glacier surges and outburst floods from ice-dammed or moraine-dammed lakes, as well as mapping large-scale geomorphological patterns of erosion and deposition beneath former ice sheets (Box 10.2). In situations where the volume change on a glacier is large enough to be noticeable photographically, it may also be possible to use repeat terrestrial photogrammetry (i.e. hand-held cameras) to monitor glacier recession.

Box 10.2 Using remotely sensed data to reconstruct the dynamics of former ice sheets

In one of the first studies of its kind, Boulton and Clark (1990) used remotely sensed data (in this case Landsat images) to map the distribution of large-scale glacial lineations (streamlined sediment ridges) across mainland Canada. These authors mapped the orientation of individual ridges and observed that in many cases sets of lineations crossed. Commonly there was a clear re-orientation of elements of pre-existing lineations into new lineation sets. Boulton and Clark (1990) therefore suggested that these changes reflected temporal changes in ice sheet flow, and that the ice divide and flow directions of the Laurentide ice sheet migrated across the landscape over time. Boulton and Clark (1990) also compared the orientations of their flow-sets with published data for fabric orientations in glacigenic deposits, bedrock striations and till stratigraphy, and found strong agreement in many areas between these separate data sets. They then proceeded to reconstruct the evolution of the Laurentide Ice Sheet through time and

---

Box 10.2 (Continued)

to construct time–distance envelopes of ice sheet extent for this part of Canada through the last glacial cycle. This work was important because it demonstrated for the first time that satellite data, as well as aerial photography, could be used to map large-scale geomorphological patterns. The study of Boulton and Clark (1990) has now been emulated by a number of other researchers interested in understanding the patterns of ice flow and dynamics of the former Laurentide, Scandinavian and British-Irish Ice Sheets.

Boulton, G.S. and Clark, C.D. 1990. A highly mobile Laurentide ice sheet revealed by satellite images of glacial lineations. *Nature*, 346, 813–817.

10.2.2 Limitations of remotely sensed data for monitoring glacier fluctuations

- Multi-temporal images are required to make inferences about glacier behaviour but such images do not always exist, or may exist only for recent years, so that commonly only short time series are available.
- The best spatial resolution is provided by aerial photography, but each photograph covers only a small geographical area (low **swath width**), so many photographs are needed to cover large areas. Conversely, satellite images provide large swath widths, but with a consequent reduction in spatial resolution. The spatial resolution of the images available therefore determines the scale of investigation. Changes in the extent of small (e.g. valley scale) glaciers are often difficult to detect at the resolution supplied by satellite data but can be monitored using aerial photography. The monitoring of large ice masses, such as the Antarctic and Greenland Ice Sheets, would be difficult using aerial photography and is best tackled using satellite data.
- Horizontal advances or recession of one part of an ice sheet margin or a single outlet glacier may not be representative of overall volume changes of a larger ice mass (Furbish and Andrews, 1984; Tangborn *et al.*, 1990).
- There are technical problems, including the delineation and quantification of changes to land-terminating glaciers in areas where the terminal zone is mantled with debris ('debris-covered glaciers'). In this situation it is difficult to identify consistently the boundary of a given glacier, let alone the trend in its behaviour.

- Glaciers of different morphology may provide contrasting evidence of fluctuations. For example, small cirque glaciers respond to changes in climate and mass balance on annual-to-decadal time-scales but large outlet glaciers often only react dynamically on decadal-to-centennial time-scales (Johannesson *et al.*, 1989; Oerlemans and Hoogendorn, 1989).

10.3 FIELDWORK MAPPING AND HISTORICAL DOCUMENTS

10.3.1 Fieldwork mapping

The use of fieldwork and ground surveys allows glacier fluctuations to be established with great precision, but only for the time period over which observations can be made. The positions of modern glacier margins can be mapped accurately using standard **theodolite-based surveys** (see section 9.8) or by using a differential global positioning system (**dGPS**) (Figure 10.1) (see section 9.8.4 and Box 10.3), then monitored by using repeat surveys over subsequent years. The fluctuations of some glaciers in relatively populated areas, such as the European Alps and Norway, have been measured over the

Figure 10.1 Using a differential global positioning system (dGPS) for mapping purposes in the field. The photographs show the base station with the static receiver (left) and the roving receiver (right) (Photographs: N.F. Glasser)

Box 10.3   Monitoring glacier changes using dGPS

For over 30 years the Okstindan Glacier project has been concerned with the glaciology of this area of Norway. One of the aims of the project was to collect data on changes in glacier volumes in response to mass-balance variations. Up to 1995 this data was collected by terrestrial photogrammetric surveys and by field surveying methods such as EDM surveys. However, Jacobsen and Theakstone (1997) reported the results of a survey of one of the Okstindan glaciers, Austre Okstindbreen, in 1995 using GPS in differential mode. These authors used two GPS receivers, one located at a fixed point (the reference receiver), the other attached to a snow scooter (the roving receiver), to map the surface of the glacier in three dimensions. The reference receiver was mounted on the roof of a hut at a known location, whilst the roving receiver was mounted on a snow scooter with the antenna at a fixed height (1.48 m) above the surface to leave the antenna clear of interference. During data collection, the scooter maintained a near-constant speed of $20 \, \text{km} \, \text{h}^{-1}$. Using the GPS in differential mode, Jacobsen and Theakstone (1997) were able to survey some 2228 points on the glacier surface in less than 6.5 hours of data collection. This compares to the survey of only 253 points in a total of 30 hours using an EDM in the previous survey carried out in 1993. Jacobsen and Theakstone (1997) also calculated that the precision of the 3D co-ordinates collected using the GPS was 0.01–0.02 m. Following the fieldwork, further processing of the GPS data allowed the authors to construct a digital elevation model (DEM) of the glacier surface, which was then compared to a DEM of the glacier surface in 1981 compiled from digital photogrammetry. The differences between the two DEMs show changes in surface elevation and surface topography over this 14-year period. This study demonstrates that GPS in differential mode is a fast, efficient technique for collecting data on glacier surface topography and that data collected in this way is an effective means of monitoring changes in glacier surface elevation and surface topography.

Jacobsen, F.M. and Theakstone, W.H. 1997. Monitoring glacier changes using a global positioning system in differential mode. *Annals of Glaciology*, 24, 314–319.

---

last century by recording the distance from one or more fixed markers on the glacier forefield.

The positions of **landforms and sediments on the glacier forefield**, such as terminal or lateral moraines, are also important clues to former glacier dimensions. A variety of landforms may be created at the ice margin by former glacier advances including glaciotectonic landforms, push moraines, dump moraines, recessional moraines, saw-tooth moraines and latero-frontal ramps and fans (Benn and Evans, 1998). Mapping the distribution of these landforms is one means of building up a picture of past glacier fluctuations, although on its own it can provide no age control on events. A related technique is to use the sedimentary evidence of former proglacial lakes as indicators of the existence of former glaciers in a catchment (Karlén, 1976; Karlén and Matthews, 1992; Leonard, 1997; Bischoff and Cummins, 2001). Turbid proglacial streams normally transport and subsequently deposit accumulations of sand or clayey silt, whereas non-glacial streams favour accumulations of lacustrine sediments with a higher organic content. Thus, the existence of former glacial lake deposits can be used to reconstruct the times when a catchment has been glacierized, especially if any of the deposits contain material suitable for dating purposes (Box 10.4). Dahl *et al.* (2003) provide a useful discussion of some of the

---

Box 10.4   Multiproxy records of glacier fluctuations

One of the most rigorous methods of investigating former glacier extent and dating these advances and recessions is with the use of multiproxy records. Multiproxy records are those that use as many lines of evidence as possible to record past events. Karlén *et al.* (1999) provide a good example of how this technique was used to map and date the former extent of rapidly receding glaciers on Mount Kenya in East Africa. These authors used geomorphological mapping to delimit the extent of former glaciers in the cirques of the mountain. They then obtained sediment cores from the bottom of a number of small moraine-dammed lakes in and around the cirques using a modified Livingstone piston corer operated from two rubber rafts. The sediment cores were sampled in the field and samples were subsequently ana-lysed in the laboratory for weight loss on ignition (by heating to 550 °C) and organic carbon content (using an ELTRA Cs 500). These methods yield information on variations in mineral content, mainly rock-flour input to the lake, which indicate the presence or absence of glaciers in a catchment. These measurements were supplemented by down-core mineral magnetic analysis of the sediment using a portable magnetic susceptibility meter (in this case a Geofyziko Brno Kappa-bridge KLY-2). X-ray photographs were taken of the cores to aid interpretation of their stratigraphy. Cores were also sampled for their microfossil content in order to investigate their palaeoecology. Finally,

---

Box 10.4   (Continued)

macrofossils in the core were submitted for ^{14}C dating by both conventional ^{14}C dating and AMS methods. When all these data were combined Karlén *et al.* (1999) concluded that there was evidence for six periods of glacier advances in the area, dated to 5700, 4500–3900, 3500–3300, 3200–2300, 1300–1200 and 600–400 cal. years BP. There was evidence for a major advance at 5700 cal. years BP, indicating a lowering of the equilibrium line altitude of around 100 m. This study illustrates how a number of methods can be combined to yield information on the timing and extent of former glacier fluctuations.

Karlén, W., Fastook, J.L., Holmgren, K., Malmström, M., Matthews, J.A., Odada, E., Risberg, J., Rosqvist, G., Sandgren, P., Shemesh, A. and Westerberg, L.O. 1999. Glacier fluctuations on Mount Kenya since ∼6000 Cal. Years BP: Implications for Holocene climatic change in Africa. *Ambio*, **28**, 409–417.

---

methodological limitations of using the non-organic/organic signature in lacustrine sediments to reconstruct former glacier extent and in the reconstruction of former glacier equilibrium line altitudes.

Another approach is to use **trimline evidence** to provide information about past glacier thickness and extent (Nesje *et al.*, 1994; McCarroll *et al.*, 1995; Ballantyne *et al.*, 1997). Trimlines in glaciated areas are the oblique or sub-horizontal lines developed on valley sides or around mountain summits that separate ice-scoured bedrock on lower slopes from weathered or frost-shattered bedrock above. Trimlines in glacierized areas can also be picked out by changes in vegetation on slopes and the extent, or lack, of soil cover on a slope. In effect, the trimline represents the former vertical extent of the glacier, and it has been argued that trimlines in glaciated areas can be used to reconstruct former ice surface altitudes (e.g. McCarroll *et al.*, 1995; Brook *et al.*, 1996; Ballantyne *et al.*, 1997; McCarroll and Ballantyne 2000). Note that unless combined with numerical-age estimates such as cosmogenic isotope dating, trimline evidence alone provides little information about the age of specific glacial events (Stone *et al.*, 1998).

10.3.2   Historical documents

Reconstructions of former glacier extent in the last 500 years are made possible by using archive material and historical documents (Lamb, 1977, 1982; Nesje and Dahl, 2000). In this category are:

- **Written documents** such as annals, chronicles, diaries, correspondence and scientific or quasi-scientific writings, government records, estate records, maritime records and commercial records that provide accounts of glacier expansion. For example, there are well-documented reports of agricultural land being abandoned in Iceland, Norway and the European Alps as a result of glacier expansion in the Little Ice Age (Grove, 1988) and information about Little Ice Age glacier damage to agricultural land in Norway from records of tax reductions (Grove and Battagel, 1983).
- **Pictorial records** such as drawings, paintings, engravings and photographs that depict the vertical and horizontal dimensions of former glacier extent. These records have been used from the sixteenth century onwards, particularly in the European Alps. For example, the positions of the Lower Grindelwald Glacier in the Swiss Alps since AD 1590 have been reconstructed from 323 illustrations provided by people visiting the glacier since this time. Over shorter time-scales (e.g. parts of the last century only) glacier-front positions can be reconstructed from time-separated terrestrial and aerial photography.

There are a number of difficulties in using this kind of information. First, the reliability and objectivity of a personal account has to be considered carefully. Drawings and written descriptions of glacier advances may be subjective or use 'poetic licence' in describing glacier-front positions. Second, it is necessary to date and interpret the information accurately in order to establish the duration of the event. The representativeness of the account must also be assessed (i.e. was the event a localized occurrence or can its spatial extent be defined by reference to other sources of information?). Finally, these data must be calibrated against recent observations and cross-referenced with instrumental records. This might be achieved by a construction of indices (e.g. the number of reported glacier advances per year or decade), which can be statistically related to analogous information derived from instrumental records of temperature or precipitation.

10.4   DATING GLACIER FLUCTUATIONS USING 'ABSOLUTE AGE' (NUMERICAL-AGE) AND 'RELATIVE AGE' ESTIMATES

Much of the evidence of past glacier fluctuations is contained in the environment record of the areas surrounding a glacier. Good accounts of the relationships between glaciers and flora and fauna can be found in Lowe and Walker (1997). Glacierized areas can often be harsh environments with deposition restricted to coarse clastic sediments rather than organic material.

Table 10.1 Useful age range of selected dating methods commonly used in glaciological and glacial geomorphological studies

Technique/method	Useful range (years)					
	0	10^2	10^3	10^4	10^5	10^6
Radiocarbon (^{14}C) dating						
Luminescence dating						
Cosmogenic nuclide dating						
Varve chronologies						
Dendrochronology						
Intact rock strength						
Rock weathering rind thickness						
Weathering pit studies						
Lichenometry						

As a result, many glaciers are surrounded by mineral (as opposed to organic) soils and although conventional dating techniques such as radiocarbon (^{14}C) are useful in some cases, this technique is not always applicable. Dating methods are traditionally divided into 'absolute age' methods, which provide a **numerical-age estimate** for a landform, or sediment and 'relative age' methods, which simply indicate the temporal sequence in which landforms or sediments were deposited. These methods can be applied across a range of time-scales (Table 10.1).

In all dating studies, the stratigraphic context of the dated samples is of vital importance. This stratigraphy provides the basic context for the material that is being dated. It is therefore good practice to document in as much detail as possible the context of the sample. The type of information recorded will vary with the techniques employed and we have attempted in the following sections to provide an indication of the type of contextual information that is required.

10.5 NUMERICAL-AGE DATING TECHNIQUES

Numerical-age dating techniques are those that give a numerical-age estimate for a landform or sediment. Three such techniques are in widespread use in glaciology and glacial geomorphology: **radiocarbon (^{14}C) dating**, **luminescence dating** and **cosmogenic nuclide ('exposure-age') dating**. The determination of numerical-age estimates involves the collection of field samples and subsequent laboratory analysis. Since this book is concerned with field techniques and not laboratory techniques we focus on the fundamental principles behind these dating techniques and the collection of field samples, rather than laboratory procedures.

10.5.1   Radiocarbon (^{14}C) dating

The radioactive decay of ^{14}C ('**radiocarbon**') is one of the most commonly used techniques in dating glacigenic deposits, especially in dating glacier advances that lie outside the limits of dendrochronological and lichenometric dating methods (Box 10.5). Carbon-14 is produced in the atmosphere by cosmic rays bombarding nitrogen, and forms a small part of the world's carbon reservoir in the atmosphere, oceans, and most living plants and animals. Carbon-14 dating can be used to obtain numerical-age

---

Box 10.5   Dating glacigenic deposits with ^{14}C methods

Some of the issues associated with dating glacier advances using ^{14}C dating are illustrated using the example of Holocene advances of Patagonian glaciers, an area where ^{14}C dating has been used extensively to date glacier advances. Here the Holocene glacial chronologies are based almost exclusively on ^{14}C dating of organic material (e.g. wood, peat and limnic sediments) in and around moraine ridges. Many authors (e.g. Mercer, 1968, 1970) collected samples for ^{14}C dating from the base of sections excavated in peat deposits ('**basal peat deposits**') within these moraine ridges. Mercer acknowledged that basal peat gives a minimal age for a feature (but not necessarily a close minimal age) because the peat did not necessarily begin to form immediately following glacier recession from the moraine. Thus there is a hiatus between glacier recession and the onset of peat formation. The start of peat growth depends on several factors, the most important of which are climate, micro-relief and composition of the surface material. Mercer (1970) considered the interval between landsurface exposure and the onset of peat formation to be short in Patagonia, of the order of 100–200 years on poorly drained sites such as depressions between moraine ridges, but potentially much longer, of the order of many centuries, on better-drained surfaces such as outwash plains. He postulated that although ^{14}C dates obtained for basal peat could be misleadingly young, they could not be misleadingly old. He therefore considered that if two or more ^{14}C dates for basal peat on the same surface give substantially different minimal ages for the same event, the younger ages could be ignored. More recently, ^{14}C dates have been obtained from better-constrained environments at other glaciers in Patagonia including organic remains in and around vegetation trimlines (e.g. Aniya, 1996), from tree remains embedded in moraines and from trees killed by glacier advances (e.g. Glasser *et al.*, 2002). These

Box 10.5 (Continued)

age-estimates for glacier advances do not necessarily agree with those originally proposed by Mercer, suggesting that either there are errors associated with the sampling strategies employed, or glacier fluctuations were indeed asynchronous in this area, reflecting not only climatic changes, but also other variables that are internal to the glacier system (e.g. Hubbard, 1997; Warren and Aniya, 1999).

Aniya, M. 1996. Holocene variations of Ameghino Glacier, southern Patagonia. *The Holocene*, 6, 247–252.

Glasser, N.F., Hambrey, M.J. and Aniya, M. 2002. An advance of Soler Glacier, North Patagonian Icefield at *c.* AD 1222–1342. *The Holocene*, 12(1), 113–120.

Hubbard, A.L. 1997. Modelling climate, topography and palaeoglacier fluctuations in the Chilean Andes. *Earth Surface Processes and Landforms*, 22, 79–92.

Mercer, J.H. 1968. Variations of some Patagonian glaciers since the Late-Glacial. *American Journal of Science*, 266, 91–109.

Mercer, J.H. 1970. Variations of some Patagonian glaciers since the Late-Glacial: II. *American Journal of Science*, 269, 1–25.

Warren, C. and Aniya, M. 1999. The calving glaciers of southern South America. *Global and Planetary Change*, 22, 59–77.

estimates for organic remains such as deformed pieces of wood (Figure 10.2), organic lake sediments and clasts of reworked peat associated with glacigenic deposits (e.g. Hormes *et al.*, 2001). Successful age-estimates have also been obtained for seeds, needles, mollusc shells, leaves, moss, roots and bones in glacigenic sediments. The basic principle behind this technique is that whilst alive, plants and animals assimilate ^{14}C from the atmosphere through photosynthesis and respiration. When an organism dies, the assimilation of ^{14}C ceases and the existing ^{14}C decays to stable nitrogen with a decay rate (the 'Libby half-life') of 5568 ± 30 years. By assuming an initial concentration of ^{14}C at death, the concentration of ^{14}C remaining can be used to determine the length of time elapsed since death.

Samples may be collected from naturally occurring sections or from bespoke core samples obtained using coring equipment such as a Russian Corer (Jowsey, 1966). Naturally occurring sections should always be cleaned prior to sampling and the stratigraphical relationships in the section recorded, together with the precise stratigraphic context of any samples collected. Some useful points to bear in mind when sampling for ^{14}C dating are:

Figure 10.2 An example of a site in the glacial environment where material could be sampled for radiocarbon (^{14}C) dating. The photograph shows a large striated boulder with a tree stump plastered around its base on the forefield of Soler Glacier, North Patagonian Icefield, Chile (Photograph: N.F. Glasser)

- Is the sample in its true stratigraphical context and is there a close relationship between the death of the sample material and the event whose age is sought?
- Is there a possibility that the sample is derived? In other words, could the sample have been re-worked into the deposit at a later date? If so, sampling will provide an erroneous age-estimate.
- Has the sample remained uncontaminated since death? Signs that a sample may be contaminated include modern root penetration and the movement of organic compounds through the deposit.
- Is it meaningful to collect more than one sample from the same horizon, given that single dates are difficult to interpret?
- Consider the sample weight required. For **conventional** ^{14}C **dating**, recommended sample weights are wood (100–200 g), peat (200–500 g), shells (75–100 g), bones (300–1000 g) and organic soil (200–1000 g). Less material is required for **accelerator mass spectrometry (AMS)** dating.

The amount of ^{14}C in a sample can be determined in two ways, both of which require specialist laboratory facilities. The first method is **conventional** ^{14}C **dating**, the second is **accelerator mass spectrometry (AMS) dating**. Both methods work on similar principles, the main difference being that

AMS dating can be performed on smaller samples and with greater precision. Most ^{14}C dates are carried out in laboratories associated with universities or government research institutions, but there are also a number of commercial laboratories that will process samples. **Radiocarbon determinations** produced by these laboratories are normally quoted as uncalibrated ^{14}C determinations and it is therefore necessary to convert these determinations into calibrated ages. The simplest method of doing this is to use the methods outlined by Stuiver *et al.* (1998a) and Stuiver *et al.* (1998b). Online access to these calibration programmes (e.g. 'CALIB') can be found at: http://www.depts.washington.edu/qil/.

The following information should always be provided when presenting radiocarbon dates.

- The individual laboratory code number, which is prefixed to radiocarbon measurements from that particular laboratory. For example, ANU-3546 refers to sample 3546 measured at the Radiocarbon Laboratory at the Australian National University.
- The Conventional Radiocarbon Age BP (with its ± error equal to ± one standard deviation, for example 5560 ± 230 BP).
- The sample isotopic fractionation (delta ^{13}C) value, whether measured or estimated.
- Any estimate of a reservoir correction (potential errors introduced by the presence of 'dead carbon' from other sources). Any radiocarbon age which possesses a reservoir correction should be termed a *Reservoir Corrected age* and this age should be given in addition to the Conventional Radiocarbon Age.
- The calibrated age of the sample, together with the calibration curve used to calculate the age. The term 'Cal AD' or 'Cal BC' is used to describe calibrated age range data.

Evaluating the reliability of published radiocarbon dates can be difficult but the following points should be considered.

- Are the sample sites clearly located on maps of the glacier forefield or on sediment logs so that the reader can evaluate their wider significance?
- Is the precise context of the sample and sample collection adequately documented? For example, is it clear from which stratigraphic horizons samples were obtained?
- Are the dates quoted as Conventional Radiocarbon Age BP or as a calibrated age? If they have been calibrated, is it stated which calibration curve was used? This is especially important for radiocarbon determinations made last millennium because calibrated data change with successive calibration curves.

- Have the authors acknowledged possible sources of error in their sample collection or analysis?
- Have the authors provided the results of all their determinations or chosen to publish only a sub-sample of the determinations?

10.5.2   Luminescence dating techniques

Luminescence dating is a technique applicable to sediment grains that have been exposed to light prior to deposition (burial). Sediments containing the long-lived radioactive isotopes of ^{238}U, ^{235}U, ^{232}Th and/or ^{40}K are continually bombarded by ionizing alpha, beta and gamma particles. As a result, metastable electrons are displaced and trapped in crystal-lattice defects. The longer the mineral is irradiated, the larger the number of metastable electrons that are trapped. These trapped electrons are released either by heating or by exposure to sunlight. The eviction of trapped electrons is known as 'zeroing' and this is required in order to reset the luminescence 'clock' during deposition. Two main types of luminescence dating are in common use: (1) **thermoluminescence (TL)** dating and (2) **optically stimulated luminescence (OSL)** dating. The main difference is in the method used to evict electrons: in TL dating this is achieved using heat, whilst in OSL dating laser light is used. Richards (2000) provided a useful guide to the theory, methods, use and limitations of OSL dating in glacial environments, using as an example the timing of glaciations in the Himalayas.

Caution is required in using this dating technique in glacial environments because a lot of the sediment transported by glaciers has not been exposed to sunlight during transportation and thus is not fully bleached (zeroed) prior to deposition (Gemmell, 1988). Material that is both subglacially derived and subglacially transported to the glacier margins, either within the basal ice or by the glacial drainage system, will not be bleached during transportation and is therefore unsuitable for the purposes of OSL dating. From experimental analysis of samples obtained from Austerdalsbreen, Norway, Gemmell (1988) concluded that zeroing of sediment does not take place at the base of valley glaciers, but englacial and supraglacial sediment does show evidence of bleaching but not total zeroing. Limited bleaching takes place as sediment passes upwards within a glacier and melts out on its surface. Benn and Owen (2002) argued that OSL dates could be obtained from glacial sediments if a careful sampling strategy is adopted. The deposits left by debris-covered glaciers, such as those of the Himalaya, are, for example, suited to this type of dating because sediment in these glaciers is predominantly supraglacially derived and transported, because the likelihood of sediment bleaching is increased during sediment cycling (e.g. by glacigenic sediment gravity flows) on the glacier surface, and because the deposition of debris in large frontal and lateral moraines by gravitational processes provides further opportunities for zeroing. Fehrentz and Radtke (2001) have successfully applied luminescence dating techniques to glaciofluvial material.

Sediments that may be sampled successfully for luminescence dating include glaciofluvial and glaciolacustrine material, as well as sandy and silty horizons at the top of debris-flow units exposed in lateral and terminal moraines (Richards, 2000). Sediment samples are normally obtained by inserting tubes into the sediment (Figure 10.3) or by using monolith tins. Normally a sample weighing around 30 g is sufficient, but it is best to take more than this in case sediment is required for other analyses (e.g. dosimetry tests or particle-size analysis). A recommended procedure is the following:

1. Ensure that the section to be sampled is a clean and fresh exposure of sediment. To do this, remove at least 0.2 m from the section face to remove all sediment that has been exposed to contemporary light.
2. Insert the sampling container or tube to its maximum depth, avoiding obvious weathering horizons and other pedogenic features.
3. Remove the sampling container or tube from the section, ensuring that the sample does not crumble and that the interior is not inadvertently exposed to light. Immediately wrap the sampling container or tube in an opaque plastic bag and seal this with tape to protect the

Figure 10.3 Sampling equipment for obtaining samples for OSL dating (Photograph: N.F. Glasser)

sample from light. Wrap the sample in a second opaque plastic bag for extra protection and to prevent water loss.

4. Label the sampling container or tube so it is clear which are the outer and the inner ends of the sample, together with a sample reference number and date. The laboratory will remove the outer face of the sample as a matter of course, as this was exposed to light during sample collection, and work only on the interior of the sediment in the sampling container or tube.

5. Store the sample in a cool, dark place prior to transportation back to the laboratory.

10.5.3   Cosmogenic nuclide dating techniques

Recent advances in **cosmogenic nuclide ('exposure-age') dating** techniques now make it possible to obtain absolute age-estimates for bare rock surfaces in glaciated areas (Brook *et al.*, 1996), to validate periglacial weathering limits (Stone *et al.*, 1998), to date erratic boulders (Jackson *et al.*, 1997) and for estimating glacial erosion rates (Briner and Swanson, 1998). Cosmogenic nuclide dating techniques have great potential for addressing unresolved chronological problems in glaciology and glacial geomorphology (Fabel and Harbor, 1999; Watchman and Twidale, 2002). In particular, these exposure-age dating techniques provide a means for establishing glacial chronologies directly from bedrock surfaces, moraine ridges and other glacial deposits over times-scales from thousands to millions of years where other techniques cannot be used (Box 10.6). The six most widely used nuclides are ^{3}He, ^{10}Be, ^{14}C, ^{21}Ne, ^{26}Al

---

Box 10.6   Cosmogenic nuclide ('exposure-age') dating techniques and former glacier fluctuations

There is considerable debate in the scientific literature as to whether or not the glacier advances of the Northern Hemisphere Younger Dryas (approximately 11 000–10 000 ^{14}C years before present) are mirrored by similar glacier advances in the Southern Hemisphere. Previous research on this topic has tended to rely on ^{14}C dates from organic material within and in front of moraine complexes to constrain the timing of glacier advances during the Younger Dryas. Cosmogenic nuclide ('exposure-age') dating techniques, however, now make it possible to obtain absolute-age estimates for bare rock surfaces and boulders in glaciated areas. This technique works by determining the amount of certain cosmogenic isotopes (^{3}He, ^{10}Be, ^{14}C, ^{21}Ne, ^{26}Al and

---

Box 10.6   (Continued)

[36]Cl) within the mineral grains of rocks exposed to cosmic rays. Ivy-Ochs *et al.* (1999) presented the result of a study designed to test whether or not glacier advances occurred simultaneously in both hemispheres during the Younger Dryas. These authors compared cosmogenic isotope concentrations (in this case [10]Be in quartz) extracted from the upper surface of boulders from two moraine complexes, one in the European Alps and one in the Southern Alps of New Zealand, in order to obtain age-estimates for the two moraine complexes. Be was extracted from quartz using standard laboratory procedures and [10]Be and [26]Al were measured with accelerator mass spectrometry (AMS). Exposure age-estimates were then calculated for each of the samples taking into account variations in the atmospheric production rates of these isotopes, together with corrections for topographic shielding and the thickness of the analysed samples. Exposure age-estimates for the New Zealand samples were 9300 ± 990, $11\,000 \pm 1360$, $11\,410 \pm 1030$, $12\,050 \pm 960$ and $12\,410 \pm 1180$ years. The date of 9300 years was considered to be an outlier, giving a mean exposure age-estimate of $11\,720 \pm 320$ years for the New Zealand moraines. This age-estimate is virtually identical to a mean exposure age of $11\,750 \pm 140$ years for the Younger Dryas in the European Alps. Ivy-Ochs *et al.* (1999) therefore suggested that, on the basis of their cosmogenic nuclide ('exposure-age') dates, glacier advances in the European and Southern Alps were in fact synchronous during the Younger Dryas.

Ivy-Ochs, S., Schlüchter, C., Kubik, P. and Denton, G.H. 1999. Moraine exposure dates imply synchronous Younger Dryas glacier advances in the European Alps and in the Southern Alps of New Zealand. *Geografiska Annaler*, **81A**, 313–323.

---

and [36]Cl. The full list of applications of terrestrial *in situ* cosmogenic nuclides includes estimating erosion rates on boulder and bedrock surfaces, fluvial incision rates, denudation rates of individual landforms or entire drainage basins, burial histories of rock surfaces and sediment, scarp retreat, fault-slip rates, palaeoseismology and palaeoaltimetry (Gosse and Phillips, 2001).

Clear field sampling strategies are fundamental to the application of cosmogenic nuclide dating since samples must be collected to reflect accurately the timing of the event in question (Figure 10.4). Samples that have

Figure 10.4 Sampling equipment for obtaining samples for cosmogenic isotope dating (Photograph: N.F. Glasser)

moved since the date of original deposition will yield dates that are too young for the date of original deposition. Care must therefore be taken to avoid sampling boulders on moraines that are not in their original position, through either sliding, slumping or upheaval by vegetation or tree roots. Gosse and Phillips (2001) have provided the following advice on sample collection. Samples should be collected from rock or landform surfaces that are sufficiently extensive, flat and horizontal. 'Sufficiently extensive' means that a sample can be collected at least 0.5 m from an edge or second face. Observations of the surrounding area are also important and, as a minimum, the following information should be recorded for each sample:

Rationale for site selection. Why was a particular piece of rock selected? What is its postulated connection to the event or process being studied? What evidence indicates that it is superior to alternative samples? What items raise doubts about its utility? How does it compare with other samples collected at the same site?

Description of object sampled. Size, shape, colour, other distinguishing features or possibly significant characteristics. Note the height of the sample above ground level. Photographs and labelled sketches are also useful.

Description of sampled material.   Note the surface textures, including presence or absence of glacial polish, striations, fluvial rounding, projection of weathering-resistant inhomogeneities, dissolution features such as weathering pits, general degree of weathering, lithology/mineralogy and percentage lichen cover.

Location of sample.   Record both the location in terms of its geomorphological setting and accurate geographic co-ordinates in latitude ($\pm0.01°$), longitude ($\pm0.01°$) and altitude (±3 m).

Orientation of sample.   Dip and dip direction can be recorded ($\pm5°$) with a compass-clinometer.

Sample thickness.   Most sample analyses are performed on samples collected within the upper 20 mm of the rock surface.

Shielding geometry.   Sample shielding must be recorded by measuring the angle ($\pm3°$) above the horizontal to the skyline at $5°$ intervals. The vegetation canopy surrounding the sample site should also be recorded.

Subsurface profile samples.   These are required in the case of depth-profile sampling. Describe the thickness of the layer sampled, the depth from the surface to the top of the layer, average bulk density above each sample, average clast size and number of clasts sampled in each layer.

Rock samples are usually removed with a hammer and chisel, although portable drills and cutoff saws can also be used if a power supply or fuel is available in the field. Sites near the edges or corners of outcrops or boulders should be avoided because of complications due to cosmic-ray penetration from multiple directions.

10.6   RELATIVE-AGE DATING TECHNIQUES

Relative-age dating techniques are those that indicate the temporal sequence in which landforms or sediments were deposited, but do not provide a numerical-age estimate for those landforms or sediments. These include a variety of dating techniques that can be applied to boulder surfaces in glacial and periglacial environments such as the measurement of intact rock strength (Schmidt hammer), measurements of weathering rind thickness and the depth of weathering pits, together with lichenometry, dendrochronology/dendroglaciology and the relationship between glaciers and surrounding landforms and sediments. Although these techniques cannot provide absolute-age estimates for events, they do have the power to discriminate between the order in which a series of events took place. Relative-age dating in glaciology and glacial geomorphology is usually based on the changes in rock surface characteristics that develop over time due to weathering (for example changes in boulder surface colour or boulder surface hardness, the development of weathering pits and weathering rinds) or by the colonization of rock surfaces by plants such as lichens. These techniques can be used as stand-alone methods or be combined with other indices to obtain absolute-age estimates for the dates of events (e.g. Chinn, 1981). Most of these techniques can be applied on the forefields of contemporary glaciers and are easily carried out at undergraduate level.

10.6.1   Intact rock strength

The use of **intact rock strength** (IRS) as a relative-age dating tool is based on the assumption that progressive weathering of rock surfaces takes place following exposure at the Earth's surface (Matthews and Shakesby, 1984). The simplest method for assessing intact rock strength or rock hardness in the field is that outlined by Selby (1980) based on the use of a standard geological hammer and penknife. In this method, the results of hammer impacts on a rock surface are compared to a published rock hardness scale (Selby, 1980) (Table 10.2). This method is not renowned for its accuracy, however, and geomorphologists seeking to produce statistically significant results have tended to favour the more accurate **Schmidt hammer** method. The Schmidt hammer works by firing a spring-loaded mass onto a rock surface and measuring the distance of its rebound. The distance of rebound depends on the mechanical strength or intact rock strength of a rock (Day, 1980). Mechanical strength declines as weathering continues, so that higher rebound values (**R-values**) refer to less weathered and thus younger surfaces (Day, 1980; McCarroll, 1989a). Intact rock strength is most accurately determined in the laboratory from rock core samples, but the Schmidt hammer (originally designed by E. Schmidt in 1948 for testing the strength of concrete) is a useful and lightweight portable field alternative. Large numbers of tests can be made quickly and data can be collected from a variety of geomorphological surfaces (e.g. bedrock exposures, weathering rinds, individual large boulders and clasts). Standard Schmidt hammers can be used to test rocks with compressive strengths in the range 20–250 MPa (Table 10.2). The potential and limitations of the Schmidt hammer are discussed by McCarroll (1989b). McCarroll and Nesje (1993) outline how Schmidt hammer measurements can be combined with rock surface roughness to estimate the vertical extent of former ice sheets.

The Schmidt hammer is extremely sensitive to discontinuities in a rock, including hairline fractures, creating difficulties in using the instrument to assess the intact rock strength of laminated and fissile rocks such as shales

Table 10.2 A strength classification for intact rock for field measurements with a standard geological hammer and Schmidt hammer (modified from Selby, 1980)

Descriptive name	Unconfined compressive strength (MPa)	Typical Schmidt hammer ('R-value')	Reaction to impact of a geological hammer	Examples of rock type
Very weak rock	1–25	10–35	Crumbles under blow from pick end; can be cut with a penknife	Chalk, lignite
Weak rock	25–50	35–40	Pick end indents deeply with a firm blow; shallow cut can be made with a penknife	Coal, siltstone, schist
Moderately strong rock	50–100	40–50	Shallow indentation with firm blow from pick end. Penknife cannot scrape surface	Slate, shale, sandstone, mudstone, ignimbrite
Strong rock	100–200	50–60	Hand-held sample breaks with one firm blow from hammer end	Marble, limestone, dolomite, andesite, granite, gneiss
Very strong rock	>200	>60	Requires many blows from hammer to break intact sample	Quartzite, dolerite, gabbro

and foliated rocks such as schists. The success (or failure) of studies using the Schmidt hammer rests on the comparison between results from sites located on surfaces of different assumed ages. The key is therefore to minimize the variation between assessments of sites. The following guidelines for data collection are recommended.

1. Suitable sites are intact, flat and horizontal rock or boulder surfaces.
2. The rock or boulder surface should be thoroughly prepared prior to testing, to produce a flat surface, preferably with a carborundum grinding stone.
3. Measurement sites should be more than 60 mm from an edge or a joint.
4. The hammer should be moved to a fresh surface for each impact.
5. A number of measurements (normally at least 15, sometimes as many as 25) should be made at each sample point in order to achieve a statistically reliable result.

6. The mean of the rebound values (R-values) is calculated and the five values deviating most from the mean are omitted (Matthews and Shakesby, 1984). A new mean Schmidt hammer reading is then calculated from the remaining 10 values and the mean can be compared with published values (Moon, 1984).

Since grinding equipment is not available to most fieldworkers, many measurements are taken on unprepared *in situ* rock surfaces. Neither have all researchers adhered to the guidelines suggested above. For example, Evans *et al.* (1999) used a Schmidt hammer to derive R-values for moraines on the forefield of a number of Icelandic glaciers. R-values were calculated from a sum of five blows per boulder on between five and seven boulders per moraine. An average of the five highest R-values for the whole moraine was then used as a representative value for the weathering of the boulders on its surface. In order to minimize variations caused by lithology they chose only massive basalt boulders. They also selected only those boulders displaying clear evidence of subglacial abrasion and therefore most likely to have been subject to weathering only since their release from glacier ice. These authors obtained reasonable R-value/lichenometric correlations where recent glacier advances had abraded debris, but weak correlations were obtained from glaciers where moraines contained re-worked material (Evans *et al.*, 1999).

10.6.2 Weathering rind thickness

Relative-age dating based on **weathering rind thickness** has been applied successfully in several glacial environments, especially in New Zealand (Chinn, 1981; Whitehouse and McSaveney, 1983; Knuepfer, 1988) and the USA (Porter, 1975; Anderson and Anderson, 1981; Colman and Pierce, 1981; Box 10.7). This technique is based on the principle that freshly exposed rock surfaces will develop a surface weathering rind over time. The longer the deposit is exposed to weathering processes, the thicker the rind that develops (Chinn, 1981). Weathering rinds are normally measured by striking a weathered rock surface with a geological hammer until the rind breaks off. The thickness of the rind is then measured with a pair of callipers at its thickest point. Digital measuring instruments, such as DigiMax electronic callipers, are preferable since these have instrument errors of only 0.1 ± 0.001 mm, and provide a reading to the nearest 0.1 mm. This level of precision can also be achieved using a $10 \times$ magnifying optical comparitor with graduations of 0.1 mm (Knuepfer, 1988), so that an accuracy of 0.1–0.2 mm is possible. The mean rind thickness in any particular locality can then be calculated from a number of closely spaced measurements. A

Box 10.7	Using weathering rinds to establish the relative ages of glacial deposits in the Cascade Range, USA

Weathering rinds develop on surface and near-surface clasts in glacial deposits as a result of the oxidation of magnesium- and iron-rich minerals. There is good reason to believe that the thickness of weathering rinds varies systematically with age – the thicker the rind, the longer the clast has been exposed to weathering. In one of the earliest studies to utilize this technique in glacial geomorphology, Porter (1975) examined the thickness of weathering rinds on drift sheets of different ages on the eastern flank of the Cascade Range, Washington, USA. Porter used only basaltic clasts (Teanaway Basalt), where the inner boundary of the weathering rinds is sharp and where the variation in rind thickness on a given clast is minimal. Weathering rinds were measured on clasts in the range of 20–100 mm in diameter collected from the surface of the drift sheets in the area. Where clasts of Teanaway Basalt were common, Porter measured the thickness of the rind of at least 50 clasts at each sample site. Where this lithology was scarce, the sample size was limited to between 20 and 30 clasts. Stones were split with a hammer and rind thickness measured (to an accuracy of the nearest 0.1 mm) with a $7 \times$ measuring magnifier. Where individual clasts showed variations in rind thickness, a number of measurements were made on each clast to obtain an average value. When all the data were assembled, Porter was able to use the mean rind thicknesses to construct time–distance curves for the former glaciers. However, because the rate of rind formation is non-linear, he could not assign absolute ages to these drift sheets without additional radiometric dating. This study does demonstrate, however, that the weathering rind measurements are a useful tool in differentiating and correlating drift sheets. The technique is low-cost and easy to carry out in the field. Furthermore, the reliability of this technique can be increased by combining weathering rind measurements with soil morphologic and chemical properties and by calibrating these measurements at sites of known ages (e.g. Knuepfer, 1988).

Knuepfer, P.L.K. 1988. Estimating ages of late Quaternary stream terraces from analysis of weathering rinds and soils. *Geological Society of America Bulletin*, 100, 1224–1236.

Porter, S.C. 1975. Weathering rinds as a relative-age criterion: Application to subdivision of glacial deposits in the Cascade Range. *Geology*, 3, 101–104.

sample size of 50 boulders from each location or geomorphological surface is normally advocated. If there are large variations in rind thickness at any one location, more than 50 measurements may be required. It is also important to sample a single lithology, since all lithologies weather in different ways and at different rates. For example, variations in rind thickness can be caused by the presence of spalling features and the large transition zone between weathered and unweathered rock (Anderson and Anderson, 1981; Whitehouse and McSaveney, 1983). Thus the simple relationship between weathering and time is complicated by variations in grain size and degree of metamorphism of rocks (Chinn, 1981). In their New Zealand studies Chinn (1981), Whitehouse *et al.* (1986) and Knuepfer (1988) sought to eliminate this possible source of error by sampling only Torlesse greywacke sandstone boulders. Knuepfer (1988) attempted to increase the reliability of this technique by combining weathering rind measurements with soil morphologic and chemical properties, and by calibrating these measurements at sites of known ages. He was able to use this technique at a resolution capable of differentiating between river terrace ages in the period 14–16 ka (thousands of years). Colman and Pierce (1981) have shown that this technique can be used back as far as 300 ka on weathering rinds of basalts and andesites in the western United States.

10.6.3   Weathering pits

Weathering pits develop primarily as a result of solution, usually controlled by differential weathering in response to lithological and structural variations in rock (Ollier, 1984). Weathering pits are not universally present on all rock surfaces and their development is favoured on certain lithologies such as granitic rocks. The basic principle is that the longer a rock or boulder surface has been exposed, the larger and deeper the weathering pits that develop. Field measurements of the diameter of the pit at its surface and the depth of the pit are sufficient to make an assessment of the relative age of weathering pits. Some indication is given by the study of weathering pits on tors in Central Otago, New Zealand (Fahey, 1986). Here, weathering pits of up to 1-m diameter were interpreted as the result of the prolonged exposure (of at least 100 000 years) of tors to weathering processes (Fahey, 1986). Other authors (e.g. McSaveney and Stirling, 1992) suggested even greater ages (between 500 000 and 1 million years) for these features based on the estimated ages of adjacent river terraces. There is no consensus as to the number of measurements required to make a statistically significant sample, since the sampling strategy will depend greatly on the aims of the research. Since, in many instances, it is unclear how long it takes for these features to form, it is also difficult to assess the significance of overall differences in weathering pit diameter. Information on weathering pit depth is commonly displayed cartographically (Figure 10.5).

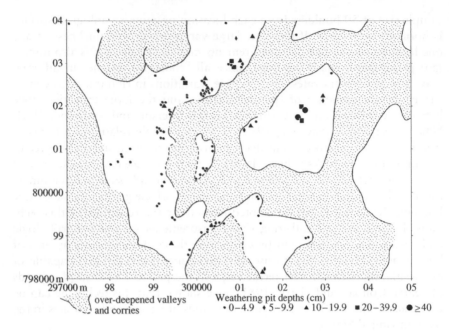

Figure 10.5 The distribution of different weathering pit depths in an area of the Cairngorm Mountains, Scotland, according to Hall and Glasser (2003). Reproduced by permission of Taylor and Francis

10.6.4 Overall assessment of geomorphological rock mass strength

Each of the above attributes (intact rock strength, weathering rind thickness and weathering pit development) can be used individually or in combination using an overall **rock mass strength rating and classification**, such as that proposed by Selby (1980). This classification provides a semi-quantitative means of estimating overall rock strength by assigning numerical values to rock or boulder surfaces according to a number of criteria. The criteria in this classification are the state of weathering of the rock, the spacing of joints, orientation of joints with respect to hillslope, width of the joints, lateral or vertical continuity of the joints, infilling of the joints and movement of water out of the rock mass (Table 10.3).

10.6.5 Lichenometry

Lichenometry is defined as the use of lichens to provide estimates of relative and absolute ages for the substrates on which they are found (Lock *et al.*,

Table 10.3 Geomorphological rock mass strength classification and scores (modified from Selby, 1980). Individual rock surfaces should be scored against the criteria listed to produce a total score out of 100 for the rock mass strength

Criterion	Decreasing rock mass strength				
1. Weathering	Unweathered	Slightly weathered	Moderately weathered	Highly weathered	Completely weathered
	$R = 10$	$R = 9$	$R = 7$	$R = 5$	$R = 3$
2. Spacing of joints	>3 m	1–3 m	0.3–1 m	50–300 mm	<50 mm
	$R = 30$	$R = 28$	$R = 21$	$R = 15$	$R = 8$
3. Joint orientations with respect to hillslope	Steep dips into slope; crossing joints interlock	Moderate dips into slope	Horizontal dips	Moderate dips out of slope	Steep dips out of slope
	$R = 20$	$R = 18$	$R = 14$	$R = 9$	$R = 5$
4. Width of joints	<0.1 mm	0.1–1 mm	1–5 mm	5–20 mm	>20 mm
	$R = 7$	$R = 6$	$R = 5$	$R = 4$	$R = 2$
5. Continuity of joints	None: continuous	Few: continuous	Continuous, no infill	Continuous, thin infill	Continuous, thick infill
	$R = 7$	$R = 6$	$R = 5$	$R = 4$	$R = 1$
6. Outflow of groundwater	None	Trace	Slight	Moderate	Great
	$R = 6$	$R = 5$	$R = 4$	$R = 3$	$R = 1$
Total score	91–100	71–90	51–70	26–50	<26

1979). Lichens are complex organisms consisting of algae and fungi living together symbiotically. Often they are the primary colonizers of a landscape, growing on both rock surfaces and boulders, and capable of survival in relatively harsh environments (Figure 10.6). The use of lichens in dating rests on the principle that there is a direct relationship between lichen size and age (the larger the lichen, the greater its age), although some authors (e.g. Caseldine and Baker, 1998) have noted that disruptions to lichen growth occur in response to climatic factors. This technique has been used successfully in glacial and periglacial environments to date the age of landforms (e.g. moraines, trimlines) and sediments (e.g. talus), which in turn provide dates for glacier fluctuations, rockfalls and debris-flow activity (Rapp and Nyberg, 1981; Innes, 1985; Jonasson *et al.*, 1991; Winchester and Harrison, 1994; McCarroll, 1995; Evans *et al.*, 1999).

Lichenometry is most commonly used in glaciology and glacial geomorphology to estimate the age of landforms such as moraine ridges in order to establish patterns of glacier advance and recession (Savoskul, 1999; Dahms, 2002; Winkler, 2003) or to differentiate between landform origins (Whalley and Palmer, 1998). The usual lichenometric measurement procedure is to measure with a set of callipers or ruler the diameter of the

Figure 10.6 Photograph of lichens growing on a boulder surface in front of Storglaciären, northern Sweden. Lens cap (lower left) for scale (Photograph: N.F. Glasser)

largest inscribed circle of nearly circular thalli (thalli is the plural of thallus, the name given to the growth area of a lichen) on boulders on moraines or adjacent bare rock surfaces (Figure 10.7). Measurements made in this way should be accurate to 0.5 mm or less (Caseldine, 1991).

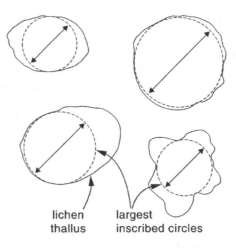

lichen   largest
thallus   inscribed circles

Figure 10.7 Suggested field measurement procedure for lichenometric studies. The diameter of the largest inscribed circle (dashed line) that fits within the lichen thallus (solid line) is used to represent lichen size. Examples are shown for different lichen shapes

Lichen-sampling strategies reported in the literature vary: some research-ers measure the single largest lichen thallus on a surface, some take the mean of the largest thallus from several sites, whilst others have used random sampling to determine the size/frequency distribution. The most common method is to measure the largest inscribed circle of the five largest thalli on a surface and then calculate the mean of these five measurements (Orombelli and Porter, 1983; Innes, 1985). On boulders where the largest measured lichen is 10 mm larger than the second largest, this measurement is often omitted from the mean calculation, especially in the context of debris flows where a single large lichen may have survived such an event (Rapp and Nyberg, 1981; Jonasson et al., 1991).

Lock et al. (1979) listed 34 species of lichens that have been used in lichenometry, but the most commonly used is the green and black lichen *Rhizocarpon geographicum*. This species is used because it is common in polar and alpine environments, it is an early colonizer of rock surfaces, it is slow-growing and it is easily identified in the field. The choice of sample site is important and sites should be chosen so that they are representative of the landform or surface being dated. On ice-cored moraines the crest is prob-ably optimal. Areas of environmental stress (e.g. exposed locations, or areas prone to inundation by snow or water) should be avoided wherever pos-sible. Some pitfalls to avoid include the presence of previously colonized rocks, 'contamination' of a deposit (for example by contemporary rockfall onto an older moraine), measuring large numbers of non-circular thalli, the time required to initiate colonization of the landform or surface by lichens and difficulties in species identification (Figure 10.8).

Lichenometry can be used as a relative dating technique (using lichen-size measurements alone) to estimate the relative ages of landforms and sediments or by using established lichen growth curve for the particular species of lichens under investigation to obtain numerical-age estimates. If such a growth curve does not exist for the field area in question, this can be established by calibrating the size of lichens against a series of surfaces of known age, for example from historical data or aerial photographs (Bradwell, 2001). Lichen growth curves can also be established using the size and ages of lichens on nearby gravestones. Establishing a lichen growth curve requires a considerable amount of time and effort. Most glacierized areas are relatively remote and, as a result, few of these areas have published growth curves. However, it is always worth checking the availability of such curves by undertaking a literature search before embarking on lichenometric studies. For example, Bradwell (2001) published lichen growth curves for Iceland, and lichen studies also exist for many areas of North America (e.g. Benedict, 1967; Osborn and Luckman, 1988), South America (Rodbell, 1992), New Zealand (Bull and Brandon, 1998; Winkler, 2000), the European Alps (Orombelli and Porter, 1983) and Scandinavia (Matthews, 1974). If it is not possible to construct growth curves for a particular area, lichenometry

Figure 10.8 Example of problems encountered when making lichenometric measurements. Fast-growing lichens, like this *Placopsis patagonica* growing in the Nef valley on the east side of the North Patagonian Icefield, Chile, often loose their centres as they grow making it hard to tell a single from a multiple growth. This specimen is almost certainly a multiple. Another problem is that exact measurement of the longest axis of the lichen is difficult on uneven surfaces (Photograph: V. Winchester)

can still be used as a relative-age technique by estimating the percentage of total lichen cover (of all present species) on a boulder or rock surface as a surrogate of exposure age. The principle behind this method is that as surface exposure of a boulder increases, the total cover of the surface with lichens will continue to increase until 100% lichen cover is reached (Nicholas and Butler, 1996). This method has the advantage that lichen-covered surfaces can be photographed in the field, and percentage cover calculated using image analysis software (McCarthy and Zaniewski, 2001).

10.6.6 Dendrochronology and dendroglaciology

Dendrochronology (tree-ring dating) makes use of the annual nature of tree growth. Each year trees develop a layer of new wood under the bark. The thickness of this layer (the tree-ring) depends on various factors, particularly climate. In conditions favourable to growth a wide ring is added, whilst in unfavourable conditions a narrow ring is added. In order to make use of the information contained in the tree-rings, long tree-ring patterns, known as

tree-ring chronologies, are constructed for use as a reference data set. These are produced by overlapping tree-ring patterns from successively older timbers, starting with living trees (Figure 10.9), buildings, and finally samples from archaeological sites and peat bogs (Schweingruber, 1987). To date a timber sample of unknown date, its rings are measured and its pattern of rings is matched against the reference chronology. Each ring on the test sample can then be given a calendar date. Cook and Kairiukstis (1990) cover all aspects of dendrochronology including data gathering, statistical analysis and environmental reconstruction. The longest tree-ring chronologies are those assembled from North American species, where dendrochronology can be used to reconstruct glacier fluctuations for several thousands of years (Osborn *et al.*, 2001).

Dendroglaciology can be used to establish recent chronologies for glacier forefields in any area where there are sufficient trees (Figure 10.10). At its simplest, dendroglaciology can be used to provide a minimum date for moraines formed during glacier advances or stillstands by sampling the oldest trees on each moraine and counting the tree-rings, thus establishing the minimum age of stabilization. For example, Luckman (1994) used this technique to reconstruct glacier fluctuations for the last ~1000 years in the Canadian Rockies. In Patagonia, Winchester and Harrison (1996) and Winchester *et al.* (2001) used tree-borings, taken with a small coring device, to date the former extent of glaciers by dating living trees at incremental

Figure 10.9 Using a coring device on a living tree to obtain an age-estimate of events, Malaya Almatinka valley, Tien Shan, Kazakhstan. Note that the corer is inserted into the tree at chest height. The core obtained suggests that the tree was partially uprooted when the boulder was tipped against it in a flood in 1878 (Photograph: V. Winchester)

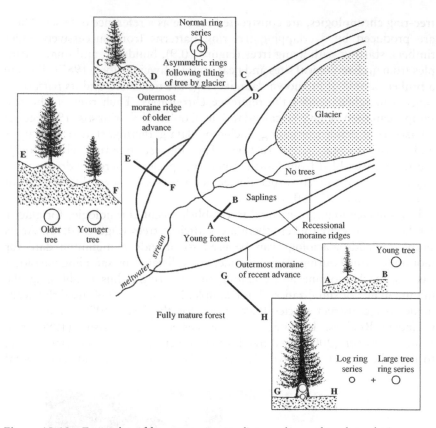

Figure 10.10 Examples of how tree-ring studies can be used to date glacier recession. The diagram depicts a forefield typical of many glaciers in mountainous areas today, with a vegetation succession from no trees near the contemporary ice margin to mature forest at some distance from the contemporary ice margin. Diagram modified from Worsley (1990) with permission from Unwin Hyman Publishers Ltd

distances from the modern glaciers. Winchester and Harrison (1996) dated 32 living trees on the outwash plain in front of, and 16 trees on the flank of, the San Quintin Glacier by taking tree cores at chest height ($\sim$1.2 m) and calculating tree age by counting the annual rings from pith to bark. These dates were then used to calculate rates of glacier recession and down-wasting for incremental time periods since the glacier's 'Little Ice Age' maximum around 1876. Winchester and Harrison (1996) pointed out that dating must take into account the number of years growth to chest height, which depends on the growth rate of the species investigated and the height from which the core is taken. Dating must also take account of the time elapsed between the exposure of a surface by glacier recession and the time for trees to become established (colonization and germination). In some cases colonization is rapid (3–10 years), whilst in other cases there is a long delay (50–70 years) between glacier recession and the start of tree growth. The time interval depends greatly on environmental factors such as aspect, exposure, soil type, rainfall and temperature.

10.6.7 The biogeography of glacier forelands

Patterns and rates of contemporary glacier recession can also be estimated by examining the biogeography of glacier forelands. The general rule is that the longer an area has been deglaciated, the more varied and better developed will be the plant communities growing there. Detailed discussion of the techniques employed in the ecology of recently deglaciated glacier fore-fields is outside the scope of this book and the reader is referred to Matthews (1992) for further details.

10.6.8 Biostratigraphy

Biostratigraphy involves the classification of sedimentary units according to observable variations in fossil content, and enables a sediment succession to be divided into biostratigraphic units or biozones according to the fossil assemblages present. This technique is commonly used for the dating and correlation of older (pre-Quaternary) glacial deposits, although biostratigraphy has also been applied to Quaternary sediments (Lowe and Walker, 1997). For example, diatom biostratigraphy (diatoms are microscopic, unicellular members of the algal kingdom) has been widely used to date and correlate Antarctic glacial sediments. Biostratigraphy is also widely used in the dating, correlation and environmental interpretation of core material. Changes in diatom species have been used for example to indicate changes in former water salinity and therefore to differentiate between glaciomarine and terrestrial sediments in cores.

10.6.9 Amino-acid geochronology

Amino-acid geochronology is a relative-age dating technique based on the premise that proteins in bones and shells are affected by a number of chemical reactions that are partly time-dependent. When an organism dies, tissues are broken down by a variety of chemical processes to produce compounds of a more simple chemical structure. The degree of alteration brought about by these chemical reactions increases with time, and therefore offers a basis for relative dating for any substance containing protein (e.g. bones, shells, teeth, tusks, hair and Foraminifera). Amino-acid measurements are made in the laboratory on samples collected in the field using either ion-exchange chromatography or gas-liquid chromatography. In the glacial context, the majority of amino-acid determinations have been made on glacigenic sediments containing shell material and on raised beach material found in close association with glacigenic sediments (Bowen *et al.*, 1985). Amino-acid ratios have also been used to rank fossils and their associated sediments according to relative age (**aminostratigraphy**) in order to correlate glacigenic sediments between different geographical areas (e.g. Bowen *et al.*, 1986). However, the application of this technique has recently been questioned by McCarroll (2002).

10.7   DATING GLACIER FLUCTUATIONS – CONCLUDING REMARKS

It is important to remember that the data obtained from any sample are only as good as the samples collected in the first place. For this reason, it is vital that all samples collected for dating glacier fluctuations are collected within a rigorous, scientific sampling strategy. The context of all samples must also be adequately documented if numerical-age estimates or relative-age estimates are to be meaningful. Figure 10.11 illustrates these points using a range of sampling sites in moraines and associated sediments and landforms. Sites where samples can be collected for radiocarbon (^{14}C) dating, OSL dating and cosmogenic nuclide ('exposure-age') dating and their possible significance are indicated.

10.8   STUDENT PROJECTS

Some of the methods for monitoring and reconstructing glacier fluctuations outlined in this chapter require expensive and specialist laboratory equipment that will not be available to the majority of undergraduates (e.g. radiocarbon (^{14}C) dating, luminescence dating and cosmogenic nuclide ('exposure-age') dating). The techniques that are probably most suitable at this level are therefore:

- Using remotely sensed data to reconstruct patterns and rates of glacier advance or recession. Time-separated aerial photographs and satellite images are ideal for this type of project work.
- Fieldwork and ground surveys using standard theodolite-based surveys and dGPS can be used to map and monitor glacier fluctuations with great precision. If measurements are only possible in 1 year, comparison with published maps or other pre-existing surveys can be used to put the contemporary measurements into their longer temporal context.

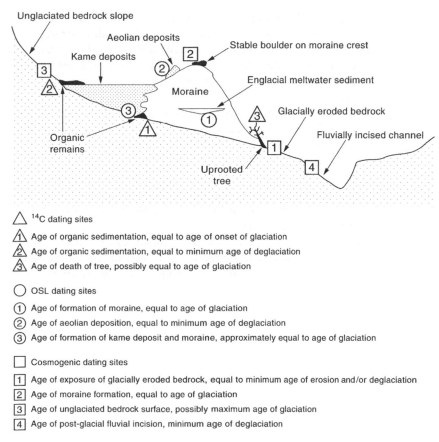

Unglaciated bedrock slope

Aeolian deposits

Kame deposits

Stable boulder on moraine crest

2

3

2

Englacial meltwater sediment

Moraine

3

Glacially eroded bedrock

1

Fluvially incised channel

Organic
remains

1

1

Uprooted
tree

4

△ ¹⁴C dating sites

⚠️1 Age of organic sedimentation, equal to age of onset of glaciation

⚠️2 Age of organic sedimentation, equal to minimum age of deglaciation

⚠️3 Age of death of tree, possibly equal to age of glaciation

○ OSL dating sites

① Age of formation of moraine, equal to age of glaciation

② Age of aeolian deposition, equal to minimum age of deglaciation

③ Age of formation of kame deposit and moraine, approximately equal to age of glaciation

☐ Cosmogenic dating sites

1 Age of exposure of glacially eroded bedrock, equal to minimum age of erosion and/or deglaciation

2 Age of moraine formation, equal to age of glaciation

3 Age of unglaciated bedrock surface, possibly maximum age of glaciation

4 Age of post-glacial fluvial incision, minimum age of deglaciation

Figure 10.11 Sampling strategies for radiocarbon (^{14}C) dating, OSL dating and cosmogenic nuclide ('exposure-age') dating of moraines and associated sediments and landforms. The possible significance of each dating site is indicated. Diagram modified from Benn and Owen (2002) with permission from Elsevier

- Field measurements of IRS are easily achieved using the geological hammer method, together with the published rock hardness scale of Selby (1980). A Schmidt hammer can also be used (if available) to improve the accuracy of this type of study.
- There are numerous possibilities for studies involving measurements of weathering rind thickness and weathering pits to investigate the relative ages of glacial landforms, in both glaciated and glacierized areas. Studies of these weathering phenomena can also be successfully combined with studies of the micro-scale roughness of rock surfaces (see Chapter 9).

- The techniques of lichenometry can be used to establish rates of ice recession on modern glacier forefields. A good sampling strategy for this type of project is to measure lichen diameters along transects from outer moraine ridges (if present) to the current glacier margin. These data can be calibrated using existing lichen growth curves, if established.
- The techniques of dendrochronology and dendroglaciology can be used to establish rates of ice recession and down-wasting for receding glaciers by taking small cores from living trees along transects across a glacier forefield. Again, a good sampling strategy for this type of project is to sample from outer moraine ridges (if present) to the current glacier margin. An alternative sampling strategy for investigating rates of glacier down-wasting is to sample inwards from lateral moraines to the modern glacier.
- There are opportunities to investigate the accuracy and limitations of these dating techniques by using multiproxy investigations (i.e. combining several techniques or dating methods at a single study site).

References

Aa, A.R. 1996. Topographic control of equilibrium-line altitude depression on reconstructed 'Little Ice Age' glaciers, Grovabreen, western Norway. *The Holocene*, 6, 82–89.

Allen, C.R., Kamb, W.B., Meier, M.F. and Sharp, R.P. 1960. Structure of the Lower Blue Glacier, Washington. *Journal of Geology*, 68, 601–625.

Alley, R.B. 2000. *The Two-Mile Time Machine*. Princeton, Princeton University Press, NJ, 229pp.

Anderson, L.W. and Anderson, D.S. 1981. Weathering rinds on quartzarenite clasts as a relative-age indicator and the glacial chronology of Mount Timpanogos, Wasatch Range, Utah. *Arctic and Alpine Research*, 13, 25–31.

Anderson, E., Harrison, S. and Passmore, D.G. 1998. Geomorphological mapping and morphostratigraphic relationships: Reconstructing the glacial history of the Macgillycuddy's Reeks, south west Ireland. *Society of Cartographers Bulletin*, 30, 9–20.

Andreassen, L.M., Elvehoy, H. and Kjollmoen, B. 2002. Using aerial photography to study glacier changes in Norway. *Annals of Glaciology*, 34, 343–348.

Aniya, M. and Welch, R. 1981. Morphometric analyses of Antarctic cirques from photogrammetric measurements. *Geografiska Annaler*, 63A, 41–53.

Aniya, M., Naruse, R. and Yamaguchi, S. 2002. Utilization of 6 × 6 cm format vertical aerial photographs for repetitive mapping of surface morphology and measurement of flow velocities of a small glacier in a remote area: Glaciar Soler, Hielo Patagónico Norte, Chile. *Annals of Glaciology*, 34, 385–390.

Annan, A.P. 1999. Ground penetrating radar workshop notes. Brevik Place, Mississauga, Ontario, ON L4W 3R7, Canada, Sensors and Software Inc., 1091.

Annan, A.P. 2001. Ground penetrating radar workshop notes. Ontario, Canada, Sensors and Software Inc., 197.

Annan, A.P. and Cosway, S.W. 1992. Ground penetrating radar survey design. Proceedings of the Symposium on the Application of Geophysics to Engineering and Environmental Problems (SAGEEP '92), Oakbrook, Illinois, USA, 329–351.

Arnold, K.C. 1966. The glaciological maps of Meighen Island, N.W.T. *Canadian Journal of Earth Sciences*, 6, 903–908.

Arrigo, K.R., Van Dijken, G.L., Ainley, D.G., Fahnestock, M.A. and Markus, T. 2002. Ecological impact of a large Antarctic iceberg. *Geophysical Research Letters*, 29, art. no. 1104.

Ashley, G.M., Shaw, J. and Smith, N.D. 1985. *Glacial Sedimentary Environments*. SEPM Short Course No. 16. Society of Paleontologists and Mineralogists, Tulsa. 246pp.

Avery, T.E. 1977. *Interpretation of Aerial Photographs*. Minneapolis, Burgess.

Avsiuk, G.A., Vinogradov, O.N. and Kravtsova, V.I. 1966. Experience in glaciological mapping of ice sheets and mountain glaciers. *Canadian Journal of Earth Sciences*, 6, 841–847.

Baines, D., Smith, D.G., Froese, D.G., Bauman, P. and Nimeck, G. 2002. Electrical resistivity ground imaging (ERGI): A new tool for mapping the lithology and geometry of channel-belts and valley-fills. *Sedimentology*, 49(3), 441–449.

Ballantyne, C.K., McCarroll, D., Nesje, A. and Dahl, S.O. 1997. Periglacial trimlines, former nunataks and the altitude of the last ice sheet in Wester Ross, northwest Scotland. *Journal of Quaternary Science*, 12, 225–238.

Bamber, J.L. 1987. Internal reflecting horizons in Spitsbergen glaciers. *Annals of Glaciology*, 9, 5–10.

Bamber, J.L., Ekholm, S. and Krabill, W.B. 2001. A new, high-resolution digital elevation model of Greenland fully validated with airborne laser altimeter data. *Journal of Geophysical Research*, 106(B4), 6733–6746.

Barnola, J.-M., Raynaud, D., Neftel, A. and Oeschger, H. 1983. Comparison of CO_2 measurements by two laboratories on air from bubbles in polar ice. *Nature*, 303, 410–413.

Barrett, P.J. 1980. The shape of rock particles: A critical review. *Sedimentology*, 27, 291–303.

Barrett, P.J. 1989. Sediment texture. In P.J. Barrett (ed.) Antarctic Cenozoic history from the CIROS-1 drillhole, McMurdo Sound. *DSIR Bulletin*, 245, 49–58.

Benedict, J.B. 1967. Recent glacial history of an alpine area in the Colorado Front Range, USA. I. Establishing a lichen growth curve. *Journal of Glaciology*, 6, 817–832.

Benn, D.I. 1994. Fabric shape and the interpretation of sedimentary fabric data. *Journal of Sedimentary Research*, 64, 910–915.

Benn, D.I. 1995. Fabric signature of till deformation, Breidamerkurjökull, Iceland. *Sedimentology*, 42, 735–747.

Benn, D.I. 2002. Discussion and reply: Clast fabric development in a shearing granular material: Implications for subglacial till and fault gouge. *Geological Society of America Bulletin*, 114(3), 382–383.

Benn, D.I. and Ballantyne, C.K. 1993. The description and representation of particle shape. *Earth Surface Processes and Landforms*, 18, 665–672.

Benn, D.I. and Ballantyne, C.K. 1994. Reconstructing the transport history of glacigenic sediments: A new approach based on the co-variance of clast form indices. *Sedimentary Geology*, 91, 215–227.

Benn, D.I. and Evans, D.J.A. 1996. The interpretation and classification of subglacially-deformed materials. *Quaternary Science Reviews*, 15, 23–52.

Benn, D.I. and Evans, D.J.A. 1998. *Glaciers and Glaciation*. London, Arnold.

Benn, D.I. and Gemmell, A.M.D. 2002. Fractal dimensions of diamictic particle-size distributions: Simulations and evaluation. *Geological Society of America Bulletin*, 114(5), 528–532.

Benn, D.I. and Owen, L.A. 2002. Himalayan glacial sedimentary environments: A framework for reconstructing and dating the former extent of glaciers in high mountains. *Quaternary International*, 97–98, 3–25.

Benn, D.I. and Ringrose, T.J. 2001. Random variation of fabric eigenvalues: Implications for the use of A-axis fabric data to differentiate till facies. *Earth Surface Processes and Landforms*, 26, 295–306.

Bennett, M.R., Hambrey, M.J. and Huddart, D. 1997. Modification of clast shape in high-arctic environments. *Journal of Sedimentary Research*, 67, 550–559.

Bennett, M.R., Huddart, D. and Waller, R.I. 2000. Glaciofluvial crevasse and conduit fills as indicators of supraglacial dewatering during a surge, Skeidararjökull, Iceland. *Journal of Glaciology*, 46, 25–34.

Bennett, M.R., Huddart, D., Glasser, N.F. and Hambrey, M.J. 2000. Resedimentation of debris on an ice-cored lateral moraine in the High-Arctic (Kongsvegen, Svalbard). *Geomorphology*, 35, 21–40.

Bennett, M.R., Waller, R.I., Glasser, N.F., Hambrey, M.J. and Huddart, D. 1999. Glacigenic clast fabrics: Genetic fingerprint or wishful thinking? *Journal of Quaternary Science*, 14(2), 125–135.

Benson, C.S. 1961. Stratigraphic studies in the snow and firn of the Greenland Ice Sheet. *Folia Geographica Danica*, 9, 13–37.

Billi, P. 1984. Quick field measurement of gravel particle size. *Journal of Sedimentary Petrology*, 54, 658–660.

Bindschadler, R. and Vornberger, P. 1998. Changes in the West Antarctic ice sheet since 1963 from declassified satellite photography. *Science*, 279(5351), 689–692.

Bindschadler, R., Scambos, T.A., Rott, H., Skvarca, P. and Vornberger, P. 2002. Ice dolines on Larsen Ice Shelf, Antarctica. *Annals of Glaciology*, 34, 283–290.

Bischoff, J.L. and Cummins, K. 2001. Wisconsin glaciation of the Sierra Nevada (79,000–15,000 yr BP) as recorded by rock flour in sediments of Owens Lake, California. *Quaternary Research*, 55, 14–24.

Björnsson, H. 1998. Hydrological characteristics of the drainage system beneath a surging glacier. *Nature*, 395, 771–774.

Blachut, T.J. and Müller, F. 1966. Some fundamental considerations on glacier mapping. *Canadian Journal of Earth Sciences*, 6, 747–759.

Blair, T.C. and MacPherson, J.G. 1999. Grain-size and textural classification of coarse sedimentary particles. *Journal of Sedimentary Research*, 69, 6–19.

Blake, E.W. and Clarke, G.K.C. 1991. Subglacial water and sediment samplers. *Journal of Glaciology*, 37(125), 188–190.

Blake, E.W. and Clarke, G.K.C. 1992. Interpretation of borehole-inclinometer data: A general-theory applied to a new instrument. *Journal of Glaciology*, 38(128), 113–124.

Blake, E.W. and Clarke, G.K.C. 1999. Subglacial electrical phenomena. *Journal of Geophysical Research-Solid Earth*, 104(B4), 7481–7495.

Blake, E.W., Clarke, G.K.C. and Gérin, M.C. 1992. Tools for examining subglacial bed deformation. *Journal of Glaciology*, 38(130), 388–396.

Blake, E.W., Fischer, U.H. and Clarke, G.K.C. 1994. Direct measurement of sliding at the glacier bed. *Journal of Glaciology*, 40(136), 595–599.

Blake, E.W., Wake, C.P. and Gerasimoff, M.D. 1998. The ECLIPSE drill: A field-portable intermediate-depth ice-coring drill. *Journal of Glaciology*, 44(146), 175–178.

Blankenship, D.D., Bentley, C.R., Rooney, S.T. and Alley, R.B. 1986. Seismic measurements reveal a saturated porous layer beneath an active Antarctic ice stream. *Nature*, 322(6074), 54–57.

Bloesch, J. and Burns, N.M. 1980. A critical review of sedimentation trap technique. *Schweizerische Zeitschrift fur Hydrologie*, 42, 15–55.

Blong, R.J. 1972. Methods of slope profile measurement in the field. *Australian Geographical Studies*, 10, 182–192.

Blott, S.J. and Pye, K. 2001. Gradistat: A grain size distribution and statistics package for the analysis of unconsolidated sediments. *Earth Surface Processes and Landforms*, 26, 1237–1248.

Boone, S.J. and Eyles, N. 2001. Geotechnical model for Great Plains hummocky moraine formed by till deformation below stagnant ice. *Geomorphology*, 38, 109–124.

Boulton, G.S. 1967. The development of a complex supraglacial moraine at the margin of Sørbreen, Ny Friesland, Vestspitsbergen. *Journal of Glaciology*, 6, 717–735.

Boulton, G.S. 1968. Flow tills and related deposits on some Vestspitsbergen glaciers. *Journal of Glaciology*, 7, 391–414.

Boulton, G.S. 1976. The development of geotechnical properties in glacial tills. In R.F. Legget (ed.) *Glacial Till: An Inter-Disciplinary Study*. The Royal Society of Canada Special Publications No. 12, Ottowa, 292–303.

Boulton, G.S. 1978a. Boulder shapes and grain-size distributions as indicators of transport paths through a glacier and till genesis. *Sedimentology*, 25, 773–799.

Boulton, G.S. 1978b. The genesis of glacial tills: A framework for geotechnical interpretation. In *The Engineering Behaviour of Glacial Materials*. GeoAbstracts, Norwich.

Boulton, G.S. and Dobbie, K.E. 1993. Consolidation of sediments by glaciers: Relationship between sediment geotechnics, soft-bed glacier dynamics and subglacial groundwater flow. *Journal of Glaciology*, 39, 26–44.

Boulton, G.S. and Hindmarsh, R.C.A. 1987. Sediment deformation beneath glaciers: Rheology and geological consequences. *Journal of Geophysical Research*, 92(B9), 9059–9082.

Boulton, G.S. and Paul, M.A. 1976. The influence of genetic properties on some geotechnical properties of till. *Journal of Engineering Geology*, 9, 159–194.

Boulton, G.S., Van der Meer, J.J.M., Beets, D.J., Hart, J.K. and Ruegg, G.H.J. 1999. The sedimentary and structural evolution of a recent push moraine complex: Holmstrømbreen, Spitsbergen. *Quaternary Science Reviews*, 18, 339–371.

Bowen, D.Q., Sykes, G.A., Reeves, A., Miller, G.H., Andrews, J.T., Brew, J.S. and Hare, P.E. 1985. Amino acid geochronology of raised beaches in south west Britain. *Quaternary Science Reviews*, 4, 279–318.

Bowen, D.Q., Rose, J., Mccabe, A.M. and Sutherland, D.G. 1986. Correlations of Quaternary glaciations in England, Ireland, Scotland and Wales. *Quaternary Science Reviews*, 5, 299–340.

Bowles, J.E. 1992. *Engineering Properties of Soils and Their Measurement*. New York, McGraw-Hill, 241pp.

Bradwell, T. 2001. A new lichenometric dating curve for southeast Iceland. *Geografiska Annaler*, 83A, 91–101.

Brandenberger, A.J. and Bull, C. 1966. Glacier surveying and mapping program of the Ohio State University. *Canadian Journal of Earth Sciences*, 6, 849–861.

Bridgland, D.R. (ed.) 1986. *Clast Lithological Analysis*. Technical Guide No. 3, London, Quaternary Research Association.

Brigham-Grette, J. 1996. Geochronology of glacial deposits. In J. Menzies (ed.) *Past Glacial Environments: Sediments, Forms and Techniques*. Oxford, Butterworth-Heinemann, 377–410.

Briner, J.P. and Swanson, T.W. 1998. Using inherited cosmogenic [36]Cl to constrain glacial erosion rates of the Cordilleran ice sheet. *Geology*, 26, 3–6.

Brockie, W.J. 1972. Field techniques in geomorphology. In F.A. Slater and T.J. Hearn (eds) *Field Techniques in Geography*. Department of University Extension, University of Otago, New Zealand, 34–48, 61pp.

Brocklehurst, S.H. and Whipple, K.X. 2004. Hypsometry of glaciated landscapes. *Earth Surface Processes and Landforms*, **29**, 907–926.

Bronge, L.B. and Bronge, C. 1999. Ice and snow-type classification in the Vestfold Hills, East Antarctica, using Landsat-TM data and ground radiometer measurements. *International Journal of Remote Sensing*, **20**(2), 225–240.

Brook, E.J., Nesje, A., Lehman, S.T., Raisbeck, G.M. and Yiou, F. 1996. Cosmogenic nuclide exposure ages a long a vertical transect in western Norway: Implications for the height of the Fennoscandian ice sheet. *Geology*, **24**, 207–210.

Brown, E., Skougstad, M.W. and Fishman, M.J. 1970. *Methods for Collection and Analysis of Water Samples for Dissolved Minerals and Gases*. Washington, DC, US Government Printing Office.

Brown, D.G., Lusch, D.P. and Duda, K.A. 1998. Supervised classification of types of glaciated landscapes using digital elevation data. *Geomorphology*, **21**, 233–250.

Brugman, M. 1986. *Water Flow at the Base of a Surging Glacier*. Pasadena, California Institute of Technology.

BS3680. 1980. Methods of measurement of liquid flow in open channels – British Part 3A: Velocity area methods. London, British Standards Institution.

Bull, W.B. and Brandon, M.T. 1998. Lichen dating of earthquake-generated regional rockfall events, Southern Alps, New Zealand. *Geological Society of America Bulletin*, **110**, 60–84.

Burger, H.R. 1992, *Exploration Geophysics of the Shallow Subsurface*. Englewood Cliffs, New Jersey, Prentice Hall.

Burkimsher, M. 1983. Investigations of glacier hydrological systems using dye tracer techniques: Observations at Pasterzengletscher, Austria. *Journal of Glaciology*, **29**(103), 403–416.

Canals, M., Casamor, J.L., Urgeles, R., Calafat, A.M., Domack, E.W., Baraza, J., Farran, M. and De Batist, M. 2002. Seafloor evidence of a subglacial sedimentary system off the northern Antarctic Peninsula. *Geology*, **30**(7), 603–606.

Carr, S. 2001. A glaciological approach for the discrimination of Loch Lomond Stadial glacial landforms in the Brecon Beacons, South Wales. *Proceedings of the Geologists' Association*, **112**, 253–262.

Carr, S.J. 2001. Micromorphological criteria for discriminating subglacial and glacimarine sediments: Evidence from a contemporary tidewater glacier, Spitsbergen. *Quaternary International*, **86**, 71–79.

Carr, S.J. and Lee, J.A. 1998. Thin section production of diamicts: Problems and solutions. *Journal of Sedimentary Research*, **68**, 217–221.

Carr, S.J., Haflidason, H. and Sejrup, H.P. 2000. Micromorphological evidence supporting Late Weichselian glaciation of the Northern North Sea. *Boreas*, **29**, 315–328.

Casassa, G. 1992. Foliation on Tyndall Glacier, southern Patagonia. *Bulletin of Glacier Research*, **10**, 75–77.

Caseldine, C. 1991. Lichenometric dating, lichen population studies and Holocene glacial history in Tröllaskagi, Northern Iceland. In J.K. Maizels and C.J. Caseldine (eds) *Environmental Change in Iceland: Past and Present*. Dordrecht, Kluwer, 219–233.

Caseldine, C. and Baker, A. 1998. Frequency distributions of *Rhizocarpon geographicum* sl, modelling, and climatic variation in Tröllaskagi, Northern Iceland. *Arctic and Alpine Research*, **30**, 175–183.

Chinn, T.J.H. 1981. Use of rock weathering-rind thickness for Holocene absolute age-dating in New Zealand. *Arctic and Alpine Research*, **13**, 33–45.

Chorley, R.J. 1959. The shape of drumlins. *Journal of Glaciology*, **3**, 339–344.

Clarke, B.G., Chen, C.C. and Aflaki, E. 1998. Intrinsic compression and swelling properties of a glacial till. *Quarterly Journal of Engineering Geology*, **31**, 235–246.

Cohen, D. 2000. Rheology of ice at the bed of Engabreen, Norway. *Journal of Glaciology*, **46**(155), 611–621.

Collins, D.N. 1978. Hydrology of an alpine glacier as indicated by the chemical composition of meltwater. *Zeitschrift für Gletscherkunde und Glazialgeologie*, **13**, 219–238.

Colman, S.M. and Pierce, K.L. 1981. Weathering rinds on andesitic and basaltic stones as a Quaternary age indicator, western United States. *US Geological Survey Professional Paper*, **1210**, 41pp.

Compton, R.R. 1985. *Geology in the Field*. New York, John Wiley & Sons.

Cook, E.R. and Kairiukstis, L.A. 1990. *Methods of Dendrochronology: Applications in the Environmental Sciences*. Dordrecht, Kluwer, 394 pp.

Cooke, R.U. and Doornkamp, J.C. 1990. *Geomorphology in Environmental Management: A New Introduction*. 2nd edition. Oxford, Clarendon.

Cooke, R.U. and Doornkamp, J.C. 1974. *Geomorphology in Environmental Management: An Introduction*. Oxford, Clarendon Press, 413pp.

Copland, L., Harbor, J. and Sharp, M. 1997a. Borehole video observation of englacial and basal ice conditions in a temperate valley glacier. *Annals of Glaciology*, **24**, 277–282.

Copland, L., Harbor, J., Minner, M. and Sharp, M. 1997b. The use of borehole inclinometry in determining basal sliding and internal deformation at Haut Glacier d'Arolla, Switzerland. *Annals of Glaciology*, **24**, 331–337.

Dahl, S.O., Bakke, J., Lie, Ø. and Nesje, A. 2003. Reconstruction of former glacier equilibrium-line altitudes based on proglacial sites: An evaluation of approaches and selection of sites. *Quaternary Science Reviews*, **22**, 275–287.

Dahms, D.E. 2002. Glacial stratigraphy of Stough Creek Basin, Wind River Range, Wyoming. *Geomorphology*, **42**, 59–83.

Davies, T.A., Bell, T., Cooper, A.K., Josenhans, H., Polyak, L., Solheim, A., Stoker, M.S. and Stravers, J.A. 1997. *Glaciated Continental Margins: An Atlas of Acoustic Images*. London, Chapman and Hall.

Day, M.J. 1980. Rock hardness: Field assessment and geomorphic importance. *Professional Geographer*, **32**, 72–81.

Day, M.J. and Goudie, A.S. 1977. Field assessment of rock hardness using the Schmidt hammer. *British Geomorphological Research Group Bulletin*, **18**, 19–29.

Demek, J. and Embleton, C. 1978. *Guide to Medium-Scale Geomorphological Mapping*, International Geographical Union Commission on Geomorphological Survey and Mapping, Brno. 348pp.

Denby, B. and Greuell, W. 2000. The use of bulk and profile methods for determining surface heat fluxes in the presence of glacier winds. *Journal of Glaciology*, **46**(154), 445–452.

Dickinson, G.C. 1979. *Maps and Air Photographs*. London, Edward Arnold, 348pp.

Dietrich, W.E. and Gallinatti, J.D. 1991. Fluvial geomorphology. In O. Slaymaker (ed.) *Field Experiments and Measurement Programs in Geomorphology*. Rotterdam, Balkema, 169–220.

Domack, E.W. and Lawson, D.E. 1985. Pebble fabric in an ice-rafted diamicton. *Journal of Geology*, 93, 577–591.

Dowdeswell, J.A. 1989. On the nature of Svalbard icebergs. *Journal of Glaciology*, 35, 224–234.

Dowdeswell, J.A. and Dowdeswell, E.K. 1989. Debris in icebergs and rates of glacimarine sedimentation: Observations from Spitsbergen and a simple model. *Journal of Geology*, 97, 221–231.

Dowdeswell, J.A. and Forsberg, C.F. 1992. The size and frequency of icebergs and bergy bits derived from tidewater glaciers in Kongsfjorden, northwest Spitsbergen. *Polar Research*, 11, 81–91.

Dowdeswell, J.A. and O'Cofaigh, C. (eds) 2002. *Glacier-Influenced Sedimentation on High-Latitude Continental Margins*. Geological Society Special Publication No. 203, London, The Geological Society.

Dowdeswell, J.A. and Powell, R.D. 1996. Submersible remotely operated vehicles (ROVs) for investigations of the glacier-ocean-sediment interface. *Journal of Glaciology*, 42, 176–183.

Dowdeswell, J.A. and Scourse, J.D. (eds) 1990. *Glacimarine Environments: Process and Sediments*. Geological Society Special Publication No. 53, London, The Geological Society.

Dowdeswell, J.A. and Sharp, M.J. 1986. Characterization of pebble fabrics in modern terrestrial glacigenic sediments. *Sedimentology*, 33, 699–710.

Dowdeswell, J.A. and Williams, M. 1997. Surge-type glaciers in the Russian High Arctic identified from digital satellite imagery. *Journal of Glaciology*, 43, 489–494.

Dowdeswell, J.A., Drewry, D.J., Liestol, O. and Orheim, O. 1984. Radio-echo sounding of Spitzbergen glaciers: Problems in the interpretation of layer and bottom returns. *Journal of Glaciology*, 30(104), 16–21.

Dowdeswell, J.A., Hambrey, M.J. and Wu, R. 1985. A comparison of clast fabric and shape in Late Precambrian and modern glacigenic sediments. *Journal of Sedimentary Petrology*, 55, 691–704.

Dowdeswell, J.A., Hamilton, G.S. and Hagen, J.O. 1991. The duration of the active phase on surge-type glaciers: Contrasts between Svalbard and other regions. *Journal of Glaciology*, 37, 388–400.

Dowdeswell, J.A., Whittington, R.J. and Hodgkins, R. 1992. The size, frequencies and freeboards of East Greenland icebergs observed using ship radar and sextant. *Journal of Geophysical Research*, 97(C3), 3515–3528.

Dowdeswell, J.A., Hodgkins, R., Nuttall, A.M., Hagen, J.O. and Hamilton, G.S. 1995. Mass-balance change as a control on the frequency and occurrence of glacier surges in Svalbard, Norwegian High Arctic. *Geophysical Research Letters*, 22, 2909–2912.

Drake, L.D. 1972. Mechanisms of clast attrition in basal till. *Geological Society of America Bulletin*, 83, 2159–2166.

Dreimanis, A. and Vagners, U.J. 1971. Bimodal distribution of rock and mineral fragments in basal tills. In R.P. Goldthwait (ed.) *Till: A Symposium*. Ohio, Ohio State University Press, 237–250.

Duck, R.W. 1994. Application of the QDA MD method of environmental discrimination to particle-size analysis of fine sediments by pipette and Sedigraph methods – a comparative study. *Earth Surface Processes and Landforms*, 19(6), 525–529.

Dunphy, P.P., Dibb, J.E. and Chupp, E.L. 1994. A gamma-ray detector for in-situ measurement of ^{137}Cs radioactivity in snowfields and glaciers. *Nuclear Instruments and Methods*, Section A, 353, 482–485.

Dwyer, J.L. 1995. Mapping tide-water glacier dynamics in East Greenland using Landsat data. *Journal of Glaciology*, 41, 584–595.

Dyurgerov, M.B. and Meier, M.F. 1997a. Mass balance of mountain and subpolar glaciers: A new global assessment for 1961–1990. *Arctic and Alpine Research*, 29(4), 379–391.

Dyurgerov, M.B. and Meier, M.F. 1997b. Year-to-year fluctuations of global mass balance of small glaciers and their contribution to sea level. *Arctic and Alpine Research*, 29(4), 392–402.

Ehlrich, R. and Weinberg, B. 1970. An exact method for the characterisation of grain shape. *Journal of Sedimentary Petrology*, 40, 205–212.

Ehrmann, W., Bloemendal, J., Hambrey, M.J., McKelvey, B. and Whitehead, J. 2003. Variations in the composition of the clay fraction of the Cenozoic Pagodroma Group, East Antarctica: Implications for determining provenance. *Sedimentary Geology*, 61, 131–152.

Engelhardt, H. and Kamb, B. 1998. Basal sliding of ice stream B, West Antarctica. *Journal of Glaciology*, 44(147), 223–230.

Engelhardt, H., Kamb, B. and Bolsey, R. 2000. A hot-water ice-coring drill. *Journal of Glaciology*, 46(153), 341–345.

Engelhardt, H.F., Harrison, W.D. and Kamb, B. 1978. Basal sliding and conditions at the glacier bed as revealed by bore-hole photography. *Journal of Glaciology*, 20, 469–508.

EPICA Community Members. 2004. Eight glacial cycles from an Antarctic ice core, *Nature*, 429(6), 623–628.

Etienne, J.L., Glasser, N.F. and Hambrey, M.J. 2003. Proglacial sediment-landform associations of a polythermal glacier: Storglaciären, northern Sweden. *Geografiska Annaler*, 85A(2), 149–164.

Etzelmüller, B. and Björnsson, H. 2000. Map analysis techniques for glaciological applications. *International Journal of Geographical Information Science*, 14, 567–581.

Evans, D.J.A., Archer, S. and Wilson, D.J.H. 1999. A comparison of the lichenometric and Schmidt hammer dating techniques based on data from the proglacial areas of some Icelandic glaciers. *Quaternary Science Reviews*, 18, 13–41.

Evin, M., Fabre, D. and Johnson, P.G. 1997. Electrical resistivity measurements on the rock glaciers of Grizzly Creek, St Elias Mountains, Yukon. *Permafrost and Periglacial Processes*, 8, 181–191.

Ewing, K.J. and Marcus, M.G. 1966. Cartographic representation and symbolization in glacier mapping. *Canadian Journal of Earth Sciences*, 6, 761–769.

Eyles, N., Eyles, C.H., Miall, A.D. 1983. Lithofacies types and vertical profile models: An alternative approach to the description and environmental interpretation of glacial diamict and diamictite sequences. *Sedimentology*, 30, 393–410.

Fabel, D. and Harbor, J. 1999. The use of in-situ produced cosmogenic radionuclides in glaciology and glacial geomorphology. *Annals of Glaciology*, 28, 103–110.

Fahey, B.D. 1986. Weathering pit development in the Central Otago mountains of southern New Zealand. *Arctic and Alpine Research*, 18, 337–348.

Fahnestock, M.A., Scambos, T.A., Bindschadler, R.A. and Kvaran, G. 2000. A millennium of variable ice flow recorded by the Ross Ice Shelf, Antarctica. *Journal of Glaciology*, 46, 652–664.

Fehrentz, M. and Radtke, U. 2001. Luminescence dating of Pleistocene outwash sediments of the Senne area (Eastern Munsterland, Germany). *Quaternary Science Reviews*, 20, 725–729.

Field, W.O. 1966. Mapping glacier termini in southern Alaska, 1931–1964. *Canadian Journal of Earth Sciences*, 6, 819–825.

Fischer, U.H. and Clarke, G.K.C. 1994. Plowing of subglacial sediment. *Journal of Glaciology*, 40(134), 97–106.

Fischer, U.H. and Clarke, G.K.C. 1997. Clast collision frequency as an indicator of glacier sliding rate. *Journal of Glaciology*, 43(145), 460–466.

Fischer, U.H., Iverson, N.R., Hanson, B., Hooke, R.L. and Jansson, P. 1998. Estimation of hydraulic properties of subglacial till from ploughmeter measurements. *Journal of Glaciology*, 44(148), 517–522.

Fischer, U.H., Clarke, G.K.C. and Blatter, H. 1999. Evidence for temporally varying 'sticky spots' at the base of Trapridge Glacier, Yukon Territory, Canada. *Journal of Glaciology*, 45(150), 352–360.

Folk, R.L. and Ward, W.C. 1957. Brazos River bar: A study in the significance of grain size parameters. *Journal of Sedimentary Petrology*, 27, 3–26.

Ford, J.P. 1984. Mapping of glacial landforms from Seasat radar images. *Quaternary Research*, 22, 314–327.

Ford, D.C. and Williams, P. 1989. *Karst Geomorphology and Hydrology*. London, Unwin Hyman, 601pp.

Fountain, A.G. 1994. Borehole water-level variations and implications for the subglacial hydraulics of South Cascade Glacier, Washington State, USA. *Journal of Glaciology*, 40(135), 293–304.

Fountain, A.G. and Jacobel, R.W. 1997. Advances in ice radar studies of a temperate alpine glacier, South Cascade Glacier, Washington, USA. *Annals of Glaciology*, 24, 303–308.

Fountain, A.G. and Vecchia, A. 1999. How many stakes are required to measure the mass balance of a glacier? *Geografiska Annaler A*, 81(4), 563–573.

Frezzotti, M., Tabacco, I.E. and Zirizzotti, A. 2000. Ice discharge of eastern Dome C drainage area, Antarctica, inferred from airborne radar survey and satellite image analysis. *Journal of Glaciology*, 46, 265–274.

Fricker, H.A., Hyland, G., Coleman, R. and Young, N.W. 2000. Digital elevation models for the Lambert Glacier-Amery Ice Shelf system, East Antarctica, from ERS-1 satellite radar altimetry. *Journal of Glaciology*, 46, 553–560.

Fricker, H.A., Young, N.W., Allison, I. and Coleman, R. 2002. Iceberg calving from the Amery ice shelf, East Antarctica. *Annals of Glaciology*, 34, 241–246.

Friedman, G.M. and Sanders, J.E. 1978. *Principles of Sedimentology*. New York, Wiley.

Fujita, K., Seko, K., Ageta, Y., Pu, J.C. and Yao, T.D. 1996. Superimposed ice in glacier mass balance on the Tibetan Plateau. *Journal of Glaciology*, 42(142), 454–460.

Fuller, S. and Murray, T. 2002. Sedimentological investigations in the forefield of an Icelandic surge-type glacier: Implications for the surge mechanism. *Quaternary Science Reviews*, 21, 1503–1520.

Furbish, D.J. and Andrews, J.T. 1984. The use of hypsometry to indicate long-term stability and response of valley glaciers to changes in mass transfer. *Journal of Glaciology*, 30(105), 199–211.

Gale, S.J. and Hoare, P.G. 1991. *Quaternary Sediments: Petrographic Methods for the Study of Unlithified Rocks*. London, Belhaven Press, 323pp.

Gale, S.J. and Hoare, P.G. 1992. Bulk sampling of coarse clastic sediments for particle-size analysis. *Earth Surface Processes and Landforms*, 17, 729–733.

Gao, J. and Liu, Y. 2001. Applications of remote sensing, GIS and GPS in glaciology: A review. *Progress in Physical Geography*, 25, 520–540.

Gardiner, V. and Dackombe, R.V. 1983. *Geomorphological Field Manual*. London, Allen and Unwin, 254pp.

Gemmell, A.M.D. 1988. Thermoluminescence dating of glacially transported sediments: Some considerations. *Quaternary Science Reviews*, 7, 277–285.

German, R., Mader, M. and Kilger, B. 1979. Glacigenic and glaciofluvial sediments, typification and sediment parameters. In C. Schlüchter (ed.) *Moraines and Varves*. Rotterdam, A.A. Balkema, 127–143.

Gillet, F. 1975. Steam, hot-water and electrical thermal drills for temperate glaciers. *Journal of Glaciology*, 14(70), 171–179.

Glasser, N.F. and Bennett, M.R. 2004. Glacial erosional landforms: Origins and significance for palaeoglaciology. *Progress in Physical Geography*, 28,43–75.

Glasser, N.F. and Hambrey, M.J. 2001. Tidewater glacier beds: Insights from iceberg debris in Kongsfjorden, Svalbard. *Journal of Glaciology*, 47(157), 295–302.

Glasser, N.F., Bennett, M.R. and Huddart, D. 1998. Ice-marginal characteristics of Fridtjovbreen (Svalbard) during its recent surge. *Polar Research*, 17, 93–100.

Glasser, N.F., Crawford, K.R., Hambrey, M.J., Bennett, M.R. and Huddart, D. 1998a. Lithological and structural controls on the surface wear characteristics of glaciated metamorphic bedrock surfaces: Ossian Sarsfjellet, Svalbard. *Journal of Geology*, 106(3), 319–329.

Glasser, N.F., Hambrey, M.J., Crawford, K.R., Bennett, M.R. and Huddart, D. 1998b. The structural glaciology of Kongsvegen, Svalbard, and its role in landform genesis. *Journal of Glaciology*, 44(146), 136–148.

Glasser, N.F., Hambrey, M.J., Etienne, J.L., Jansson, P. and Pettersson, R. 2003. The origin and significance of debris-charged ridges at the surface of Storglaciären, northern Sweden. *Geografiska Annaler*.

Glen, J.W. 1955. The creep of polycrystalline ice. *Proceedings of the Royal Society of London*, Series A, 228, 519–538.

Goldthwait, R.P. 1988. Classification of glacial morphological features. In R.P. Goldthwait and C.L. Matsch (eds) *Genetic Classification of Glacigenic Deposits*. Rotterdam, Balkema, 267–277.

Gomez, B., Dowdeswell, J.A. and Sharp, M.J. 1988. Microstructural control of quartz sand grain shape and texture: Implication for the discrimination of debris transport pathways through glaciers. *Sedimentary Geology*, 57, 119–129.

Goodman, D.J. 1975. Radio-echo sounding on temperate glaciers. *Journal of Glaciology*, 14, 57–69.

Gordon, S., Sharp, M., Hubbard, B., Smart, C., Ketterling, B. and Willis, I. 1998. Seasonal reorganization of subglacial drainage inferred from measurements in boreholes. *Hydrological Processes*, 12(1), 105–133.

Gordon, S., Sharp, M., Hubbard, B., Willis, I., Smart, C., Copland, L., Harbor, J. and Ketterling, B. 2001. Borehole drainage and its implications for the investigation of glacier hydrology: Experiences from Haut Glacier d'Arolla, Switzerland. *Hydrological Processes*, 15(5), 797–813.

Gosse, J.C. and Phillips, F.M. 2001. Terrestrial in situ cosmogenic nuclides: Theory and application. *Quaternary Science Reviews*, 20, 1475–1560.

Goudie, A.S. 1990. *Geomorphological Techniques*. 2nd edition. London and Boston, Unwin Hyman.

Goudie, A.S. 1994. *Geomorphological Techniques*. London and New York, Routledge, 570pp.

Graham, J. 1988. Collection and analysis of field data. In M.E. Tucker (ed.) *Techniques in Sedimentology*. Oxford, Blackwell, 5–62.

Graham, D.J. and Midgley, N.G. 2000. Graphical representation of particle shape using triangular diagrams: An Excel spreadsheet method. *Earth Surface Processes and Landforms*, 25, 1473–1477.

Gray, D.M. and Male, D.H. 1981. *Handbook of Snow: Principles, Processes, Management and Use*. Toronto, Pergamon Press, 776pp.

Greenwood, B. 1969. Sediment parameters and environmental discrimination: An application of multivariate statistics. *Canadian Journal of Earth Sciences*, 6, 1347–1358.

Grisak, G.E., Cherry, J.A., Vonhof, J.A. and Blumele, J.P. 1976. Hydrogeologic and hydrochemical properties of fractured till in the Interior Plains Region. In R.F. Legget (ed.) *Glacial Till: An Inter-Disciplinary Study*. The Royal Society of Canada Special Publications No. 12, Ottawa, 304–333.

Grove, J.M. 1988. *The Little Ice Age*. London, Methuen.

Grove, J.M. and Battagel, A. 1983. Tax records from western Norway as an index of Little Ice Age environmental and economic deterioration. *Climatic Change*, 5, 265–282.

Gudmundsson, G.H. 1999. A three-dimensional numerical model of the confluence area of Unteraargletscher, Bernese Alps, Switzerland. *Journal of Glaciology*, 45(150), 219–230.

Gudmundsson, G.H., Bauder, A., Luthi, M., Fischer, U.H. and Funk, M. 1999. Estimating rates of basal motion and internal ice deformation from continuous tilt measurements. *Annals of Glaciology*, 28, 247–252.

Haefeli, R. 1966. Some notes on glacier mapping and ice movement. *Canadian Journal of Earth Sciences*, 3, 863–876.

Hagen, J.O. 1987. Glacier surge at Usherbreen, Svalbard. *Polar Research*, 5, 239–252.

Hagen, J.O. and Saetrang, A. 1991. Radio-echo soundings of sub-polar glaciers with low-frequency radar. *Polar Research*, 26, 15–57.

Haldorsen, S. 1981. Particle-size distribution of subglacial till and its relation to glacial crushing and abrasion. *Boreas*, 10(1), 91–105.

Hall, A.M. and Glasser, N.F. 2003. Reconstructing the basal thermal regime of an ice stream in a landscape of selective linear erosion: Glen Avon, Cairngorm Mountains, Scotland. *Boreas*, 32, 191–207.

Hall, D.K. and Martinec, J. 1985. *Remote Sensing of Ice and Snow*. London, Chapman and Hall, 189pp.

Hall, D.K., Williams, R.S., Jr and Sigurðsson, O. 1995. Glaciological observations of Brúarjökull, Iceland, using synthetic aperture radar and thematic mapper satellite data. *Annals of Glaciology*, 21, 271–276.

Hambrey, M.J. 1975. The origin of foliation in glaciers: Evidence from some Norwegian examples. *Journal of Glaciology*, 14(70), 181–185.

Hambrey, M.J. 1977. Foliation, minor faults and strain in glacier ice. *Tectonophysics*, 39(1–3), 397–416.

Hambrey, M.J. 1994. *Glacial Environments*. London, UCL Press.

Hambrey, M.J. 1999. The record of Earth's glacial climate during the last 3000 Ma. *Terra Antartica Reports*, 3, 73–107.

Hambrey, M.J. and Dowdeswell, J.A. 1994. Flow regime of the Lambert Glacier-Amery Ice Shelf system, Antarctica: Structural evidence from Landsat imagery. *Annals of Glaciology*, 20, 401–406.

Hambrey, M.J. and Dowdeswell, J.A. 1997. Structural evolution of a surge-type glacier: Hessbreen, Svalbard. *Annals of Glaciology*, 24, 375–381.

Hambrey, M.J. and Glasser, N.F. 2003. Glacial sediments: Processes, environments and facies. In G.V. Middleton (ed.) *Encyclopedia of Sediments and Sedimentary Rocks*. Dordrecht, Kluwer, 316–331.

Hambrey, M.J. and Lawson, W.J. 2000. Structural styles and deformation fields in glaciers: A review. In A.J. Maltman, B.P. Hubbard and M.J. Mambrey (eds) *Deformation of Glacial Materials*. Geological Society of London Special Publication, 176, 59–83.

Hambrey M.J. and Milnes, A.G. 1975. Boudinage in glacier ice – some examples. *Journal of Glaciology*, 14(72), 383–393.

Hambrey, M.J. and Müller, F. 1978. Structures and ice deformation in the White Glacier, Axel Heiberg Island, Northwest Territories, Canada. *Journal of Glaciology*, 20(82), 41–66.

Hambrey, M.J., Dowdeswell, J.A., Murray, T. and Porter, P.R. 1996. Thrusting and debris entrainment in a surging glacier: Bakaninbreen, Svalbard. *Annals of Glaciology*, 22, 241–248.

Hambrey, M.J., Huddart, D., Bennett, M.R. and Glasser, N.F. 1997. Genesis of 'hummocky moraines' by thrusting in glacier ice: Evidence from Svalbard and Britain. *Journal of the Geological Society*, London, 154, 623–632.

Hambrey, M.J., Bennett, M.R., Dowdeswell, J.A., Glasser, N.F. and Huddart, D. 1999. Debris entrainment and transfer in polythermal valley glaciers. *Journal of Glaciology*, 45(149), 69–86.

Hamilton, G.S. 2002. Mass balance and accumulation rate across Siple Dome, West Antarctica, *Annals of Glaciology*, 35, 102–106.

Hamilton, G.S. and Dowdeswell, J.A. 1996. Controls on glacier surging in Svalbard. *Journal of Glaciology*, 42, 157–168.

Hamilton, G.S., Whillans, I.M. and Morgan, P.J. 1998. First point measurements of ice-sheet thickness change in Antarctica. *Annals of Glaciology*, 27, 125–129.

Hansen, D.P. and Wilen, L.A. 2002. Performance and applications of an automated *c*-axis ice-fabric sampler. *Journal of Glaciology*, 48(160), 159–170.

Hanson, B., Hooke, R.L. and Grace, E.M. 1998. Short-term velocity and water-pressure variations down-glacier from a riegel, Storglaciären, Sweden. *Journal of Glaciology*, 44(147), 359–367.

Harbor, J., Sharp, M., Copland, L., Hubbard, B., Nienow, P. and Mair, D. 1997. Influence of subglacial drainage conditions on the velocity distribution within a glacier cross section. *Geology*, 25(8), 739–742.

Harper, J.T., Humphrey, N.F. and Pfeffer, W.T. 1998. Crevasse patterns and the strain-rate tensor: A high resolution comparison. *Journal of Glaciology*, 44, 68–76.

Harris, C., Williams, G., Brabham, P., Eaton, G. and McCarroll, D. 1997. Glaciotectonized Quaternary sediments at Dinas Dinlle, Gwynedd, North Wales, and their bearing on the style of deglaciation in the eastern Irish Sea. *Quaternary Science Reviews*, 16, 109–127.

Harris, S.E. 1943. Friction cracks and the direction of glacial movement. *Journal of Geology*, 51, 244–258.

Harrison, W.D., Echelmeyer, K.A., Cosgrove, D.M. and Raymond, C.F. 1992. The determination of glacier speed by time-lapse photography under unfavourable conditions. *Journal of Glaciology*, 38(129), 257–265.

Hart, J.K. 1994. Till fabric associated with deformable beds. *Earth Surface Processes and Landforms*, 19, 15–32.

Hart, J.K. and Roberts, D. 1994. Criteria to distinguish between subglacial glaciotectonic and glacimarine sedimentation. *Sedimentary Geology*, 91, 191–213.

Hay, I. 1995. Writing a review. *Journal of Geography in Higher Education*, 19, 357–363.

Helk, J.V. 1966. Glacier mapping in Greenland. *Canadian Journal of Earth Sciences*, 6, 771–774.

Heucke, E. 1999. A light portable steam-driven ice drill suitable for drilling holes in ice and firn. *Geografiska Annaler A*, 81(4), 603–609.

Hicock, S.R., Goff, J.R., Lian, O.B. and Little, E.C. 1996. On the interpretation of subglacial till fabric. *Journal of Sedimentary Research*, 66, 928–934.

Hodge, S.M. 1976. Direct measurement of basal water pressures: A pilot study. *Journal of Glaciology*, 16(74), 205–218.

Hodge, S.M. 1979. Direct measurement of basal water pressures: Progress and problems. *Journal of Glaciology*, 23(89), 309–319.

Hodgkins, R. 1996. Seasonal trend in suspended-sediment transport from an Arctic glacier, and implications for drainage-system structure. *Annals of Glaciology*, 22, 147–151.

Hodson, A.J. and Ferguson, R.I. 1999. Fluvial suspended sediment transport from cold and warm-based glaciers in Svalbard. *Earth Surface Processes and Landforms*, 24, 957–974.

Hodson, A.J., Tranter, M., Dowdeswell, J.A., Gurnell, A.M. and Hagen, J.O. 1997. Glacier thermal regime and suspended-sediment yield: A comparison of two high-Arctic glaciers. *Annals of Glaciology*, 24, 32–37.

Holmes, C.D. 1960. Evolution of till-stones shapes, central New York. *Geological Society of America Bulletin*, 71, 1645–1660.

Holmlund, P., Burman, H. and Rost, T. 1996. Sediment mass exchange between turbid meltwater streams and proglacial deposits of Storglaciären, Northern Sweden. *Annals of Glaciology*, 22, 63–67.

Hooke, R.L. 1981. Flow law for polycrystalline ice in glaciers: Comparison of theoretical predictions, laboratory data, and field measurements. *Reviews of Geophysics and Space Physics*, 19(4), 664–672.

Hooke, L. and Hudleston, P.J. 1978. Origin of foliation in glaciers. *Journal of Glaciology*, 20(83), 285–299.

Hooke, R.L. and Iverson, N.R. 1995. Grain size distribution in deforming subglacial tills: Role of grain fracture. *Geology*, 23, 57–60..

Hooke, R.L., Hanson, B., Iverson, N.R., Jansson, P. and Fischer, U.H. 1997. Rheology of till beneath Storglaciären, Sweden. *Journal of Glaciology*, 43(143), 172–179.

Hooyer, T.S. and Iverson, N.R. 2000. Clast-fabric development in a shearing granular material: Implications for subglacial till and fault gouge. *GSA Bulletin*, 112(5), 683–692.

Hooyer, T.S. and Iverson, N.R. 2002. Flow mechanism of the Des Moines lobe of the Laurentide ice sheet. *Journal of Glaciology*, 48(163), 575–586.

Hormes, A., Müller, B.U. and Schlüchter, C. 2001. The Alps with little ice: Evidence for eight Holocene phases of reduced glacier extent in the Central Swiss Alps. *The Holocene*, 11, 255–265.

Hotzel, I.S. and Miller, J.D. 1983. Icebergs: Their physical dimensions and the presentation and application of measured data. *Annals of Glaciology*, 4, 116–123.

Hubbard, A. 2000. The verification and significance of three approaches to longitudinal stresses in high-resolution models of glacier flow. *Geografiska Annaler Series A, Physical Geography*, 82A(4), 471–487.

Hubbard, A. and Hubbard, B. 2000. The potential contribution of high-resolution glacier flow modelling to structural glaciology. In A.J. Maltman, B.P. Hubbard and M.J. Hambrey (eds) *Deformation of Glacial Materials. Geological Society of London Special Publication*, 176, 135–146.

Hubbard, B. and Sharp, M. 1995. Basal ice facies and their formation in the western Alps. *Arctic and Alpine Research*, 27(4), 301–310.

Hubbard, B., Sharp, M.J., Willis, I.C., Nielsen, M.K. and Smart, C.C. 1995. Borehole water-level variations and the structure of the subglacial hydrological system of Haut Glacier d'Arolla, Valais, Switzerland. *Journal of Glaciology*, 41(139), 572–583.

Hubbard, A., Blatter, H., Nienow, P., Mair, D. and Hubbard, B. 1998. Comparison of a three-dimensional model for glacier flow with field data from Haut Glacier d'Arolla, Switzerland. *Journal of Glaciology*, 44(147), 368–378.

Hubbard, B., Siegert, M.J. and McCarroll, D. 2000. Spectral roughness of glaciated bedrock geomorphic surfaces: Implications for glacier sliding. *Journal of Geophysical Research*, 105(B9), 21295–21303.

Hubbard, B., Tison, J.L., Janssens, L. and Spiro, B. 2000. Ice-core evidence of the thickness and character of clear-facies basal ice: Glacier de Tsanfleuron, Switzerland. *Journal of Glaciology*, 46(152), 140–150.

Huddart, D. 1994. Rock-type controls on downstream changes in clast parameters in sandur systems in southeast Iceland. *Journal of Sedimentary Research*, 64, 215–225.

Huddart, D., Bennett, M.R., Hambrey, M.J., Glasser, N.F. and Crawford, K.R. 1998. Origin of well rounded gravels in glacial deposits from Brøggerhalvøya, northwest Spitsbergen: Potential problems caused by sediment reworking in the glacial environment. *Polar Research*, 17(1), 61–69.

Huggel, C. and Kääb, A. 2002. Remote-sensing based assessment of hazards from glacier lake outbursts: A case study in the Swiss Alps. *Journal of Canadian Geotechnics*, 39, 1–15.

Humlum, O. 1985. Changes in texture and fabric of particles in glacial traction with distance from source, Mýrdalsjökull, Iceland. *Journal of Glaciology*, 31, 150–156.

Humphrey, N.F., Engelhardt, H.F., Fahnestock, M. and Kamb, B. 1993. Characteristics of the bed of the Lower Columbia Glacier, Alaska. *Journal of Geophysical Research*, 98(B1), 837–846.

Hunter, L.E., Powell, R.D. and Lawson, D.E. 1996. Flux of debris transported by ice at three Alaskan tidewater glaciers. *Journal of Glaciology*, 42, 123–135.

Huybrechts, P. 1993. Glaciological modelling of the Late Cenozoic East Antarctic ice sheet: Stability or dynamism? *Geografiska Annaler*, 75A, 221–238.

Icefield_Instruments_Inc. 2002. Icefield Instruments Inc. http//www.icefield.yk.ca/index.htm.

Iken, A., Echelmeyer, K. and Harrison, W.D. 1989. A light-weight hot-water drill for large depth: Experiences with drilling on Jakobshavns glacier, Greenland. *Ice Core Drilling*. Proceedings of the third international workshop on ice drilling technology, 10–14 October 1988, France, Grenoble, 123–136.

Illenberger, W.K. 1991. Pebble shape (and size!). *Journal of Sedimentary Petrology*, **61**, 756–767.

Innes, J.L. 1985. Short Communications. Lichenometric dating of debris-flow deposits on alpine colluvial fans in southwest Norway. *Earth Surface Processes and Landforms*, **10**, 519–524.

Iverson, N.R. 1991. Morphology of glacial striae: Implications for abrasion of glacier beds and fault surfaces. *Geological Society of America Bulletin*, **103**, 1308–1316.

Iverson, N.R., Hooyer, T.S. and Hooke, R. Le, B. 1996. A laboratory study of sediment deformation: Stress heterogeneity and grain-size evolution. *Annals of Glaciology*, **22**, 167–175.

Jacka, T.H. 1984. Laboratory studies on relationships between ice crystal size and flow-rate. *Cold Regions Science and Technology*, **10**(1), 31–42.

Jackson, L.E., Phillips, F.M., Shimamura, K. and Little, E.C. 1997. Cosmogenic ^{36}Cl dating of the Foothills erratics train, Alberta, Canada. *Geology*, **25**, 195–198.

Jacobs, S.S. 1992. Is the Antarctic Ice Sheet growing? *Nature*, **360**, 29–33.

Jansson, K.N., Kleman, J. and Marchant, D.R. 2002. The succession of ice-flow patterns in North-Central Québec-Labrador, Canada. *Quaternary Science Reviews*, **21**, 503–523.

Jensen, J.R. 1996. *Introductory Image Processing: A Remote Sensing Perspective*. London, Prentice Hall.

Jiskoot, H., Boyle, P. and Murray, T. 1998. The incidence of glacier surging in Svalbard: Evidence from multivariate statistics. *Computers and Geoscience*, **24**, 387–399.

Jiskoot, H., Murray, T. and Boyle, P. 2000. Controls on the distribution of surge-type glaciers in Svalbard. *Journal of Glaciology*, **46**, 412–422.

Jiskoot, H., Pedersen, A.K. and Murray, T. 2001. Multi-model photogrammetric analysis of the 1990s surge of Sortebrae, East Greenland. *Journal of Glaciology*, **47**(159), 677–687.

Johannesson, T., Raymond, C.F. and Waddington, E.D. 1989. Time-scale for adjustment of glaciers to changes in mass balance. *Journal of Glaciology*, **35**(121), 355–369.

Johnsen, S.J., Dansgaard, W., Gundestrup, N., Hansen, S.B.B., Nielsen, J.O. and Reeh, N. 1980. A fast light-weight core drill. *Journal of Glaciology*, **25**(91), 169–174.

Jonasson, C., Kot, M. and Kotarba, A. 1991. Lichenometrical studies and dating of debris flow deposits in the High Tatra Mountains, Poland, *Geografiska Annaler*, **73A**, 141–146.

Jones, A.P., Tucker, M.E. and Hart, J.K. 1999. *The Description and Analysis of Quaternary Stratigraphic Field Sections*. Technical Guide No. 7, London, Quaternary Research Association.

Jowsey, P.C. 1966. An improved peat sampler. *New Phytologist*, **65**, 245–248.

Kääb, A. and Funk, M. 1999. Modelling mass balance using photogrammetric and geophysical data: A pilot study at Griesgletscher, Swiss Alps. *Journal of Glaciology*, **45**, 575–583.

Kääb, A., Paul, F., Maisch, M., Hoelzle, M. and Haeberli, W. 2002. The new remote-sensing-derived Swiss glacier inventory: II. First results. *Annals of Glaciology*, **34**, 362–366.

Karlén, W. 1976. Lacustrine sediments and tree-limit variations as indicators of Holocene climatic fluctuations in Lappland, Northern Sweden. *Geografiska Annaler*, **58A**, 1–34.

Karlén, W. and Matthews, J.A. 1992. Reconstructing Holocene glacier variations from glacial lake sediments: Studies from Nordvestlandet and Jostedalsbreen-Jotunheimen, southern Norway. *Geografiska Annaler*, **74A**, 327–348.

Karrow, P.F. 1976. The texture, mineralogy and petrography of North American tills. In R.F. Legget (ed.) *Glacial Till: An Inter-Disciplinary Study*. The Royal Society of Canada Special Publications No. 12, Ottowa, 83–98.

Kaser, G., Fountain, A. and Jansson, P. 2002. *A Manual for Monitoring the Mass Balance of Mountain Glaciers*, 59, UNESCO.

Kasser, P. and Röethlisberger, H. 1966. Some problems of glacier mapping experienced with the 1:10000 map of the Aletsch Glacier. *Canadian Journal of Earth Sciences*, 6, 799–809.

Kennedy, B.A. 1992. First catch your hare . . . Research design for individual projects. In A. Rogers, H. Viles and A. Goudie (eds) *The Student's Companion to Geography*. Oxford, Blackwell.

Kennett, J.P. 1982. *Marine Geology*. New Jersey, Prentice Hall.

Khatwa, A. and Tulaczyk, S. 2001. Microstructural interpretations of modern and Pleistocene subglacially deformed sediments: The relative role of parent material and subglacial processes. *Journal of Quaternary Science*, 16, 507–517.

Khatwa, A., Hart, J.K. and Payne, A.J. 1999. Grain textural analysis across a range of glacial facies. *Annals of Glaciology*, 28, 111–117.

Kick, W. 1966. Measuring and mapping of glacier variations. *Canadian Journal of Earth Sciences*, 6, 775–781.

Kilpatrick, F.A. 1970. Dosage requirements for slug injections of Rhodamine BA and WT dyes, *US Geological Survey Professional Paper 700-B*, 250–253.

King, C.A.M. 1966. *Techniques in Geomorphology*. London, Edward Arnold, 342pp.

Kirby, R.P. 1969. Till fabric analyses from the Lothians, central Scotland. *Geografiska Annaler*, 51A, 48–60.

Kirkbride, M.P. 1995. Ice flow vectors on the debris-mantled Tasman Glacier, 1957–1986. *Geografiska Annaler*, 77A, 147–157.

Kirkbride, M.P. and Warren, C.R. 1997. Calving processes at a grounded ice cliff. *Annals of Glaciology*, 24, 116–121.

Kirkbride, M.P., Duck, R.W., Dunlop, A., Drummond, J., Mason, M., Rowan, J.S. and Taylor, D. 2001. Development of a geomorphological database and geographical information system for the North West Seaboard: Pilot study. *Scottish Natural Heritage Commissioned Report BAT/98/99/137*. Scottish Natural Heritage, Edinburgh.

Kite, G. 1993. Computerized streamflow measurement using slug injection. *Hydrological Processes*, 7, 227–233.

Kjaer, K.H. 1999. Mode of subglacial transport deduced from till properties: Mýrdalsjökull, Iceland. *Sedimentary Geology*, 128, 271–292.

Kjaer, K.H. and Krüger, J. 1998. Does clast size influence fabric strength? *Journal of Sedimentary Research*, 68, 746–749.

Kjaer, K.H. and Krüger, J. 2001. The final phase of dead-ice moraine development: Process and sediment architecture, Kötlujökull, Iceland. *Sedimentology*, 48, 935–952.

Kleman, J. 1990. On the use of glacial striae for reconstruction of palaeo-ice sheet flow patterns. *Geografiska Annaler*, 72A, 217–36.

Kleman, J. 1994. Preservation of landforms under ice sheets and ice caps. *Geomorphology*, 9, 19–32.

Kleman, J. and Hattestrand, C. 1999. Frozen-bed Fennoscandian and Laurentide ice sheets during the last glacial maximum. *Nature*, 402, 63–66.

Kleman, J. and Stroeven, A.P. 1997. Preglacial surface remnants and Quaternary glacial regimes in northwestern Sweden. *Geomorphology*, 19, 35–54.

Klohn, E.J. 1965. The elastic properties of a dense glacial till deposit. *Canadian Geotechnical Journal*, 2, 90–98.

Knap, W.H., Brock, B.W., Oerlemans, J. and Willis, I.C. 1999. Comparison of Landsat TM-derived and ground-based albedos of Haut Glacier d'Arolla, Switzerland. *International Journal of Remote Sensing*, 20, 3293–3310.

Knight, P.G. 1987, *Observations at the Edge of the Greenland Ice Sheet: Boundary Condition Implications for Modellers*. International Association of Hydrological Sciences Publication, 170, 359–366.

Knight, P.G. and Hubbard, B. 1999. Ice facies: A case study from the basal ice facies of the Russell Glacier, Greenland Ice Sheet. In A.P. Jones, M.E. Tucker and J.K. Hart (eds) *The Description and Analysis of Quaternary Stratigraphic Field Sections*. London, Quaternary Research Association, 9.1–9.4.

Knuepfer, P.L.K. 1988. Estimating ages of late Quaternary stream terraces from analysis of weathering rinds and soils. *Geological Society of America Bulletin*, 100, 1224–1236.

Kohler, J. 1995. Determining the extent of pressurized flow beneath Storglaciaren, Sweden, using results of tracer experiments and measurements of input and output discharge. *Journal of Glaciology*, 41(138), 217–231.

Kohler, J., Moore, J., Kennett, M., Engeset, R. and Elvehøy, H. 1997. Using ground-penetrating radar to image previous years' summer surfaces for mass-balance measurements. *Annals of Glaciology*, 24, 355–360.

Konecny, G. 1966. Application of photogrammetry to surveys of glaciers in Canada and Alaska. *Canadian Journal of Earth Science*, 3, 783–798.

König, M., Winther, J.-G. and Isaksson, E. 2001a. Measuring snow and ice properties from satellite. *Reviews of Geophysics*, 39, 1–28.

König, M., Winther, J.-G., Knudsen, N.T. and Guneriussen, T. 2001b. Firn-line detection on Austre Okstindbreen, Norway, with airborne multipolarization SAR. *Journal of Glaciology*, 47, 251–257.

König, M., Wadham, J., Winther, J.-G., Kohler, J. and Nuttall, A.-M. 2002. Detection of superimposed ice on the glaciers Kongsvegen and midre Lovénbreen, Svalbard, using SAR satellite imagery. *Annals of Glaciology*, 34, 335–342.

Krimmel, R.M. and Rasmussen, L.A. 1986. Using sequential photography to estimate ice velocity at the terminus of Columbia Glacier, Alaska. *Annals of Glaciology*, 8, 117–123.

Krüger, J. 1984. Clasts with stoss-lee form in lodgement tills: A discussion. *Journal of Glaciology*, 30, 241–243.

Krüger, J. and Kjaer, K.H. 1999. A data chart for field description and genetic interpretation of glacial diamicts and associated sediments – with examples from Greenland, Iceland and Denmark. *Boreas*, 28, 386–402.

Krumbein, W.C. 1941. Measurement and geological significance of shape and roundness of sedimentary particles. *Journal of Sedimentary Petrology*, 11, 64–72.

Kuhn, M., Dreiseitl, E., Hofinger, S., Kaser, G., Markl, G. and Span, N. 1999. Measurements and models of the mass balance of Hintereisferner. *Geografiska Annaler*, 81A(4), 659–670.

Lachniet, M.S., Larson, G.J., Lawson, D.E., Evenson, E.B. and Alley, R.B. 2001. Microstructures of sediment flow deposits and subglacial sediments: A comparison. *Boreas*, 30, 254–262.

Lamb, H.H. 1977. *Climate: Present, Past and Future*. London, Methuen.

Lamb, H.H. 1982. *Climate, History and the Modern World*. London, Methuen.

Landim, P.M.B. and Frakes, L.A. 1968. Distinction between tills and other diamictons based on textural characteristics. *Journal of Sedimentary Petrology*, 38, 1213–1223.

Langway, C.C.J. 1958. Ice fabrics and the Universal stage, *Cold Regions Research and Engineering Laboratory*, 16.

Laverdiere, C., Guimont, P. and Pharand, M. 1979. Marks and forms on glacier beds: Formation and classification. *Journal of Glaciology*, 23, 414–416.

Lawson, D.E. 1979. Sedimentological analysis of the western terminus of the Matanuska Glacier, Alaska. *Cold Regions Research Engineering Laboratory Report*, 79(9), 1–112.

Lawson, T.J. 1996. Glacial striae and former ice movement: The evidence from Assynt, Sutherland. *Scottish Journal of Geology*, 32, 59–65.

Lawson, W.J., Sharp, M.J. and Hambrey, M.J. 1994. The structural geology of a surge-type glacier. *Journal of Structural Geology*, 16, 1447–1462.

Lee, J.R., Rose, J., Riding, J.B., Moorlock, B.S.P. and Hamblin, R.J.O. 2002. Testing the case for a Middle Pleistocene Scandinavian glaciation in Eastern England: Evidence for a Scottish ice source for tills within the Corton Formation of east Anglia, UK. *Boreas*, 31, 345–355.

Leeder, M. 1983. *Sedimentology, Process and Product*. London, Allen and Unwin.

Lees, G. 1964. A new method for determining the angularity of particles. *Sedimentology*, 3, 2–21.

Lefauconnier, B. and Hagen, J.O. 1991. Surging and calving glaciers in eastern Svalbard. *Norsk Polarinstitutt Meddelser*, 116.

Leonard, E.M. 1997. The relationship between glacial activity and sediment production: Evidence from a 4450-year varve record of Neoglacial sedimentation in Hector Lake, Alberta, Canada. *Journal of Paleolimnology*, 17, 319–330.

Leonard, K.C. and Fountain, A.G. 2003. Map-based methods for estimating glacier equilibrium-line altitudes, *Journal of Glaciology*, 49, 329–336.

Lewis, T., Gilbert, R. and Lamoureux, S.F. 2002. Spatial and temporal changes in sedimentary processes at Proglacial Bear Lake, Devon Island, Nunavut, Canada. *Arctic, Antarctic and Alpine Research*, 34(2), 119–129.

Li, Z., Su, W.X. and Zeng, Q.Z. 1998. Measurements of glacier variation in the Tibetan Plateau using Landsat data. *Remote Sensing of Environment*, 63(3), 258–264.

Lillesand, T.M. and Kiefer, R.W. (eds) 2000. *Remote Sensing and Image Interpretation*. New York, John Wiley, 736pp.

Lindner, L., Marks, L., Ostaficzuk, S., Pekala, K. and Szczesny, R. 1985. Application of photogeological mapping to studies of glacial history of south Spitsbergen. *Earth Surface Processes and Landforms*, 10, 387–399.

Lliboutry, L. 2002. Overthrusts due to easy-slip/poor slip transitions at the bed: The mathematical singularity with non-linear isotropic viscosity. *Journal of Glaciology*, 48(160), 109–117.

Lock, W.W., Andrews, J.T. and Webber, P.J. 1979. A manual for lichenometry. *British Geomorphological Research Group Technical Bulletin*, 26, 47pp.

Loiselle, A.A. and Hurtubise, J.E. 1976. Properties and behaviour of till as construction material. In R.F. Legget (ed.) *Glacial Till: An Inter-Disciplinary Study*. The Royal Society of Canada Special Publications No. 12, Ottowa, 346–363.

Lowe, J.J. and Walker, M.J.C. 1997. *Reconstructing Quaternary Environments*. 2nd edition. Essex, Addison Wesley Longman Limited.

Luckman, B.H. 1994. Glacier fluctuation and tree-ring records for the last millennium in the Canadian Rockies. *Quaternary Science Reviews*, 12, 441–450.

Macheret, Y.Y. and Zhuravlev, A.B. 1982. Radio-echo sounding of Svalbard glaciers. *Journal of Glaciology*, 28(99), 295–314.

Macheret, Y.Y., Moskalevsky, M.Y. and Vasilenko, E.V. 1993. Velocity of radio-waves in glaciers as an indicator of their hydrothermal state, structure and regime. *Journal of Glaciology*, 39(132), 373–384.

Mahaney, W.C. and Kalm, V. 2000. Comparative scanning electron microscopy study of oriented till blocks, glacial grains and Devonian sands in Estonia and Latvia. *Boreas*, 29, 35–51.

Mair, D., Nienow, P., Willis, I. and Sharp, M. 2001. Spatial patterns of glacier motion during a high-velocity event: Haut Glacier d'Arolla, Switzerland. *Journal of Glaciology*, 47(156), 9–20.

Mair, D., Nienow, P., Sharp, M.J., Wohlleben, T. and Willis, I. 2002. Influence of subglacial drainage system evolution on glacier surface motion: Haut Glacier d'Arolla, Switzerland. *Journal of Geophysical Research-Solid Earth*, 107(B8), art. no. 2175.

Makinson, K. 1994. The BAS hot water drill. In H. Oerter (ed.) *Filchner Ronne Ice Shelf Programme*, Report No. 7. Alfred-Wegener-Institute for polar and marine research, Germany, Bremerhaven, 20–26.

Mark, D.M. 1974. On the interpretation of till fabrics. *Geology*, 2, 101–104.

Massonnet, D. and Feigl, K.L. 1998. Radar interferometry and its application to changes in the Earth's surface. *Reviews of Geophysics*, 36, 441–500.

Matthews, J.A. 1974. Families of lichenometric dating curves from Storbreen Gletscher-vorfeld, Jotunheim, Norway. *Norsk Geografisk Tidsskrift*, 28, 215–235.

Matthews, J.A. 1992. *The Ecology of Recently-Deglaciated Terrain: A Geoecological Approach to Glacier Forelands and Primary Succession.* Cambridge, Cambridge University Press.

Matthews, J.A. and Shakesby, R.A. 1984. The status of the 'Little Ice Age' in southern Norway: Relative-age dating of Neoglacial moraines with Schmidt hammer and lichenometry. *Boreas*, 13, 333–346.

McCarroll, D. 1989a. Potential and limitations of the Schmidt hammer for relative-age dating: Field tests on Neoglacial moraines, Jotunheimen, southern Norway. *Arctic and Alpine Research*, 21, 268–275.

McCarroll, D. 1989b. Schmidt hammer relative-age evaluation of a possible pre-'Little Ice Age' Neoglacial moraine, Leirbreen, southern Norway. *Norsk Geologisk Tidsskrift*, 69, 125–130.

McCarroll, D. 1991. The Schmidt hammer, weathering and rock surface roughness. *Earth Surface Processes and Landforms*, 16, 477–480.

McCarroll, D. 1992. A new instrument and techniques for the field measurement of rock surface roughness. *Zeitschrift für Geomorphologie*, 36, 69–79.

McCarroll, D. 1995. A new approach to lichenometry: Dating single-age and diachronous surfaces. *The Holocene*, 4, 383–396.

McCarroll, D. 1997. A template for calculating rock surface roughness. *Earth Surface Processes and Landforms*, 22, 1229–1230.

McCarroll, D. 2002. Amino-acid geochronology and the British Pleistocene: Secure stratigraphical framework or a case of circular reasoning? *Journal of Quaternary Science*, 17, 647–651.

McCarroll, D. and Ballantyne, C.K. 2000. The last ice sheet in Snowdonia. *Journal of Quaternary Science*, 15, 765–778

McCarroll, D. and Nesje, A. 1993. Vertical extent of ice sheets in Nordfjord, western Norway: Measuring degree of rock surface weathering using Schmidt hammer and rock surface roughness. *Boreas*, 22, 255–265.

McCarroll, D., Ballantyne, C.K., Nesje, A. and Dahl, S.O. 1995. Nunataks of the last ice sheet in Scotland. *Boreas*, 24, 305–323.

McCarthy, D.P. and Zaniewski, K. 2001. Digital analysis of lichen cover: A technique for use in lichenometry and lichenology. *Arctic, Antarctic and Alpine Research*, 33, 107–113.

McManus, J. 1988. Grain size determination and interpretation. In M.E. Tucker (ed.) *Techniques in Sedimentology*. Oxford, Blackwell, 63–85.

McSaveney, M.J. and Stirling, M.W. 1992. Central Otago: Basin and Range Country. In J.M. Soons and M.J. Selby (eds) *Landforms of New Zealand*. 2nd edition. Auckland, Longman Paul.

Meier, M. 1966. Some glaciological interpretations of remapping programs of South Cascade, Nisqually, and Klawatti Glaciers, Washington. *Canadian Journal of Earth Sciences*, 6, 811–818.

Menzies, J. 2000. Micromorphological analyses of microfabrics and microstructures indicative of deformation processes in glacial sediments. In A.J. Maltman, B. Hubbard and M.J. Hambrey (eds) *Deformation of Glacial Materials*. Geological Society Special Publication, 176, 245–257.

Menzies, J. and Maltman, A.J. 1992 Microstructures in diamictons: Evidence of subglacial bed conditions. *Geomorphology*, 6, 27–40.

Miall, A.D. 1978. Lithofacies types and vertical profile models in braided river deposits: A summary. *Memoir of the Canadian Society of Petroleum Geologists*, 5, 597–604.

Middleton, R.T. 2000. Hydrogeological characterisation using high-resolution electrical resistivity and radar tomographic imaging. Unpublished PhD Thesis, Lancaster University.

Milligan, V. 1976. Geotechnical aspects of glacial tills. In R.F. Legget (ed.) *Glacial Till: An Inter-Disciplinary Study*. The Royal Society of Canada Special Publications No. 12, Ottowa, 269–291.

Mills, H.H. 1979. Downstream rounding of pebbles: A quantitative review. *Journal of Sedimentary Petrology*, 49, 295–302.

Milsom, J. 2002. *Field Geophysics*. 3rd edition. Chichester, Wiley, 208pp.

Moncrieff, A.C.M. 1989. Classification of poorly-sorted sedimentary rocks. *Sedimentary Geology*, 65, 191–194.

Moon, B.P. 1984. Refinement of a technique for determining rock mass strength for geomorphological purposes. *Earth Surface Processes and Landforms*, 9, 189–193.

Moore, J.C., Mulvaney, R. and Paren, J.G. 1989. Dielectric stratigraphy of ice: A new technique for determining total ionic concentrations in polar ice cores. *Geophysical Research Letters*, 16, 1177–1180.

Murray, T. and Clarke, G.K.C. 1995. Black-box modeling of the subglacial water-system. *Journal of Geophysical Research – Solid Earth*, 100(B6), 10231–10245.

Murray, T., Dowdeswell, J.A., Drewry, D.J. and Frearson, I. 1998. Geometric evolution and ice dynamics during a surge of Bakaninbreen, Svalbard. *Journal of Glaciology*, 44, 263–272.

Murray, T., Stuart, G.W., Fry, M., Gamble, N.H. and Crabtree, M.D. 2000a. Englacial water distribution in a temperate glacier from surface and borehole radar velocity analysis. *Journal of Glaciology*, 46(154), 389–398.

Murray, T., Stuart, G.W., Miller, P.J., Woodward, J., Smith, A.M., Porter, P.R. and Jiskoot, H. 2000b. Glacier surge propagation by thermal evolution at the bed. *Journal of Geophysical Research*, **105**, 13491–13507.

Narod, B.B. and Clarke, G.K.C. 1980. Airborne UHF radio-echo sounding of three Yukon glaciers. *Journal of Glaciology*, **25**, 23–31.

Narod, B.B. and Clarke, G.K.C. 1994. Miniature high-power impulse transmitter for radio-echo sounding. *Journal of Glaciology*, **40**(134), 190–194.

Naslund, J.O., Fastook, J.L. and Holmlund, P. 2000. Numerical modelling of the ice sheet in western Dronning Maud Land, East Antarctica: Impacts of present, past and future climates. *Journal of Glaciology*, **46**, 54–66.

Neal, A. 2004. Ground-penetrating radar and its use in sedimentology: Principles, problems and progress. *Earth-Science Reviews*, **66**, 261–330.

Nesje, A. and Dahl, S.O. 2000. *Glaciers and Environmental Change*. London, Arnold.

Nesje, A., McCarroll, D. and Dahl, S.O. 1994. Degree of rock surface weathering as an indicator of ice-sheet thickness along an east-west transect across southern Norway. *Journal of Quaternary Science*, **9**, 337–347.

Nicholas, J.W. and Butler, D.R. 1996. Application of relative-age dating techniques on rock glaciers of the La Sal Mountains, Utah: An interpretation of Holocene paleoclimates, *Geografiska Annaler*, **78A**, 1–18.

Nienow, P., Sharp, M. and Willis, I. 1998. Seasonal changes in the morphology of the subglacial drainage system, Haut Glacier d'Arolla, Switzerland. *Earth Surface Processes and Landforms*, **23**(9), 825–843.

Nye, J.F. 1953. The flow law of ice from measurements in glacier tunnels, laboratory experiments, and the Jungfraufirn borehole experiment. *Proceedings of the Royal Society of London*, Series A, **219**, 477–489.

Oerlemans, J. 1988. Simulation of historic glacier variations with a simple climate-glacier model. *Journal of Glaciology*, **34**(118), 333–341.

Oerlemans, J. 1994. Quantifying global warming from the retreat of glaciers. *Science*, **264**, 243–245.

Oerlemans, J. 1997a. Climate sensitivity of Franz Joseph Glacier, New Zealand, as revealed by numerical modelling. *Arctic and Alpine Research*, **29**(2), 233–239.

Oerlemans, J. 1997b. A flowline model for Nigardsbreen, Norway: Projection of future glacier length based on dynamic calibration with historic record. *Annals of Glaciology*, **24**, 382–389.

Oerlemans, J. and Fortuin, J.P.F. 1992. Sensitivity of glaciers and small ice caps to greenhouse warming. *Science*, **258**, 115–118.

Oerlemans, J. and Hoogendorn, N.C. 1989. Mass balance gradients and climatic change. *Journal of Glaciology*, **35**(121), 399–405.

Oke, T.R. 1987. *Boundary Layer Climates*. London and New York, Routledge, 435pp.

Okuda, S. 1991. Rapid mass movement. In O. Slaymaker (ed.) *Field Experiments and Measurement Programs in Geomorphology*. Rotterdam, Balkema, 61–105.

Ollier, C.D. 1984. *Weathering*. 2nd edition. Edinburgh, Oliver & Boyd.

Olsen, L. 1983. A method for determining total clast roundness in sediments. *Boreas*, **12**, 17–21.

Orombelli, G. and Porter, S.C. 1983. Lichen growth curves for the southern flank of the Mont Blanc Massif, Western Italian Alps. *Arctic and Alpine Research*, **15**, 193–200.

Osborn, G.D. and Luckman, B.H. 1988. Holocene glacier fluctuations in the Canadian Cordillera (Alberta and British Columbia). *Quaternary Science Reviews*, **7**, 115–128.

Osborn, G.D., Robinson, B.J. and Luckman, B.H. 2001. Holocene and latest Pleistocene fluctuations of Stutfield Glacier, Canadian Rockies. *Canadian Journal of Earth Sciences* 38(8), 1141–1155.

Østrem, G. 1966. Surface coloring of glaciers for air photography. *Canadian Journal of Earth Sciences*, 6, 877–880.

Østrem, G. and Brugman, M. 1991. *Glacier Mass-Balance Measurements*. NHRI Science Report No. 4, National Hydrology Research Institute, Saskatoon, Canada.

Park, R.G. 1997. *Foundations of Structural Geology*. London, Chapman and Hall.

Parsons, A.J. and Knight, P. 1995. *How to Do Your Dissertation in Geography and Related Disciplines*. London, Chapman and Hall.

Paterson, W.S.B. 1966. Test of contour accuracy on a photogrammetric map of Athabasca Glacier. *Canadian Journal of Earth Sciences*, 6, 909–915.

Paterson, W.S.B. 1994. *The Physics of Glaciers*. Pergamon, Elsevier, 480pp.

Paul, F. 2002. Changes in glacier area in Tyrol, Austria, between 1969 and 1992 derived from Lansat TM and Austrian glacier inventory data. *International Journal of Remote Sensing*, 23, 787–799.

Paul, F., Kääb, A., Maisch, M., Kellenberger, T. and Haeberli, W. 2002. The new remote-sensing-derived Swiss glacier inventory: I. Methods. *Annals of Glaciology*, 34, 355–361.

Peach, P.A. and Perrie, L.A. 1975. Grain-size distribution within glacial varves. *Geology*, 3, 43–46.

Peterson, J.A. and Robinson, G. 1969. Trend-surface mapping of cirque floor levels. *Nature*, 222(5188), 75.

Petit, J.R., Jouzel, J., Raynaud, D., Barkov, N.I., Barnola, J.M., Basile, I., Bender, M., Chappellaz, J., Davis, M., Delaygue, G., Delmotte, M., Kotlyakov, V.M., Legrand, M., Lipenkov, V.Y., Lorius, C., Pepin, L., Ritz, C., Saltzman, E. and Stievenard, M. 1999. Climate and atmospheric history of the past 420,000 years from the Vostok ice core, Antarctica. *Nature*, 399(6735), 429–436.

Petrie, G. and Price, R.J. 1966. Photogrammetric measurements of the ice wastage and morphological changes near the Casement Glacier, Alaska. *Canadian Journal of Earth Sciences*, 6, 827–840.

Phillips, E.R. and Auton, C.R. 2000. Micromorphological evidence for polyphase deformation of glaciolacustrine sediments from Strathspey, Scotland. In A.J. Maltman, B. Hubbard and M.J. Hambrey (eds) *Deformation of Glacial Materials*. Geological Society Special Publication, 176, 279–292

Plewes, L.A. and Hubbard, B. 2001. A review of the use of radio-echo sounding in glaciology. *Progress in Physical Geography*, 25(2), 203–236.

Pohjola, V.A. 1993. TV-video observations of bed and basal sliding on Storglaciären, Sweden. *Journal of Glaciology*, 39(131), 111–118.

Pohjola, V.A. 1994. TV-video observations of englacial voids in Storglaciären, Sweden. *Journal of Glaciology*, 40(135), 231–240.

Porter, S.C. 1975. Weathering rinds as a relative-age criterion: Application to subdivision of glacial deposits in the Cascade Range. *Geology*, 3, 101–104.

Porter, S.C. 2001. Snowline depression in the tropics during the Last Glaciation. *Quaternary Science Reviews*, 20, 1067–1091.

Porter, P.R. and Murray, T. 2001. Mechanical and hydraulic properties of till beneath Bakaninbreen, Svalbard. *Journal of Glaciology*, 47, 167–175.

Porter, P.R., Murray, T. and Dowdeswell, J.A. 1997. Sediment deformation and basal dynamics beneath a glacier surge front: Bakaninbreen, Svalbard. *Annals of Glaciology*, **24**, 21–26.

Poulin, A.O. and Harwood, T.A. 1966. Infrared mapping of thermal anomalies in glaciers. *Canadian Journal of Earth Sciences*, **6**, 881–885.

Powers, M.C. 1953. A new roundness scale for sedimentary particles. *Journal of Sedimentary Petrology*, **23**, 117–119.

Quigley, R.M. and Ogunbadejo, T.A. 1976. Till geology, mineralogy and geotechnical behaviour, Sarnia, Ontario. In R.F. Legget (ed.) *Glacial Till: An Inter-Disciplinary Study*. The Royal Society of Canada Special Publications No. 12, Ottawa, 336–345.

Radhakrishna, H.S. and Klym, T.W. 1974. Geotechnical properties of a very dense glacial till. *Canadian Geotechnical Journal*, **11**, 396–409.

Ragan, D.M. 1969. Structures at the base of an icefall. *Journal of Geology*, **77**, 647–667.

Raiswell, R. 1984. Chemical-models of solute acquisition in glacial melt waters. *Journal of Glaciology*, **30**(104), 49–57.

Raper, S.C.B., Briffa, K.R. and Wigley, T.M.L. 1996. Glacier change in Northern Sweden from AD 500: A simple geometric model of Storglaciären. *Journal of Glaciology*, **42**(141), 341–351.

Rapp, A. and Nyberg, R. 1981. Alpine debris flows in Northern Scandinavia. Morphology and dating by lichenometry. *Geografiska Annaler*, **63A**, 183–196.

Raymond, C.F. 1971. Flow in a transverse section of Athabasca Glacier, Alberta, Canada. *Journal of Glaciology*, **10**(58), 55–84.

Reeh, N. 1984. *Antitorque Leaf Springs: A Design Guide for Ice-drill and Torque Leaf Springs*. Hanover, US Army Cold Regions Research and Engineering Laboratory, 69–72.

Rees, W.G. 2001. *Physical Principles of Remote Sensing*. 2nd edition. Cambridge, Cambridge University Press, 372pp.

Reineck, H.E. and Singh, R. 1980. *Depositional Sedimentary Environments*. New York, Springer-Verlag.

Rémy, F., Shaeffer, P. and Legrésy, B. 1999. Ice flow physical processes derived from ERS-1 high-resolution map of Antarctica and Greenland ice sheet. *Geophysical Journal International*, **139**, 645–656.

Reynolds, J.M. 1997. *An Introduction to Applied and Environmental Geophysics*. Chichester, John Wiley & Sons.

Richards, B.W.M. 2000. Luminescence dating of Quaternary sediments in the Himalayas and High Asia: A practical guide to its use and limitations for constraining the timing of glaciation. *Quaternary International*, **65/66**, 49–61.

Rigsby, G.P. 1951. Crystal fabric studies on Emmons Glacier, Mount Rainer, Washington. *Journal of Geology*, **49**, 590–598.

Robin, G.de.Q. 1966. Mapping the Antarctic ice sheet by satellite altimetry. *Canadian Journal of Earth Sciences*, **6**, 893–901.

Robin, G.de.Q., Evans, S. and Bailey, J.T. 1969. Interpretation of radio-echo sounding in polar ice sheets. *Philosophical Transactions of the Royal Society of London*, Series A, **265**(1166), 437–505.

Robinson, G., Peterson, J.A. and Anderson, P.A. 1971. Trend surface analysis of corrie altitudes in Scotland. *Scottish Geographical Magazine*, **87**, 142–146.

Rodbell, D.T. 1992. Lichenometric and radiocarbon dating of Holocene glaciation, Cordillera Blanca, Peru. *The Holocene*, **2**, 1–10.

Rolstad, C., Amlien, J., Hagen, J.O. and Lundén, B. 1997. Visible and near-infrared digital images for determination of ice velocities and surface elevation during a surge on Osbornebreen, a tidewater glacier in Svalbard. *Annals of Glaciology*, 24, 255–261.

Rose, J. and Letzer, J.M. 1975. Drumlin measurements: A test of the reliability of data derived from 1:25,000 scale topographic maps. *Geological Magazine*, 112, 361–371.

Röthlisberger, H. and Lang, H. 1987. Glacial hydrology. In A.M. Gurnell and M.J. Clark (eds) *Glacial Hydrology*. Chichester, John Wiley & Sons, 207–256.

Salem, H.S. 2001. The influence of clay conductivity on electric measurements of glacial aquifers. *Energy Sources*, 23, 225–234.

Savoskul, O.S. 1999. Holocene glacier advances in the headwaters of Sredniaya Avacha, Kamchatka, Russia. *Quaternary Research*, 52, 14–26.

Schlosser, E., Van Lipzig, N. and Oerter, H. 2002. Temporal variability of accumulation at Neumayer station, Antarctica, from stake array measurements and a regional atmospheric model. *Journal of Glaciology*, 48, 87–94.

Schmeits, M.J. and Oerlemans, J. 1997. Simulation of historical variations in length of Unterer Grindelwaldgletscher, Switzerland. *Journal of Glaciology*, 43(143), 152–164.

Schweingruber, F.H. 1987. *Tree Rings: Basics and Applications of Dendrochronology*. Dordrecht, Reidel, 276pp.

Schytt, V. 1966. The purpose of glacier mapping. *Canadian Journal of Earth Sciences*, 6, 743–746.

Seaberg, S.Z., Seaberg, J.Z., Hooke, R.L. and Wiberg, D.W. 1988. Character of the englacial and subglacial drainage system in the lower part of the ablation area of Storglaciären, Sweden, as revealed by dye-trace studies. *Journal of Glaciology*, 34(117), 217–227.

Selby, M.J. 1980. A rock mass strength classification for geomorphological purposes: With tests from Antarctica and New Zealand, *Zeitschrift für Geomorphologie*, 24, 31–51.

Sharp, M. and Richards, K. 1996. Integrated studies of hydrology and water quality in glacierized catchments – Preface. *Hydrological Processes*, 10(4), 475–478.

Sharp, M.J. and Gomez, B. 1986. Processes of debris comminution in the glacial environment and implications for quartz sand-grain micromorphology. *Sedimentary Geology*, 46, 33–47.

Sharp, M., Dowdeswell, J.A. and Gemmell, J.C. 1989a. Reconstructing past glacier dynamics and erosion from glacial geomorphic evidence: Snowdon, North Wales. *Journal of Quaternary Science* 4, 115–130.

Sharp, M., Gemmell, J.C. and Tison, J.-L. 1989b. Structure and stability of the former subglacial drainage system of the Glacier de Tsanfleuron, Switzerland. *Earth Surface Processes and Landforms*, 14, 119–134.

Sharp, M., Richards, K., Willis, I., Arnold, N., Nienow, P., Lawson, W. and Tison, J.-L. 1993. Geometry, bed topography and drainage system structure of the Haut Glacier d'Arolla, Switzerland. *Earth Surface Processes and Landforms*, 18, 119–134.

Shoshany, M. 1989. A digitizer for field measurements of microtopography and sediments texture. *Journal of Sedimentary Petrology*, 59, 628–629.

Shreve, R.L. 1972. Movement of water in glaciers. *Journal of Glaciology*, 11, 205–214.

Shumskii, P.A. 1964. *Principles of Structural Glaciology*. New York, Dover Publications, 497pp.

Sidjak, R.W. and Wheate, R.D. 1999. Glacier mapping of the Illecillewaet icefield, British Columbia, Canada, using Landsat TM and digital elevation data. *International Journal of Remote Sensing*, 20, 273–284.

Siegert, M.J., Dowdeswell, J.A., Gorman, M.R. and McIntyre, N.F. 1996. An inventory of Antarctic sub-glacial lakes. *Antarctic Science*, 8(3), 281–286.

Siegert, M.J., Kwok, R., Mayer, C. and Hubbard, B. 2000. Water exchange between the subglacial Lake Vostok and the overlying ice sheet. *Nature*, 403(6770), 643–646.

Slaymaker, O. (ed.) 1991. *Field Experiments and Measurement Programs in Geomorphology*. Rotterdam, Balkema, p. 224.

Smart, C.C. 1996. Statistical evaluation of glacier boreholes as indicators of basal drainage systems. *Hydrological Processes*, 10(4), 599–613.

Smart, C.C. and Ketterling, D.B. 1997. A low-cost electrical conductivity profiler for glacier boreholes. *Journal of Glaciology*, 43(144), 365–369.

Smart, P.L. and Laidlaw, I.M.S. 1976. An evaluation of some fluorescent dyes for water tracing. *Water Resources Research*, 13(1), 15–33.

Smeets, C., Duynkerke, P.G. and Vugts, H.F. 1998. Turbulence characteristics of the stable boundary layer over a mid-latitude glacier. Part 1: A combination of katabatic and large scale forcing. *Boundary-Layer Meteorology*, 87, 117–145.

Sneed, E.D. and Folk, R.L. 1958. Pebbles in the lower Colorado River, Texas: A study in particle morphogenesis. *Journal of Geology*, 66, 114–150.

Souchez, R., Lemmens, M., Lorrain, R., Tison, J.L., Jouzel, J. and Sugden, D. 1990. Influence of Hydroxyl-bearing minerals on the isotopic composition of ice from the basal zone of an ice-sheet. *Nature*, 345(6272), 244–246.

Stenni, B., Serra, F., Frezzotti, M., Maggi, V., Traversi, R., Becagli, S. and Udisti, R. 2000. Snow accumulation rates in northern Victoria Land, Antarctica, by firn-core analysis. *Journal of Glaciology*, 46, 541–552.

Stoker, M.S., Pheasant, J.B. and Josenhans, H. 1997. Seismic methods and interpretation. In T.A. Davies, T. Bell, A.K. Cooper, H. Josenhans, L. Polyak, A. Solheim, M.S. Stoker and J.A. Stravers (eds) *Glaciated Continental Margins: An Atlas of Acoustic Images*. London, Chapman and Hall, 9–26.

Stone, D.B. and Clarke, G.K.C. 1996. In situ measurements of basal water quality and pressure as an indicator of the character of subglacial drainage systems. *Hydrological Processes*, 10(4), 615–628.

Stone, D.B., Clarke, G.K.C. and Blake, E.W. 1993. Subglacial measurement of turbidity and electrical-conductivity. *Journal of Glaciology*, 39(132), 415–420.

Stone, D.B., Clarke, G.K.C. and Ellis, R.G. 1997. Inversion of borehole-response test data for estimation of subglacial hydraulic properties. *Journal of Glaciology*, 43(143), 103–113.

Stone, J.O., Ballantyne, C.K. and Fifield, L.K. 1998. Exposure dating and validation of periglacial weathering limits, northwest Scotland. *Geology*, 26, 587–590.

Strahler, A.N. 1952. Hypsometric (area-altitude) analysis of erosional topography. *Geological Society of America Bulletin*, 63, 1117–1142.

Stroeven, A.P., Fabel, D., Hattestrand, C. and Harbor, J. 2002. A relict landscape in the centre of Fennoscandian glaciation: Cosmogenic radionuclide evidence of tors preserved through multiple glacial cycles. *Geomorphology*, 44, 145–154.

Studinger, M., Bell, R.E. and Tikku, A.A. 2004. Estimating the depth and shape of subglacial Lake Vostok's water cavity from aerogravity data. *Geophysical Research Letters*, 31, L12401, doi:10.1029/2004GL019801.

Stuiver, M., Reimer, P.J., Bard, E., Beck, J.W., Burr, G.S., Hughen, K.A., Kromer, B., McCormac, F.G., van der Plicht, J. and Spurk, M. 1998a. INTCAL98 radiocarbon age calibration, 24000-0 cal BP. *Radiocarbon*, 40, 1041–1083.

Stuiver, M., Reimer, P.J. and Braziunas, T.F. 1998b. High-precision radiocarbon age calibration for terrestrial and marine samples. *Radiocarbon*, 40, 1127–1151.

Sugden, D.E., Denton, G.H. and Marchant, D.R. 1995. Landscape evolution of the dry valleys, transantarctic mountains: Tectonic implications. *Journal of Geophysical Research*, 100(B7), 9949–9967.

Sugden, D.E., Summerfield, M.A., Denton, G.H., Wilch, T.I., McIntosh, W.C., Marchant, D.R. and Rutford, R.H. 1999. Landscape development in the Royal Society Range, southern Victoria Land, Antarctica: Stability since the mid-Miocene. *Geomorphology*, 28, 181–200.

Suppe, J. 1985. *Principles of Structural Geology*. New Jersey, Prentice Hall.

Tangborn, W.V., Fountain, A.G. and Sikonia, W.G. 1990. Effect of area distribution with altitude on glacier mass balance: A comparison of North and South Klawatti Glaciers, Washington State, U.S.A. *Annals of Glaciology*, 14, 278–282.

Taylor, J.R. 1997. *An Introduction to Error Analysis: The Study of Uncertainties in Physical Measurements*. University Science Books, Sausalito 327pp. ISBN: 0-935702-42-3.

Taylor, M.P. and Brewer, P.A. 2001. A study of Holocene floodplain particle size characteristics with special reference to palaeochannel infills from the upper Severn basin, Wales, UK. *Geological Journal*, 36, 143–157.

Theakstone, W.H. 1966. Deformed ice at the bottom of Osterdalsisen, Norway. *Journal of Glaciology*, 6(43), 19–21.

Thorp, P.W. 1981. An analysis of the spatial variability of glacial striae and friction cracks in part of the Western Grampians of Scotland. *Quaternary Studies*, 1, 71–94.

Tison, J.-L. and Hubbard, B. 2000. Ice crystallographic evolution at a temperate glacier: Glacier de Tsanfleuron, Switzerland. In A.J. Maltman, B. Hubbard and M.J. Hambrey (eds) *Deformation of Glacial Materials*, London, 23–38.

Tison, J.-L., Souchez, R. and Lorrain, R. 1989. On the incorporation of unconsolidated sediments in basal ice: Present-day examples. *Zeitschrift für Geomorphologie*, 72, 173–183.

Tranter, M., Brown, G.H., Raiswell, R., Sharp, M.J. and Gurnell, A.M. 1993. A conceptual model of solute acquisition by Alpine glacial meltwaters. *Journal of Glaciology*, 39, 573–581.

Tranter, M., Sharp, M.J., Brown, G.H., Willis, I.C., Hubbard, B., Nielsen, M.K., Smart, C.C., Gordon, S., Tulley, M. and Lamb, H.R. 1997. Variability in the chemical composition of in situ subglacial meltwaters. *Hydrological Processes*, 11(1), 59–77.

Troeh, F.R. 1965. Landform equations fitted to contour maps. *American Journal of Science*, 263, 616–627.

Tucker, M. 1988. *Techniques in Sedimentology*. Oxford, Blackwell Scientific Publications.

Tucker, M.E. 1996. Sedimentary rocks in the field. 2nd edition. London, John Wiley, 153pp.

Tulaczyk, S., Kamb, B., Scherer, R. and Engelhardt, H.F. 1998. Sedimentary processes at the base of a West Antarctic ice stream: Constraints from textural and compositional properties of subglacial debris. *Journal of Sedimentary Research*, 68(3), 487–496.

Tulaczyk, S., Kamb, B. and Engelhardt, H. 2000. Basal mechanics of ice stream B: I. Till mechanics. *Journal of Geophysical Research*, 105, 463–481.

Tulaczyk, S., Kamb, B. and Engelhardt, H. 2001. Estimates of effective stress beneath a beneath a modern West Antarctic Ice Stream from till preconsolidation and void ratio. *Boreas*, 30, 101–114.

Udden, J.A. 1914. Mechanical composition of clastic sediments. *Bulletin of the Geological Society of America*, 25, 655–744.

Urick, R.J. 1983. *Principles of Underwater Sound*. New York, McGraw-Hill.

Vallon, M., Petit, J.-R. and Fabre, B. 1976. Study of an ice core to the bedrock in the accumulation zone of an alpine glacier. *Journal of Glaciology*, 17(75), 13–28.

Van der Meer, J.J.M. 1993. Microscopic evidence of subglacial deformation. *Quaternary Science Reviews*, 12, 553–587.

Van der Meer, J.J.M. 1996. Micromorphology. In J. Menzies (ed.) *Past Glacial Environments: Sediments, Forms, Techniques*. Oxford, Pergamon, 335–356.

Van der Wateren, F.M., Kluiving, S.J. and Bartek, L.R. 2000. Kinematic indicators of subglacial shearing. In A.J. Maltman, B. Hubbard and M.J. Hambrey (eds) *Deformation of Glacial Materials*. Geological Society Special Publication, 176, 259–278.

Waddington, B.S. and Clarke, G.K.C. 1995. Hydraulic-properties of subglacial sediment determined from the mechanical response of water-filled boreholes. *Journal of Glaciology*, 41(137), 112–124.

Wadell, H. 1932. Volume, shape and roundness of rock particles. *Journal of Geology*, 40, 443–451.

Wadell, H. 1933. Sphericity and roundness of rock particles. *Journal of Geology*, 41, 310–331.

Wadell, H. 1935. Volume, shape and roundness of quartz particles. *Journal of Geology*, 43, 250–280.

Wadham, J.L., Hodson, A.J., Tranter, M. and Dowdeswell, J.A. 1997. The rate of chemical weathering beneath a quiescent, surge-type, polythermal-based glacier, southern Spitsbergen, Svalbard. *Annals of Glaciology*, 24, 27–31.

Wadham, J.L., Hodgkins, R., Cooper, R.J. and Tranter, M. 2001. Evidence for seasonal subglacial outburst events at a polythermal glacier, Finsterwalderbreen, Svalbard. *Hydrological Processes*, 15(12), 2259–2280.

Walden, J., Smith, J.P. and Dackombe, R.V. 1987. The use of mineral magnetic analyses in the study of glacial diamicts: A pilot study. *Journal of Quaternary Science*, 2, 73–80.

Walden, J., Smith, J.P. and Dackombe, R.V. 1996. A comparison of mineral magnetic, geochemical and mineralogical techniques for compositional studies of glacial diamicts. *Boreas*, 25, 115–130.

Walden, J., Oldfield, F. and Smith, J.P. (eds) 1999. *Environmental Magnetism: A Practical Guide*. Technical Guide No. 6, London, Quaternary Research Association.

Walder, J.S. and Hallet, B. 1979. Geometry of former subglacial water channels and cavities. *Journal of Glaciology*, 23, 335–346.

Walling, D.E. 1991. Drainage basin studies. In O. Slaymaker (ed.) *Field Experiments and Measurement Programs in Geomorphology*. Rotterdam, Balkema, 17–59.

Wang, Y. and Azuma, N. 1999. A new automatic ice-fabric analyzer which uses image-analysis techniques. *Annals of Glaciology*, 29, 155–162.

Warburton, J. and Beecroft, I. 1993. Use of meltwater stream material loads in the estimation of glacial erosion rates. *Zeitschrift für Geomorphologie*, 37, 19–28.

Warren, C.R., Glasser, N.F., Kerr, A.R., Harrison, S., Winchester, V. and Rivera, A. 1995. Characteristics of tidewater calving activity at Glaciar San Rafael, Chile. *Journal of Glaciology*, 41(138), 273–289.

Watchman, A.L. and Twidale, C.R. 2002. Relative and 'absolute' dating of land surfaces. *Earth Science Reviews*, 58, 1–49.

Waters, R.S. 1958. Morphological mapping. *Geography*, 43, 10–18.

Watts, R.D. and England, A.W. 1976. Radio-echo sounding of temperate glaciers: Ice properties and sounder design criteria. *Journal of Glaciology*, 17(75), 39–48.

Weertman, J. 1957. On the sliding of glaciers. *Journal of Glaciology*, 3(21), 33–38.

Weertman, J. 1964. The theory of glacier sliding. *Journal of Glaciology*, 5(39), 287–303.

Welch, R. and Howarth, P.J. 1968. Photogrammetric measurements of glacial landforms. *Photogrammetric Record*, 7, 75–96.

Welch, B.C., Pfeffer, W.T., Harper, J.T. and Humphrey, N.F. 1998. Mapping subglacial surfaces of temperate valley glaciers by two-pass migration of a radio-echo sounding survey. *Journal of Glaciology*, 44(146), 164–170.

Wellner, J.S., Lowe, A.L., Shipp, S.S. and Anderson, J.B. 2001. Distribution of glacial geomorphic features on the Antarctic continental shelf and correlation with substrate: Implications for ice behavior. *Journal of Glaciology*, 47, 397–411.

Wentworth, C.K. 1922. A scale of grade and class terms of clastic sediments. *Journal of Geology*, 377–390.

Wentworth, C.K. 1935. An analysis of the shapes of glacial cobbles. *Journal of Sedimentary Petrology*, 6, 85–96.

Wessels, R.L., Kargel, J.S. and Kieffer, H.H. 2002. ASTER measurements of supraglacial lakes in the Mount Everest region of the Himalaya. *Annals of Glaciology*, 34, 399–408.

Whalley, W.B. and Krinsley, D.H. 1974. A scanning electron microscope study of surface textures of quartz grains from glacial environments. *Sedimentology*, 21, 87–105.

Whalley, W.B. and Palmer, C.F. 1998. A glacial interpretation for the origin and formation of the Marinet Rock Glacier, Alpes Maritimes, France. *Geografiska Annaler*, 80A, 221–236.

Whitehouse, I.E. and McSaveney, M.J. 1983. Diachronous talus surfaces in the Southern Alps, New Zealand, and their implications to talus accumulation. *Arctic and Alpine Research*, 15, 53–64.

Whitehouse, I.E., McSaveney, M.J., Knuepfer, P.L.K. and Chinn, T.J. 1986. Growth of weathering rinds on Torlesse sandstone, Southern Alps, New Zealand. In S.M. Colman and D.P. Dethier (eds) *Rates of Chemical Weathering of Rocks and Minerals*. Orlando, Academic Press, 419–435.

Wilhelms, F., Kipfstuhl, J., Miller, H., Firestone, J. and Heinloth, K. 1998. Precise dielectric profiling of ice cores: A new device with improved guarding and its theory. *Journal of Glaciology*, 44(146), 171–174.

Williams, R.S., Hall, D.K. and Benson, C.S. 1991. Analysis of glacier facies using satellite techniques. *Journal of Glaciology*, 37, 120–128.

Williams, R.S., Jr, Hall, D., Sigurðsson, O. and Chien, J.Y.L. 1997. Comparison of satellite-derived with ground-based measurements of the fluctuations of the margins of Vatnajökull, Iceland, 1973–92. *Annals of Glaciology*, 24, 72–80.

Wilson, C.J.L. 2004. *Fabric Analyser & Investigator*. School of Earth Sciences, University of Melbourne.

Wilson, C.J.L. and Sim, H.M. 2002. The localization of strain and *c*-axis evolution in anisotropic ice. *Journal of Glaciology*, 48(163), 601–610.

Winchester, V. and Harrison, S. 1994. A development of the lichenometric method applied to the dating of glacially influenced debris flows in Southern Chile. *Earth Surface Processes and Landforms*, 19, 137–151.

Winchester, V. and Harrison, S. 1996. Recent oscillations of the San Quintin and San Rafael Glaciers, Patagonian Chile. *Geografiska Annaler*, 78A, 35–49.

Winchester, V., Harrison, H. and Warren, C.R. 2001. Recent retreat Glaciar Nef, Chilean Patagonia, dated by lichenometry and dendrochronology. *Arctic, Antarctic and Alpine Research*, 33, 266–273.

Wingtes, T. 1985. Studies on crescentic fractures and crescentic gouges with the help of close-range photography. *Journal of Glaciology*, 31, 340–349.

Winkler, S. 2000. The 'Little Ice Age' maximum in the Southern Alps, New Zealand: Preliminary results at Mueller Glacier. *The Holocene*, 10, 643–647.

Winkler, S. 2003. A new interpretation of the date of the 'Little Ice Age' glacier maximum at Svartisen and Okstindan, northern Norway. *The Holocene*, 13, 83–95.

Wolfe, P.R. 1983. *Elements of Photogrammetry*. 2nd edition. New York, McGraw-Hill.

Woodruff, J.F. and Evenden, L.J. 1962. Geomorphic measurement from aerial photos. *Professional Geographer*, 14, 23–26.

Worsley, P. 1990. Tree-ring dating (dendrochronology). In A.S. Goudie (ed.) *Geomorphological Techniques*. London, Unwin Hyman, 415–421.

Yingkui, L., Gengnian, L. and Zhijiu, C. 2001. Longitudinal variations in cross-section morphology along a glacial valley: A case-study from the Tien Shan, China. *Journal of Glaciology*, 47, 243–250.

Young, J.A.T. 1969. Variations in till macrofabric over very short distances. *Geological Society of America Bulletin*, 80, 2343–2352.

Zingg, T. 1935. Beitrag zur Schotteranalyse. *Schweizerische Mineralogische und Petrographische Mitteilungen*, 15, 39–140.

Index

Field Techniques in Glaciology and Glacial Geomorphology   Bryn Hubbard and Neil Glasser
© 2005 John Wiley & Sons, Ltd

Printed and bound by CPI Group (UK) Ltd, Croydon, CR0 4YY

Printed and bound by CPI Group (UK) Ltd, Croydon, CR0 4YY

27/10/2024

14580210-0005